AF330102

Computational Quantum Chemistry
Molecular Structure and Properties *in Silico*

RSC Theoretical and Computational Chemistry Series

Editor-in-Chief
Jonathan Hirst, *University of Nottingham, Nottingham, UK*

Series Editors:
Kenneth Jordan, *University of Pittsburgh, Pittsburgh, USA*
Carmay Lim, *Academia Sinica, Taipei, Taiwan*
Walter Thiel, *Max Planck Institute for Coal Research, Mülheim an der Ruhr, Germany*

Titles in the Series:
1: Knowledge-based Expert Systems in Chemistry: Not Counting on Computers
2: Non-Covalent Interactions: Theory and Experiment
3: Single-Ion Solvation: Experimental and Theoretical Approaches to Elusive Thermodynamic Quantities
4: Computational Nanoscience
5: Computational Quantum Chemistry: Molecular Structure and Properties *in Silico*

How to obtain future titles on publication:
A standing order plan is available for this series. A standing order will bring delivery of each new volume immediately on publication.

For further information please contact:
Book Sales Department, Royal Society of Chemistry, Thomas Graham House, Science Park, Milton Road, Cambridge, CB4 0WF, UK
Telephone: +44 (0)1223 420066, Fax: +44 (0)1223 420247
Email: booksales@rsc.org
Visit our website at www.rsc.org/books

Computational Quantum Chemistry
Molecular Structure and Properties *in Silico*

Joseph J W McDouall
School of Chemistry, University of Manchester, UK
Email: joe.mcdouall@manchester.ac.uk

RSC Publishing

RSC Theoretical and Computational Chemistry Series No. 5

ISBN: 978-1-84973-608-4
ISSN: 2041-3181

A catalogue record for this book is available from the British Library

Published by The Royal Society of Chemistry,
Thomas Graham House, Science Park, Milton Road,
Cambridge CB4 0WF, UK

Registered Charity Number 207890

For further information see our web site at www.rsc.org

Printed in the United Kingdom by Henry Ling Limited, Dorchester, DT1 1HD, UK

Preface

In writing this book I have had in mind the needs of a postgraduate student, or other researcher of a similar level, starting to work in the area of computational quantum chemistry. The standard methods of electronic structure theory are presented, as are the principles necessary for the evaluation of molecular properties. The output of quantum chemical studies, in the form of wavefunctions, densities and orbitals, allow connections to be made with underlying chemical concepts. This can be done in many ways and a number of methods of analysis are described. An increasingly important area of research is the quantum chemical treatment of molecules containing heavy elements. This is the realm of relativistic quantum chemistry and a chapter is devoted to introducing the background theory to some of the commonly used relativistic electronic structure methods.

I have assumed no greater acquaintance with quantum mechanics or mathematical techniques other than might be encountered in a good under-graduate course in chemistry. Where necessary I have introduced numerical methods and alluded to their computational implementation. The book begins with an overview in Chapter 1 that provides an introduction to how chemistry can be studied by theoretical and computational techniques. This chapter might also find use at an undergraduate level in providing some connections between quantum mechanics, computing and chemistry. The rest of the book aims to fill in some of the details associated with the ideas introduced in Chapter 1. At the risk of being old-fashioned, I have attempted to include enough detail so that the reader can see how a computational method is arrived at and how it might be implemented. The very nature of the subject is technical and requires a, relatively straightforward, mathematical language. I have made no attempt to use a sentence when an equation would do instead.

Computational quantum chemistry is a large and mature subject, with many specialist sub-fields. In a general introduction such as this there must, of

RSC Theoretical and Computational Chemistry Series No. 5
Computational Quantum Chemistry: Molecular Structure and Properties *in Silico*
By Joseph J W McDouall
© Joseph J W McDouall 2013
Published by the Royal Society of Chemistry, www.rsc.org

necessity, be omissions in the material covered. Some obvious omissions related to electronic structure theory include semi-empirical methods, effective core potentials and explicitly correlated techniques. I have also omitted any discussion of dynamics on potential energy surfaces.

Finally I must acknowledge that the preparation of this manuscript owes much to, and would not have been possible without, the huge amount of typing undertaken by my wife, Jacky.

Joseph J. W. McDouall

To all my family, past and present.

Contents

RSC Theoretical and Computational Chemistry Series No. 5

Computational Quantum Chemistry: Molecular Structure and Properties *in Silico*

By Joseph J W McDouall

© Joseph J W McDouall 2013

Published by the Royal Society of Chemistry, www.rsc.org

Chapter 3   The Computation of Molecular Properties

Chapter 4   Understanding Molecular Wavefunctions, Orbitals and Densities

Chapter 5   Relativistic Effects and Electronic Structure Theory

CHAPTER 1
Computational Quantum Chemistry

1.1 What Does Computational Quantum Chemistry Offer?

Computational quantum chemistry has been in development for almost nine decades. Its progress has been intimately linked to developments in computing hardware and technology. Today computational quantum chemistry provides a complementary way of investigating a wide range of chemistry. In particular it provides reliable information on molecular structures, molecular properties, reactions mechanisms and energetics. Detailed mechanistic questions can be addressed using the techniques of computational quantum chemistry. An advantage over traditional experimental techniques is that it provides a route to the study of chemical questions which may be experimentally difficult, or expensive, or dangerous. The purpose is always to answer a chemical question and in that sense computational quantum chemistry is the complement to experiment, either approach on its own is much less convincing. This complementarity of techniques is very familiar to chemists. For example, to determine a molecular structure a range of spectroscopies must be used and each provides a component of the overall picture. Now to these spectroscopies are added quantum chemical techniques that can provide further information.

Computational quantum chemistry is an elegant conjunction of chemistry, physics, mathematics and computer science. Chemistry defines the question. Physics defines the laws that are obeyed by the chemical system. Mathematics formulates a numerical representation of the problem. Computer science solves

RSC Theoretical and Computational Chemistry Series No. 5
Computational Quantum Chemistry: Molecular Structure and Properties *in Silico*
By Joseph J W McDouall
© Joseph J W McDouall 2013
Published by the Royal Society of Chemistry, www.rsc.org

the mathematical model, yielding numbers that encapsulate physical significance. For example, does a particular alkylation reaction proceed more efficiently with the alkyl chloride or the corresponding iodide? To answer this question, at the simplest level, we could compute the geometries of the transition structures and reactants, from which we would obtain the activation energies and so determine which reaction should be more efficient. The insight gained from such numerical answers can lead to further questions. This often results in an iterative refinement of questions, answers and models, see Figure 1.1. By such a process our understanding of a chemical question deepens.

The historical development of quantum chemistry can be categorised into a number of eras. The earliest, first age of quantum chemistry, was characterised by computational results of a qualitative nature. These did much to help develop understanding of potential energy surfaces, geometries of molecules at equilibrium, reactive transition structures, and molecular orbital concepts. These insights were able to explain the physical origins of experimentally measured properties. The second age of quantum chemistry came about through the development of computer technology and accompanying developments in numerical algorithms. This enabled much more elaborate computations to be performed. In this second era, semi-quantitative agreement with experiment was already obtained for some measured quantities. Despite this improved accuracy, quantum chemical techniques were still not able to displace experimental measurements, but had become sufficiently reliable that

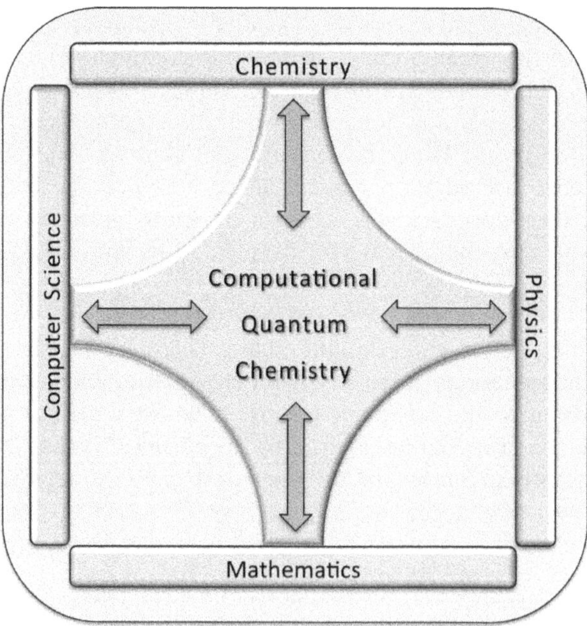

Figure 1.1   Interplay of disciplines that make up computational quantum chemistry.

they could be applied to situations in which experiments were not yet feasible. For example, the study of very short-lived molecular species, or the study of the properties of postulated molecules that had yet to be synthesised. The third age of quantum chemistry is best summarised by Graham Richards in his influential article of 1979:[1] *"The work represents perhaps a near perfect instance of theory being in harmony with experiment, each aspect vital to the other and the combination much more than the sum of the separate parts"*. Experimental measurements cannot be interpreted or understood in the absence of a reliable theoretical framework. The studies referred to by Richards showed the computational work to be an equal partner to the experiments. There have also been cases where computational studies have preceded experiments through the correct predictions of measured quantities, which have subsequently been confirmed by experiment.[2] Since the dawn of this new age of quantum chemistry, alluded to above, rapid developments have continued and their success has made computational quantum chemistry an essential component of many modern chemical investigations.

Historically, computational quantum chemistry was restricted to the realm of specialists who had access to high performance computing facilities, a good knowledge of software construction and numerical methods, as well as a good understanding of the underlying quantum mechanical models. There is still a strong need for this type of specialist who can push the subject forward by developing new methods, or providing very efficient computer implementations of established techniques. However, the standard models that we shall discuss in this book have been developed to the point that serious molecular questions can be tackled by any good scientist, not just the computational specialist. Today elaborate quantum chemical calculations can be carried out using fast desktop machines, and readily available software, by non-specialists. The same care and rigour must be applied to the design and execution of such calculations as would be applied to the design of any scientific investigation. A poorly thought out study, whether computational or experimental, cannot produce useful results.

1.2   The Model: Quantum Mechanics

An interesting experiment, which the author has carried out on numerous occasions, is to ask a room full of 200 undergraduate chemistry students: *What is chemistry?* Invariably, one obtains a fascinating range of answers. Many will tell you that chemistry is about "making things", for example materials, medicines, or fuels. Others may tell you that chemistry is about understanding the physical processes that govern chemical properties, for example the rate of reaction between two molecules, or the colour of a molecule. All these answers, and many others beside, are equally valid. Yet the overriding answer is: *Chemistry is a game that electrons play!* In a sense this answer encapsulates all the other answers, since everything chemical is under the control of the electrons that participate in the chemical process. There are no chemical phenomena that cannot be traced back to the behaviour of electrons.

So chemistry is about electrons. To understand chemistry we need to understand the behaviour of electrons. We are familiar with electrons being negatively charged particles with mass. Additionally we know from the experiments of Davisson and Germer in 1925, involving the diffraction of electrons by a crystal, that electrons can behave as waves. This wave-particle duality is quantified in the de Broglie relation

$$\lambda = \frac{h}{p} \tag{1.1}$$

where λ is the wavelength associated with a particle of mass, m, moving with velocity, v. The linear momentum is, $p = mv$, and h is Planck's constant. For slow-moving macroscopic objects the wavelength given by eqn (1.1) is undetectably small. However electrons confined within atoms and molecules are very light and fast-moving with comparatively large de Broglie wavelengths. This is the realm of quantum mechanics and the correct description of quantum mechanical particles, such as the electron, is provided by the Schrödinger equation. The electronic structure and properties of any molecule, in any of its available stationary states may be determined, in principle, by solution of Schrödinger's (time-independent) equation.

$$\hat{H}\Psi_A = E_A\Psi_A \tag{1.2}$$

In eqn (1.2), A labels the state of interest. For example, the ground state or the first electronically excited state. To begin we shall concern ourselves with the ground state only and suppress the state label. At the simplest level we want to find the energy, E, and the wavefunction, Ψ, based on the hamiltonian operator, $\hat{H}$, for the molecular system of interest. The Schrödinger equation can be solved exactly only for one-electron systems. Hence much of the apparatus of computational quantum chemistry is concerned with finding increasingly accurate approximations to the Schrödinger equation for many-electron molecular systems. As we shall see in due course, the accuracy of the approximations is intimately related to the computational cost of the underlying numerical algorithms.

The first chemical application of the Schrödinger equation was undertaken by Heitler and London in 1927. In their landmark paper they calculated the potential energy curve of the hydrogen molecule. Today we are able to perform calculations on much larger systems, perhaps including up to 1000 atoms, and the methods we use are very different from those used by Heitler and London. Developments in computational quantum chemistry have been closely allied to developments in computational hardware as well as algorithmic developments (Figure 1.1). This endeavour shows no sign of abating and the demand for computational studies to complement experimental work grows continually.

This is easily understood since, as we have asserted, chemistry is about the behaviour of electrons. The Schrödinger equation furnishes us, in principle, with all information about the behaviour of electrons in molecules and in turn, all information about chemistry. As we have stated already, approximations are key and it emerges that there is in practice no "best" method in computational quantum chemistry. Studies on real chemical problems always involve a trade-off between accuracy and computational cost. For certain methods we can make formal statements about their relative merits as approximations to the Schrödinger equation. However if such methods are too computationally demanding to be applicable to a problem of interest then describing them as "better" is, at best, vague.

1.2.1   The Schrödinger Equation and the Born-Oppenheimer Approximation

Before proceeding to some details, it is useful to briefly describe a very powerful notational expedience, introduced by Paul Dirac in 1939, which we shall use throughout this book. We shall write the many equations and integrals that appear using Dirac notation. For example, consider how we can obtain the energy, E, from the Schrödinger equation as shown below (the conventional notation will be shown on the left hand side and the equivalent in Dirac notation on the right).

$$\hat{H}\Psi = E\Psi \quad \equiv \quad \hat{H}|\Psi\rangle = E|\Psi\rangle \tag{1.3}$$

Now pre-multiply by Ψ^* (the complex conjugate of Ψ) and integrate over all variables, call them τ,

$$\int \Psi^* \hat{H}\Psi \, d\tau = E \int \Psi^* \Psi \, d\tau \quad \equiv \quad \langle\Psi|H|\Psi\rangle = E\langle\Psi|\Psi\rangle \tag{1.4}$$

Now rearrange to obtain E:

$$E = \frac{\int \Psi^* \hat{H}\Psi \, d\tau}{\int \Psi^* \Psi \, d\tau} \quad \equiv \quad E = \frac{\langle\Psi|H|\Psi\rangle}{\langle\Psi|\Psi\rangle} \tag{1.5}$$

A quantity denoted in $|\rangle$ is termed a "ket" and here represents the wave-function. $\langle|$ is called a "bra" and represents the complex conjugate of $|\rangle$. For real quantities, $\langle|$ and $|\rangle$ are the same. When an operator is pre- and post-multiplied by a *bra* and *ket*, integration is assumed implicitly. Denoting a general operator as $\hat{C}$, we form

$$\langle bra|C|ket\rangle \quad e.g. \langle\Psi|H|\Psi\rangle \tag{1.6}$$

Accordingly this notation is often referred to as "bracket" notation. It is very widely used. There are many other subtle features, related to the description of vector spaces, which are implicit in the Dirac notation, but they will not concern us here. In the rest of this text we shall use Dirac notation and conventional notations as suits the discussion.

It turns out that despite the simple form in which the Schrödinger equation can be written, its solutions are far from simple to obtain. In fact the Schrödinger equation can only be solved for one-electron systems. To deal with more complex atoms and molecules we must introduce a number of approximations. There are three key ideas which we shall adopt. To motivate the first of these, let us look in more detail at the quantities that enter the Schrödinger equation. The model of the atom that we shall use consists of a set of protons positioned at the atomic nucleus and surrounded by a number of electrons. The number of protons is given by the atomic number, Z, which tells us the number of protons carrying a unit positive charge, e, in the atomic nucleus. For neutral atoms, Z also gives the number of electrons surrounding the nucleus, each with unit negative charge, $-e$. In the absence of electric or magnetic fields, the hamiltonian operator, $\hat{H}$, then includes terms which specify the kinetic and potential energies of the electrons and nuclei. $\hat{H}$ includes (i) the kinetic energy of motion for electrons and nuclei; (ii) the potential energy of attraction between electrons and nuclei; (iii) the potential energy of repulsion between electrons and similarly the potential energy of repulsion between nuclei. These terms have the following form:

$$-\frac{\hbar^2}{2m_e} \sum_i^{\text{electrons}} \nabla_i^2 - \frac{\hbar^2}{2} \sum_A^{\text{nuclei}} \frac{1}{M_A} \nabla_A^2 \tag{1.7}$$

$$-\frac{e^2}{4\pi\varepsilon_0} \sum_A^{\text{nuclei}} \sum_i^{\text{electrons}} \frac{Z_A}{r_{iA}} \tag{1.8}$$

$$\frac{e^2}{4\pi\varepsilon_0} \sum_{i<j}^{\text{electrons}} \frac{1}{r_{ij}} + \frac{e^2}{4\pi\varepsilon_0} \sum_{A<B}^{\text{nuclei}} \frac{Z_A Z_B}{R_{AB}} \tag{1.9}$$

The quantities that enter eqns (1.7) – (1.9) are:
  $\hbar$: Planck's constant divided by 2π,
  m_e: the rest mass of the electron
  M_A: the mass of nucleus A
  e: the charge on the proton
  ε_0: the permittivity of free-space
∇_i^2 and ∇_A^2: are the kinetic energy operators for electron i and nucleus A, respectively. ∇^2 is known as the "laplacian operator" and has the general form

(in cartesian coordinates) $\nabla^2 = \left(\dfrac{\partial^2}{\partial x^2}, \dfrac{\partial^2}{\partial y^2}, \dfrac{\partial^2}{\partial z^2} \right)$.

r_{iA}, r_{ij} and R_{AB}: are the distance vector between electron i and nucleus A; the distance vector between electron i and electron j; the distance vector between nucleus A and nucleus B, respectively. For example, if A and B have cartesian coordinates, (x_A, y_A, z_A) and (x_B, y_B, z_B), respectively, the magnitude of the distance vector between A and B can be written as $R_{AB} = \sqrt{(x_B - x_A)^2 + (y_B - y_A)^2 + (z_B - z_A)^2}$. All other distances are defined similarly.

Before proceeding any further we can simplify the forms of eqns (1.7) – (1.9) by introducing atomic units (au). In this system of units a number of fundamental constants take the value of unity, hence

$$e = 1$$

$$m_e = 1$$

$$\hbar = 1$$

$$4\pi\varepsilon_0 = 1$$

Table 1.1 shows these, and other key quantities assigned a value of 1 au, with their SI equivalents. A more complete list of quantities is given in Appendix 1A.

Using these definitions we can write the kinetic energy terms in au as:

$$-\frac{1}{2} \sum_{i}^{\text{electrons}} \nabla_i^2 - \frac{1}{2} \sum_{A}^{\text{nuclei}} \frac{1}{\bar{M}_A} \nabla_A^2 \tag{1.10}$$

where $\bar{M}_A$ is the ratio of the mass of nucleus A to the mass of the electron, $\bar{M}_A = \dfrac{M_A}{m_e}$. Similarly the potential energy of attraction between electrons and nuclei may be written in au as:

Table 1.1   Essential atomic units and their SI equivalents.

Quantity	*Atomic Unit*	*Equivalent in SI units*
Charge	$m_e = 1$	1.602176×10^{-19} C
Mass	$e = 1$	9.109382×10^{-31} kg
Angular Momentum	$\hbar = \dfrac{h}{2\pi} = 1$	1.054571×10^{-34} J s
Permittivity of Free-Space	$4\pi\varepsilon_0 = 1$	1.112650×10^{-10} C^2 J^{-1} m^{-1}
Length	$a_0 = 1$	5.291772×10^{-11} m
Energy	$E_h = 1$	4.359744×10^{-18} J

$$-\sum_{A}^{\text{nuclei}}\sum_{i}^{\text{electrons}}\frac{Z_A}{r_{iA}} \tag{1.11}$$

and the repulsive potential energy terms as:

$$\sum_{i<j}^{\text{electrons}}\frac{1}{r_{ij}}+\sum_{A<B}^{\text{nuclei}}\frac{Z_A Z_B}{R_{AB}} \tag{1.12}$$

The full molecular hamiltonian now becomes:

$$\hat{H}=-\frac{1}{2}\sum_{i}^{\text{electrons}}\nabla_i^2-\frac{1}{2}\sum_{A}^{\text{nuclei}}\frac{1}{\bar{M}_A}\nabla_A^2-$$
$$\sum_{A}^{\text{nuclei}}\sum_{i}^{\text{electrons}}\frac{Z_A}{r_{iA}}+\sum_{i<j}^{\text{electrons}}\frac{1}{r_{ij}}+\sum_{A<B}^{\text{nuclei}}\frac{Z_A Z_B}{R_{AB}} \tag{1.13}$$

The ratio of the mass of the proton to that of the electron is $\frac{m_P}{m_e}\approx 1836$. Therefore even the lightest nucleus, hydrogen, with its single proton is three orders of magnitude heavier than the electron. This large disparity in mass means that the relatively light electron will move much more quickly than the nucleus to which it is attached. This implies a difference in the time scales governing the motion of electrons and nuclei. The electrons will be able to execute several periods of their motion before the nuclei have moved to any significant degree. This means that we can quantise the motion of the electrons for a fixed position of the nuclei. If the nuclei are fixed then their kinetic energy is zero. Hence we can remove the term $-\frac{1}{2}\sum_{A}^{\text{nuclei}}\frac{1}{\bar{M}_A}\nabla_A^2$ from the molecular hamiltonian in eqn (1.13), to yield the simpler "electronic" hamiltonian

$$\hat{H}_{\text{Electronic}}=-\frac{1}{2}\sum_{i}^{\text{electrons}}\nabla_i^2-\sum_{A}^{\text{nuclei}}\sum_{i}^{\text{electrons}}\frac{Z_A}{r_{iA}}+\sum_{i<j}^{\text{electrons}}\frac{1}{r_{ij}} \tag{1.14}$$

Since the kinetic energy of the nuclei are assumed to be zero, the potential energy of repulsion between nuclei assumes a constant value (for a given position of the nuclei). This is termed the "nuclear repulsion" energy and is given by

$$V_{AB}=\sum_{A>B}^{\text{nuclei}}\frac{Z_A Z_B}{R_{AB}} \tag{1.15}$$

Using the electronic hamiltonian in eqn (1.14) the electronic Schrödinger equation may be solved

$$\hat{H}_{\text{Electronic}}|\Psi_{\text{Electronic}}\rangle = E_{\text{Electronic}}|\Psi_{\text{Electronic}}\rangle \tag{1.16}$$

The solution of this equation is the electronic wavefunction, $|\Psi_{\text{Electronic}}\rangle$, which depends explicitly on the set of all electronic coordinates $\{\mathbf{r}_i\}$ and parametrically on the set of all nuclear coordinates, $\{\mathbf{R}_A\}$. Formally, we should write the electronic wavefunction as $|\Psi(\{\mathbf{r}_i\},\{\mathbf{R}_A\})_{\text{Electronic}}\rangle$ to show this dependence. We shall omit writing the explicit dependence to simplify the notation. The parametric dependence on $\{\mathbf{R}_A\}$ implies that for every different arrangement of the nuclei, the electronic wavefunction will be different and accordingly the electronic energy obtained from eqn (1.16) will vary with the nuclear arrangement (geometry) of the system. The total molecular potential energy is given by the sum of the electronic energy, obtained from eqn (1.16), and the nuclear repulsion energy, obtained from eqn (1.15):

$$E_{\text{Total}} = E_{\text{Electronic}} + V_{AB} \tag{1.17}$$

E_{Total} is only a "potential" energy in as much as the motion of the nuclei are being considered (since $E_{\text{Electronic}}$ contains the kinetic energy of the electrons). E_{Total} depends on $\{\mathbf{R}_A\}$ and plotting E_{Total} as a function of $\{\mathbf{R}_A\}$ yields a potential energy curve or surface. This is the Born-Oppenheimer approximation. The simplest example of this is the familiar plot of potential energy against internuclear distance for a diatomic molecule, for example hydrogen chloride as shown in Figure 1.2.

We now have a form for $\hat{H}_{\text{Electronic}}$ from which we can obtain E_{Total} provided we know $|\Psi_{\text{Electronic}}\rangle$, the electronic wavefunction. The Born-Oppenheimer approximation is the first of our key approximations in dealing with the Schrödinger equation. It is the central starting point to nearly all quantum chemical methods.

We shall not consider the nuclear problem in any depth but note that if we wished to consider the nuclear motion, for example to study the effects of isotopic substitution, we would introduce the appropriate nuclear Schrödinger equation

$$\hat{H}_{\text{Nuclear}}|\Psi_{\text{Nuclear}}\rangle = E_{\text{Nuclear}}|\Psi_{\text{Nuclear}}\rangle \tag{1.18}$$

This nuclear equation governs the vibrations, rotations and translations of a molecule. These different types of nuclear motion often operate on different time and energy scales and, provided there is no strong coupling between the different types of motion, we can decompose eqn (1.18) as

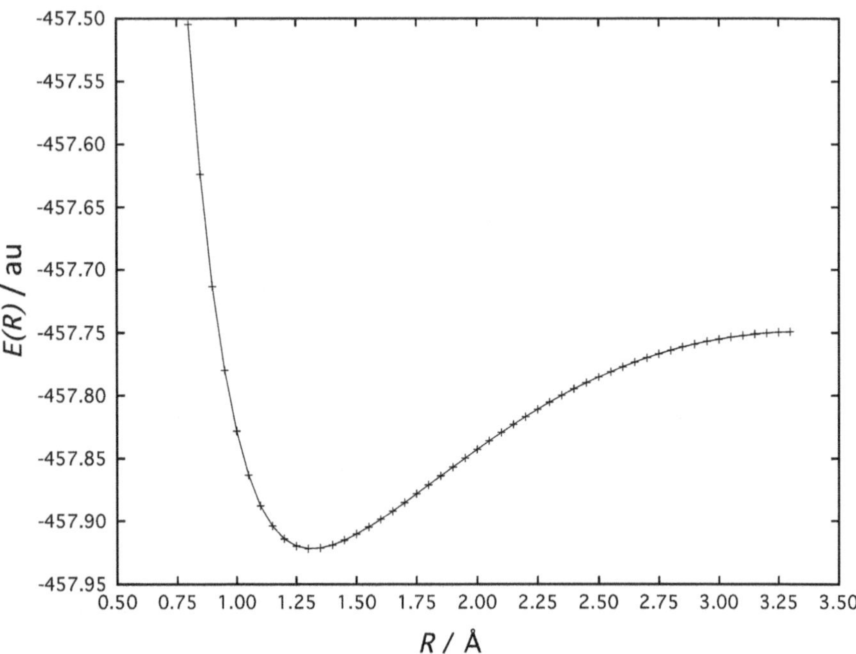

Figure 1.2   Potential energy curve for hydrogen chloride.

$$\hat{H}_{\text{Nuclear}} = \hat{H}_{\text{Vibration}} + \hat{H}_{\text{Rotation}} + \hat{H}_{\text{Translation}}$$

$$|\Psi_{\text{Nuclear}}\rangle = |\Psi_{\text{Nuclear}}\rangle|\Psi_{\text{Rotation}}\rangle|\Psi_{\text{Translation}}\rangle \tag{1.19}$$

$$E_{\text{Nuclear}} = E_{\text{Vibration}} + E_{\text{Rotation}} + E_{\text{Translation}}$$

Since we have assumed there to be no strong coupling terms, we can proceed more efficiently by solving independent Schrödinger equations for vibration, rotation and translation:

$$\hat{H}_{\text{Vibration}}|\Psi_{\text{Vibration}}\rangle = E_{\text{Vibration}}|\Psi_{\text{Vibration}}\rangle$$

$$\hat{H}_{\text{Rotation}}|\Psi_{\text{Rotation}}\rangle = E_{\text{Rotation}}|\Psi_{\text{Rotation}}\rangle \tag{1.20}$$

$$\hat{H}_{\text{Translation}}|\Psi_{\text{Translation}}\rangle = E_{\text{Translation}}|\Psi_{\text{Translation}}\rangle$$

Adding in the separation of the electronic problem, as discussed above, we can summarise the outcome of applying the Born-Oppenheimer approximation to eqn (1.3) as yielding the energy, for all internal and external motions of a molecule, which can be written as

$$E = E_{\text{Electronic}} + E_{\text{Vibration}} + E_{\text{Rotation}} + E_{\text{Translation}} \tag{1.21}$$

In all that follows, unless we explicitly state otherwise, we shall be concerned with the electronic problem and the energy, E_{Total}, as given in eqn (1.17). Note that E_{Total} depends on atom type through the nuclear charge Z, but is independent of nuclear mass. In referring to the electronic hamiltonian and wavefunction we shall now omit the subscript "Electronic" to simplify the notation.

1.2.2　Electronic Wavefunctions and the Antisymmetry Principle

The next key idea necessary to develop our approach is based on experimental evidence from the electronic spectroscopy of atoms, which led Wolfgang Pauli in 1925 to introduce the concept of spin for all fundamental particles. Spin is quantified by the spin quantum number, s. For example, electrons and protons are particles with spin value, $s = \frac{1}{2}$. Non-integer values of s characterise particles known as "fermions". Helium nuclei have, $s = 0$, and the photon, $s = 1$. Particles with integer values of s are called "bosons". Returning to the case of electrons with $s = \frac{1}{2}$, and introducing a magnetic field, the electron can align itself either parallel or antiparallel to the field. These orientations are described by the spin magnetic quantum number, $m_s = +\frac{1}{2}$ (parallel) and $m_s = -\frac{1}{2}$ (antiparallel). To satisfy the known physical evidence of electronic states in atoms Pauli formulated, as an independent postulate of quantum mechanics, the antisymmetry principle which states that:

"The wavefunction describing any state of an N-electron system must be antisymmetric under any permutation of the electronic coordinates."

The consequence of this principle is that no two identical fermions (in our case electrons) can occupy the same quantum state simultaneously. More familiarly, for electrons in atoms, we encounter the Pauli exclusion principle as the requirement that no two electrons can have the same four quantum numbers. If the principal quantum number, n, the angular quantum number, l, and the magnetic quantum number, m_l, are the same for two electrons, then the spin magnetic quantum number, m_s, must be different. The outcome of this is the requirement that if two electrons enter the same atomic orbital their spins must be antiparallel. The antisymmetry principle is a generalisation of this requirement that extends it to any number of electrons in any type of system.

We have referred to electrons entering orbitals in an atom and at this stage we must elaborate a little on what we mean by an orbital. An orbital is a solution to a one-electron Schrödinger equation. For example, s, p, d, f ... orbitals are all solutions of the Schrödinger equation for the hydrogen atom. The electronic hamiltonian given in eqn (1.14) and the electronic Schrödinger equation given eqn (1.16), contain no reference to spin. To accommodate the rôle of spin we must distinguish between spatial orbitals, $\phi(\mathbf{r})$, which depend solely on spatial (positional) coordinates $\mathbf{r}$, and spin-orbitals, $\phi(\mathbf{x})$, in which the variable $\mathbf{x}$ includes the space ($\mathbf{r}$) and spin ($\mathbf{s}$) variables, $\mathbf{x} = \{\mathbf{r}, \mathbf{s}\}$. The spin variable has associated with it the spin operators $\hat{s}_x, \hat{s}_y, \hat{s}_z$. The hamiltonian given in eqn (1.13) works on functions of $\mathbf{r}$, and the spin operators work on functions of $\mathbf{s}$. Writing the spin-orbital $\phi(\mathbf{x})$ as a product of space and spin-functions

$$\phi(\mathbf{x}) = \phi(\mathbf{r})\omega(\mathbf{s}) \tag{1.22}$$

we find that $\phi(\mathbf{x})$ is simultaneously a solution of (in atomic units)

$$\hat{s}_z\,\phi(\mathbf{x}) = m_s\,\phi(\mathbf{x})$$
$$\hat{h}\,\phi(\mathbf{x}) = \varepsilon\,\phi(\mathbf{x}) \tag{1.23}$$

In the second equation above, $\hat{h}$ refers to a one-electron hamiltonian operator. Observation, in the form of the Stern-Gerlach experiment of 1921, showed that $\hat{s}_z\omega(\mathbf{s}) = m_s\omega(\mathbf{s})$ has only two solutions, which are usually denoted α and β, or $\uparrow$ and $\downarrow$, respectively:

$$\hat{s}_z\,\alpha(\mathbf{s}) = +\frac{1}{2}\alpha(\mathbf{s})$$
$$\hat{s}_z\,\beta(\mathbf{s}) = -\frac{1}{2}\beta(\mathbf{s}) \tag{1.24}$$

Hence each spatial orbital $\phi(\mathbf{r})$ yields two possible spin-orbitals

$$\phi(\mathbf{x}) = \phi(\mathbf{r})\alpha(\mathbf{s})$$
$$\bar{\phi}(\mathbf{x}) = \phi(\mathbf{r})\beta(\mathbf{s}) \tag{1.25}$$

The antisymmetry principle stated above, refers to the interchange of the coordinates, $\mathbf{x_1}$ and $\mathbf{x_2}$, of two electrons, that is the wavefunction must be antisymmetric under the permutation of both space and spin coordinates.

There are many ways to construct wavefunctions which satisfy the antisymmetry principle. Here we shall consider only one simple scheme. The electronic wavefunction, $|\Psi(\mathbf{x}_1,\mathbf{x}_2,\mathbf{x}_3\cdots\mathbf{x}_N)\rangle$, for an N-electron atom or molecule is written as a determinant of N spin-orbitals, called a "Slater determinant":

$$|\Psi(\mathbf{x}_1,\mathbf{x}_2,\mathbf{x}_3...\mathbf{x}_N)\rangle = \frac{1}{\sqrt{N!}}\begin{vmatrix} \phi_i(\mathbf{x}_1) & \phi_j(\mathbf{x}_1) & \cdots & \phi_k(\mathbf{x}_1) \\ \phi_i(\mathbf{x}_2) & \phi_j(\mathbf{x}_2) & \cdots & \phi_k(\mathbf{x}_2) \\ \vdots & \vdots & \vdots & \vdots \\ \phi_i(\mathbf{x}_N) & \phi_j(\mathbf{x}_N) & \cdots & \phi_k(\mathbf{x}_N) \end{vmatrix} \tag{1.26}$$

The $\dfrac{1}{\sqrt{N!}}$ term is a normalisation factor, since expanding an $N \times N$ determinant will produce $N!$ terms, see Appendix 1B. The Slater determinant is completely specified by the spin-orbitals from which it is constructed. We note the following properties of this form for $|\Psi(\mathbf{x}_1,\mathbf{x}_2,\mathbf{x}_3\cdots\mathbf{x}_N)\rangle$:

(a) N electrons are distributed in N spin–orbitals, but with no specification of which electron is associated with which orbital. Electrons are physically indistinguishable particles and here they are associated equally with all orbitals in the Slater determinant.

(b) It is a property of determinants that interchanging two rows or columns changes the sign of the determinant (Appendix 1B). Swapping two rows in eqn (1.26) corresponds to swapping the coordinates of two electrons, and this is reflected in a change of sign in $|\Psi(\mathbf{x}_1,\mathbf{x}_2,\mathbf{x}_3\cdots\mathbf{x}_N)\rangle$ satisfying the antisymmetry requirement.

(c) If two electrons occupy the same spin-orbital, that is possess the same $\mathbf{x}$ coordinates, then two rows of $|\Psi(\mathbf{x}_1,\mathbf{x}_2,\mathbf{x}_3\cdots\mathbf{x}_N)\rangle$ will be equivalent. Any determinant with two equivalent rows, or two equivalent columns, when evaluated will yield zero. Hence $|\Psi(\mathbf{x}_1,\mathbf{x}_2,\mathbf{x}_3\cdots\mathbf{x}_N)\rangle$ vanishes and is not allowed. This satisfies the Pauli exclusion principle.

For many electronically simple atoms and molecules we can adopt a single Slater determinant as a satisfactory first approximation to the electronic wavefunction. Later on we shall see that a single Slater determinant is also a satisfactory starting point for developing the exact wavefunction. It is often useful to adopt a shorthand notation for the Slater determinant in eqn (1.26) in which only the diagonal elements are specified and the electron order is implicit

$$|\Psi(\mathbf{x}_1,\mathbf{x}_2,\mathbf{x}_3\cdots\mathbf{x}_N)\rangle = |\phi_i(\mathbf{x}_1)\phi_j(\mathbf{x}_2)\ldots\phi_k(\mathbf{x}_N)\rangle = |\phi_i\phi_j\ldots\phi_k\rangle \qquad (1.27)$$

1.2.3 Molecular Orbitals and Basis Set Expansions

At this stage we have a form for $\hat{H}$ given in eqn (1.14) and a form for $|\Psi\rangle$ given in eqn (1.26). To proceed further we must add a little more detail to the specification of $|\Psi\rangle$. We have stated that the Slater determinant is completely specified by the spin-orbitals, $\phi(\mathbf{x})$, from which it is built. Some functional form must be chosen for the set of atomic or molecular orbitals $\{\phi\}$ that enter eqn (1.26). The usual choice is to approximate each $\phi(\mathbf{r})$ as a linear combination of atomic orbitals (LCAO), with the atomic orbitals being located on the nuclei. In practice, each spatial orbital, $\phi(\mathbf{r})$, is written as

$$\phi_i(\mathbf{r}) = \sum_{\mu=1}^{m} c_{\mu i}\chi_\mu(\mathbf{r}) \qquad (1.28)$$

where the set $\{\chi\}$ is a set of m 'atomic basis functions' usually referred to as a basis set. The $c_{\mu i}$ are the mixing coefficients of the LCAO, and they are determined by minimising the energy. The process of optimising the $c_{\mu i}$ is called the "self-consistent field" (SCF) method. We shall say much more about the SCF process in chapter 2.

In principle, the set of functions $\{\chi\}$ can be of any type, provided that they are able to represent the molecular orbitals accurately and efficiently. If a great many functions are required to obtain accuracy, then the basis set will incur a computational cost that makes it inefficient. Equally, if a particular type of basis function leads to very compact basis sets, but the resultant integrals are computationally demanding, again this will be inefficient. As is often the case in computational quantum chemistry it is necessary to find a balance between accuracy and efficiency. The use of true atomic orbitals, solutions of the Schrödinger equation for the hydrogen atom, is accurate but inefficient. In analogy with atomic orbitals it is possible to use Slater basis functions that have a radial dependence of $e^{-\zeta r}$, in which r is the distance from the nucleus and ζ is an exponent that determines the radial distribution. Large values of ζ contract the function nearer to the nucleus, whereas small values of ζ expand the function away from the nucleus. Slater orbitals have been found to be computationally less efficient than the gaussian type orbitals with a radial dependence of $e^{-\alpha r^2}$, where α is the exponent of the gaussian type orbital. This subtle difference in the radial form has a very significant effect on the speed with which integrals may be evaluated in molecular systems. While both Slater and gaussian type orbitals are used, the gaussian type orbitals are used much more widely and dominate in most quantum chemical applications and developments. These two types of function show very different radial behaviour, see Figure 1.3. The radial dependence of the Slater type orbital possesses a cusp at the nucleus: an infinite gradient. In contrast,

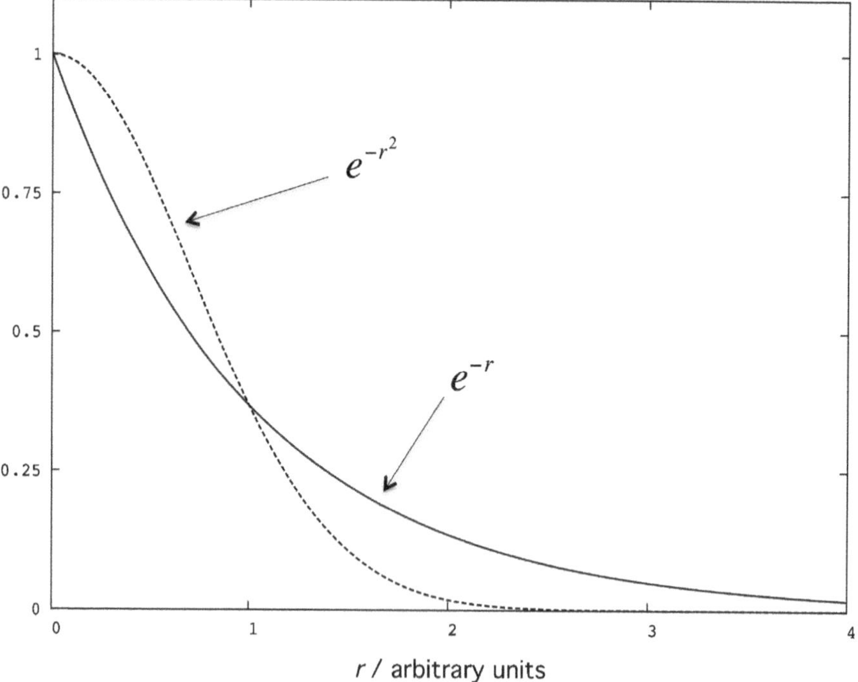

Figure 1.3 Comparison of the radial dependence of Slater type (solid line) and gaussian type orbitals (dashed line). Nucleus is located at the origin.

the gaussian type orbital has a zero gradient at the nucleus and decays more rapidly at larger distances from the nucleus. Nevertheless it turns out that the gaussian type orbitals are so much more efficient, computationally, that it is preferable to use larger sets of gaussian type orbitals than to deal with more compact Slater type orbital sets.

In all that follows we shall only concern ourselves with gaussian type orbitals. The choice of basis set, $\{\chi\}$, is critical to the success of any proposed calculation and we shall return to this subject in detail in Chapter 2.

The three approximations we have discussed so far: the Born-Oppenheimer approximation; representing the electronic wavefunction by a single Slater determinant; and the expansion of orbitals in atomic basis functions, constitute the most widely adopted approaches in computational quantum chemistry. While each approximation lends itself to improvement and refinements, a great deal of chemistry can be studied, with some reliability, using these simple ideas.

1.3 Chemistry *in Silico*: Where Do You Start?

From what has been presented so far we can assume that we are able to compute the potential energy of a molecule for any choice of the coordinates (geometry) of the constituent atoms. How does this enable us to study chemical problems? The discussion of chemical processes is often formulated in terms of the properties of reactants, products and transition states, from which information about thermochemistry, rates of reaction and molecular structure may be derived. By defining the notions of reactant, product, transition state and thermodynamic functions in terms of quantities that we can compute, we can begin to formulate a complete computational quantum chemistry.

1.3.1 Potential Energy Curves, Forces and Force Constants

Let us return to the familiar potential energy curve for a diatomic molecule, Figure 1.4, we have only one coordinate (bond distance) to consider and associate the minimum energy point on the curve with the equilibrium bond length, R_e. The minimum energy point is characterised by being a stationary point on the potential energy curve. A stationary point is defined by a zero first derivative:

$$\frac{dE(R)}{dR} = 0 \tag{1.29}$$

Physically, this quantity can be related to the force acting on the atoms

$$force = -\frac{dE(R)}{dR} \tag{1.30}$$

The negative sign shows that the force acts in the direction that will lower the

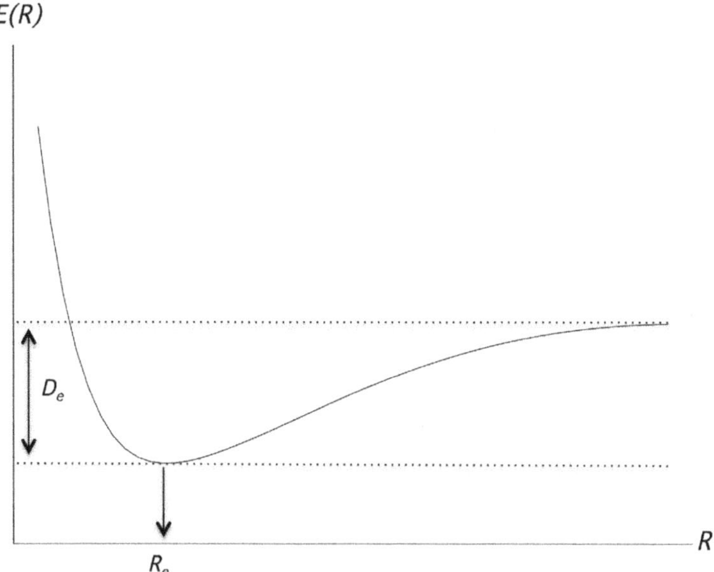

Figure 1.4   Using a potential energy curve to find the equilibrium bond length, R_e, and the bond dissociation energy, D_e.

potential energy. So any point beyond R_e has a positive slope, $dE(R)/dR$, and the force is negative. Beyond R_e, the bond distance is greater than at equilibrium and the force acts in the direction that would lower the potential energy, corresponding to contraction of the bond distance. At distances less than R_e, the slope of the curve is negative and the force acts in the direction of bond elongation.

We can also obtain the bond dissociation energy (in the absence of vibrational zero point energy) as the difference between the minimum energy point and the energy of the potential energy curve when the bond distance is large enough that any further change does not appreciably affect the potential energy (this is the plateau region on the far right hand side of the curve). This gives us two pieces of chemically useful information: the bond length and the bond strength. Another useful piece of information we can obtain is the vibrational wavenumber (frequency), which for a diatomic molecule is given by

$$\tilde{v} = \frac{1}{2\pi c}\sqrt{\frac{k}{\mu}} \tag{1.31}$$

where c is the speed of light, μ is the reduced mass which for a diatomic molecule A–B is written in terms of the constituent atomic masses as

$$\mu = \frac{M_A M_B}{M_A + M_B} \tag{1.32}$$

In eqn (1.31), k is the force constant which is the second derivative of the potential energy with respect to the bond distance, R, evaluated at the minimum energy point, R_e,

$$k = \left(\frac{d^2 E(R)}{dR^2}\right)_{R_e} \tag{1.33}$$

1.3.2 Potential Energy Surfaces, Stationary Points and Reactivity

How do these ideas change when we move to molecules with more than one geometric coordinate? Consider the water molecule and its potential energy as a function of the valence angle, θ, and the symmetric stretching of the O–H bonds, R, as shown in Figure 1.5. We can now construct the potential energy as a function of both coordinates, $E(R,\theta)$. The two geometric coordinates and the energy produce a three dimensional surface shown in Figure 1.6. At the minimum energy point we can read off the equilibrium values of the two geometric coordinates. As in the case of the diatomic potential energy curve, Figure 1.4, the minimum energy point is characterised as a stationary point which implies

$$\frac{dE(R,\theta)}{dR} = 0$$
$$\frac{dE(R,\theta)}{d\theta} = 0 \tag{1.34}$$

The force is no longer a single number but a vector, $-\mathbf{g}$, containing two components:

$$\mathbf{g} = \begin{pmatrix} \dfrac{dE(R,\theta)}{dR} \\ \dfrac{dE(R,\theta)}{d\theta} \end{pmatrix} \tag{1.35}$$

Accordingly, the second derivative is no longer a single number but a 2×2

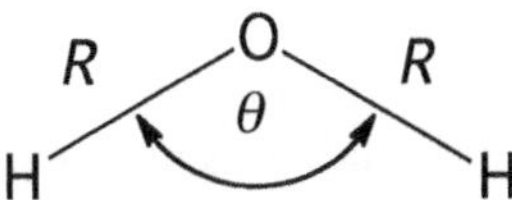

Figure 1.5 Coordinates for the symmetric stretching and bending of water.

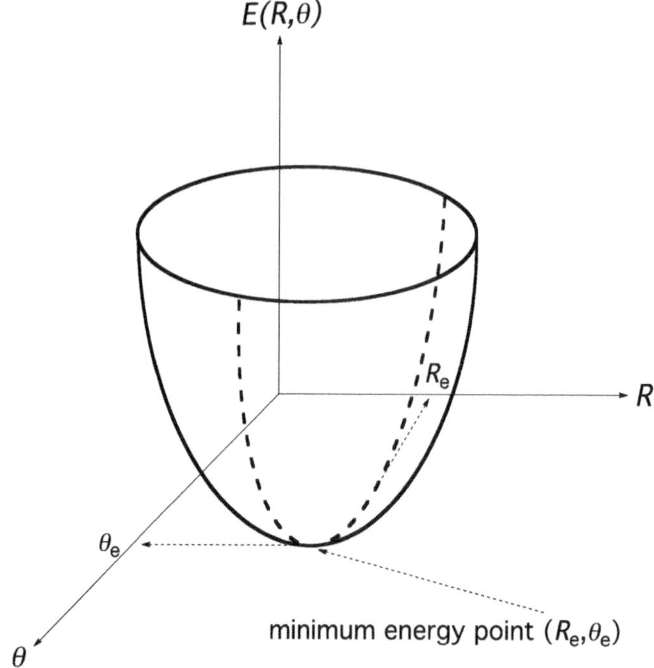

Figure 1.6 Schematic three-dimensional potential energy surface for water, the coordinates correspond to those in **Figure 1.5**.

matrix, **F**:

$$\mathbf{F} = \begin{pmatrix} \dfrac{d^2 E(R,\theta)}{dR^2} & \dfrac{d^2 E(R,\theta)}{dRd\theta} \\[2ex] \dfrac{d^2 E(R,\theta)}{d\theta dR} & \dfrac{d^2 E(R,\theta)}{d\theta^2} \end{pmatrix} \tag{1.36}$$

If we had considered all the possible internal motions of the water molecule we would need to allow the two O–H bonds to move independently, each having its own internal coordinate. The potential energy would then depend on three coordinates, for example $E(R_1,R_2,\theta)$, yielding a four dimensional *hypersurface*. The gradient vector, **g**, would now contain three terms and the force constant matrix, **F**, would be a 3×3 matrix. Beyond three dimensions we should formally refer to a *hypersurface* but, in practice, potential energy functions all tend to be referred to as *"potential energy surfaces."*

In general, for a molecule consisting of N_{atoms} atoms there will be $3N_{\text{atoms}} - 6$ independent geometric variables (bond lengths, bond angles and torsional angles). The set of independent variables is not unique. For example, the water molecule can be represented equally by either of the two sets of variables shown in Figure 1.7.

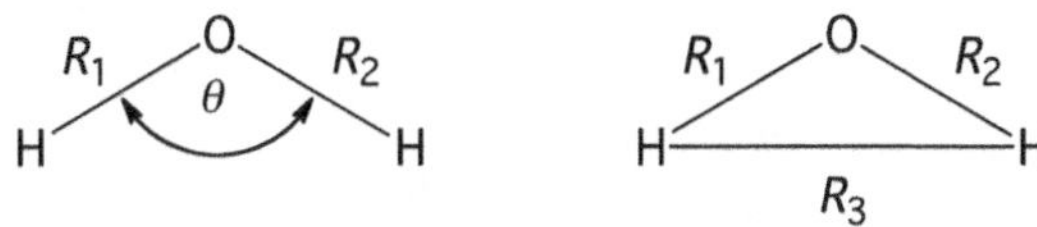

Figure 1.7   Two different sets of $3N_{atoms} - 6 = 3$ variables for describing the water molecule.

The sum, $3N_{atoms} - 6$, specifies the number of internal motions that a molecule may possess. If we consider the general set of cartesian coordinates, $\{x,y,z\}$, then each atom in a molecule may be specified by three such cartesian coordinates relative to an origin. This provides a total of $3N_{atoms}$ cartesian coordinates. This coordinate set describes not only the internal motions, one atom moving relative to another, but also the external motions. The external motions are the global translations and rotations of the molecule. We can translate an isolated molecule as a whole along any of the cartesian axes without changing the energy. To see this consider that in eqn (1.13) all the terms depend only on the relative position of one atom to another. Equally we can rotate an isolated molecule about any of the cartesian axes, again without changing the energy. Hence our set of coordinates should include $3N_{atoms}$, -3 coordinates for global translations and -3 coordinates for global rotations giving a total of $3N_{atoms} - 6$. In the case of a linear molecule there are only two axes for rotation, since rotation along the bond axis does not displace the molecule in space. Hence for linear molecules the number of internal coordinates is $3N_{atoms} - 5$.

A stationary point is defined for a general potential energy surface by

$$\frac{dE(\mathbf{R})}{dR_i} = 0 \quad i = 1,2, \cdots 3N_{atoms} - 6 \tag{1.37}$$

where $\mathbf{R}$ refers to the set of all nuclear coordinates and R_i to a specific member of the set. So provided we can locate a geometry that satisfies eqn (1.37), then we can find a minimum energy configuration for any molecular system. How we go about locating stationary points on surfaces of arbitrary complexity we shall discuss in chapter 3. Eqn (1.37) is a necessary condition for finding a minimum energy configuration, but it is not a sufficient condition. When we deal with chemical reactions, we are no longer just interested in minimum energy configurations. Let us continue with the simple example of water in the set of two coordinates shown in Figure 1.5 and consider the inversion of the angle as an example of a very simple reaction, as shown in Figure 1.8. We know that the minimum energy structure of water has a bond angle of $\theta_e = 104.5°$, and this will be the value assumed for this variable in the reactant

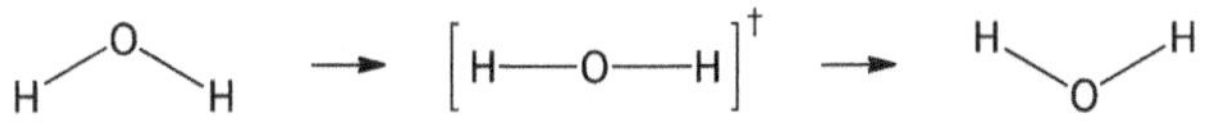

Figure 1.8   Inversion of the bond angle in water as an example of a chemical reaction.

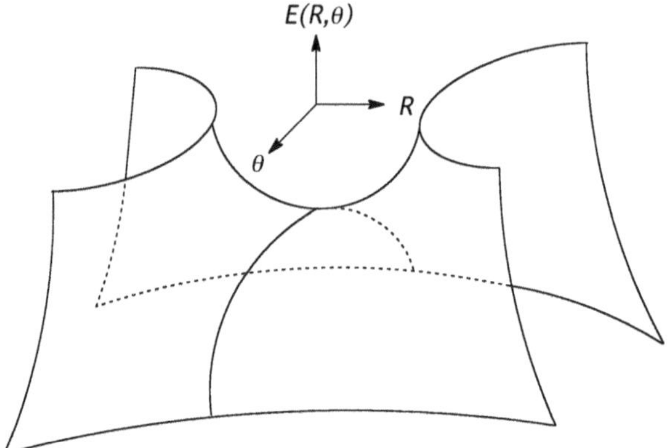

Figure 1.9 Saddle point structure around the transition state corresponding to the reaction shown in **Figure 1.8**.

and (equivalent) product shown in Figure 1.8. The linear structure will correspond to a stationary point, but it is a transition structure rather than a minimum on the potential energy surface. The potential energy surface around the linear structure is shown schematically in Figure 1.9 and corresponds to a saddle point structure. In the stationary saddle point structure one unique coordinate lies at a maximum, and all other coordinates (only one in this example) correspond to minima. Along the R coordinate the energy contours are those of a minimum, $d^2E(R,\theta)/dR^2 > 0$. Conversely, the energy contours along the θ coordinate correspond to a maximum, $d^2E(R,\theta)/d\theta^2 < 0$. If we plot the potential energy as a function of θ (at the corresponding minimum value along the R coordinate) we obtain the reaction energy profile shown in Figure 1.10. From the shape of the energy profile we can associate the coordinate, θ, with the reaction coordinate. The difference in energy between $\theta = \theta_e$ and $\theta = 180°$ gives the activation barrier for this process. This is a

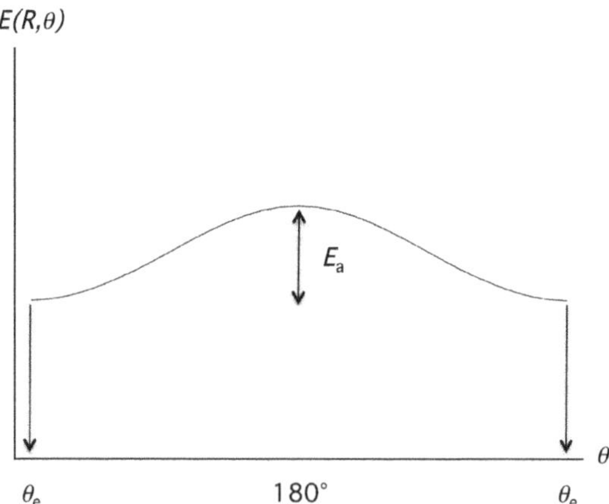

Figure 1.10 Energy profile for the reaction shown in **Figure 1.8**.

particularly simple example in which the reaction coordinate can be described in terms of just one internal coordinate, θ. In most reactions, the reaction coordinate will depend on many internal coordinates. In fact the reaction coordinate is, in principle, a function of all $3N_{\text{atoms}} - 6$ internal coordinates.

We have seen how a stationary point may correspond to a minimum or a saddle point transition structure. What distinguishes them is the curvature of the potential energy surface around the stationary point. So in general when studying chemical processes of arbitrary complexity we must first of all find stationary points that satisfy eqn (1.37) and then characterise these stationary points by computing the force constant (second derivative) matrix. The second derivatives contain the information about the curvature of the surface. Having computed the force constant matrix (eqn (1.36)), we must go one step further and transform it to diagonal form:

$$\mathbf{Q}^{-1}\mathbf{F}\mathbf{Q} = \begin{pmatrix} \dfrac{d^2 E(R,\theta)}{dq_1^2} & 0 \\ 0 & \dfrac{d^2 E(R,\theta)}{dq_2^2} \end{pmatrix} \tag{1.38}$$

Given the matrix $\mathbf{F}$, the matrix $\mathbf{Q}$ can be found using standard numerical methods, see Appendix 1C. The columns of $\mathbf{Q}$ give linear combinations of the initial set of variables $\{R,\theta\}$. These linear combinations are called normal coordinates, and denoted $\{q_1,q_2\}$, and can be written as

$$q_1 = Q_{11}R + Q_{21}\theta$$
$$q_2 = Q_{12}R + Q_{22}\theta \tag{1.39}$$

where Q_{ij} refers to the elements of the transformation matrix $\mathbf{Q}$. For the simple case considered, q_1 and q_2 will correspond to symmetric (positive) and antisymmetric (negative) combinations of R and θ. Each normal coordinate will be dominated by either R or θ, so $Q_{11} >> Q_{21}$ and $Q_{22} >> Q_{12}$ in eqn (1.39), since these two coordinates do not couple very strongly. By contrast, in a reaction such as the collinear exchange shown in Figure 1.11, the reaction proceeds via the antisymmetric combination of R_1 and R_2. The transition structure being formed from the reactants by the shortening of R_1 and the lengthening of R_2,

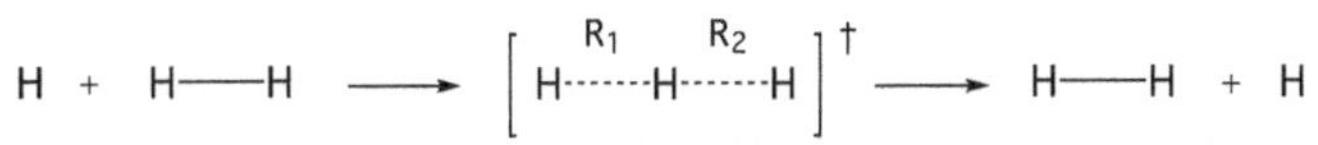

$$\text{H} + \text{H}\!\!-\!\!\text{H} \longrightarrow \left[\overset{R_1 \quad R_2}{\text{H}\cdots\text{H}\cdots\text{H}} \right]^{\dagger} \longrightarrow \text{H}\!\!-\!\!\text{H} + \text{H}$$

Figure 1.11   Collinear exchange reaction involving two coordinates R_1 and R_2.

$$q_1 = Q_{11}R_1 + Q_{21}R_2$$
$$q_2 = Q_{12}R_1 + Q_{22}R_2$$

$$(1.40)$$

At the transition structure geometry, the normal coordinate corresponding to the maximum in the energy profile will be an equal mixture of R_1 and R_2 ($Q_{11} = -Q_{21}$), while on the reactant side the reaction coordinate will be dominated by R_2 ($Q_{21} > Q_{11}$) and on the product side by R_1 ($Q_{11} > Q_{21}$). So we can see that the reaction coordinate is not a fixed linear combination of internal coordinates but changes as a reaction proceeds.

Chemical reactants, products and stable intermediates are characterised as minima and must have

$$\frac{d^2E(\mathbf{R})}{dq_i^2} > 0 \quad i = 1, 2, \cdots 3N_{atoms} - 6 \qquad (1.41)$$

For transition structures there will be one unique normal coordinate which corresponds to the reaction coordinate. Hence a transition staructure is characterised by

$$\frac{d^2E(\mathbf{R})}{dq_i^2} > 0 \quad i = 1, 2, \cdots 3N_{atoms} - 7 \qquad (1.42)$$

and one unique coordinate, q_{RC}, corresponding to the reaction coordinate for which

$$\frac{d^2E(\mathbf{R})}{dq_{RC}^2} < 0 \qquad (1.43)$$

Stationary points with more than one coordinate, or direction, in which the potential energy surface is a maximum are not of any chemical significance since there will always be a path of lower energy by which a reaction may proceed.

The conditions given in eqn (1.37), (1.41), (1.42) and (1.43) are sufficient to determine whether a stationary point refers to a stable structure or a transition structure, but they cannot tell us how the stationary point relates to the global structure of the potential energy surface. To illustrate this, consider the torsional conformations of 1,2-dichloroethane shown in Figure 1.12. The torsion angle, τ, is taken as the angle between the two C–Cl bonds looking along the C–C bond. The lowest energy conformation corresponds to $\tau = 180°$ and is indicated in Figure 1.12 as **I**. Each $60°$ rotation in τ leads from conformation **I** to another stationary point. The first of these corresponds to an eclipsed conformation, **II**. **II** is a transition structure connecting conformation **I** to conformation **III**. **III** is a minimum energy structure but lies higher in energy than conformation **I**. **III** is called a *local minimum*, while **I**

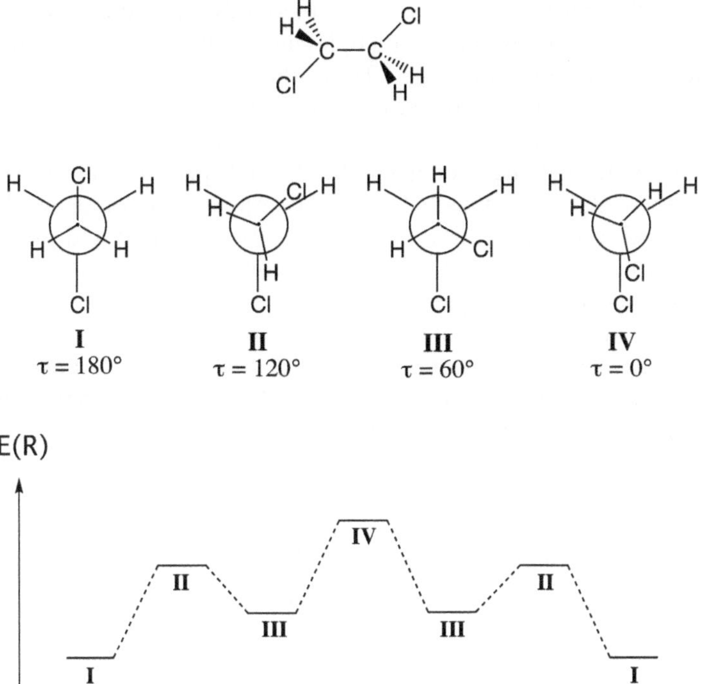

Figure 1.12   Conformations and torsional energies of 1,2-dichloroethane.

is the *global minimum*. Rotation through another 60° from **III** takes us to transition structure **IV**. Both **II** and **IV** are characterised by a maximum in the potential energy along the τ coordinate and will satisfy the conditions in eqns (1.42) and (1.43). There are two rotational barriers, the first corresponding to $E(\mathbf{II}) - E(\mathbf{I})$ and the second to $E(\mathbf{IV}) - E(\mathbf{III})$. At non-zero temperatures, the equilibrium distribution of 1,2-dichloroethane will consist of conformers **I** and **III**. The relative fractions of each component will depend on the energy difference, $E(\mathbf{III}) - E(\mathbf{I})$, as given by the Boltzmann distribution.

1.3.3   Linking the Electronic Energy with Thermodynamic State Functions

So now, in principle, we can study systems of arbitrary complexity by applying these simple ideas. We must use $dE(R)/dR_i$ and $d^2E(R)/dq_i^2$ to enable us to think about chemical processes on high-dimensional potential energy surfaces. The difference in energy between different stationary points provides us with energies of reaction or activation energies. Activation energies govern the kinetics of all activated processes, while energies of reaction determine reaction equilibrium and other thermochemical features of a reaction. The energies we have discussed so far refer to an isolated molecule at the absolute zero of temperature. We must link these with the thermodynamic quantities familiar from experiments, for example enthalpies and Gibbs energies, which are usually defined for molar quantities. To do this we use statistical mechanical expressions to relate our computed internal energies at 0 K, $U(0)_{\text{Electronic}}$, to enthalpies or free energies at non-zero temperatures, $H(T)$ and $G(T)$, respectively. Note that the experimentally derived quantities are usually the *changes* in these state functions, ΔU, ΔH, ΔG. The computations we describe will yield the absolute quantities, U, H, G, for each reactant and product we consider, and from these ΔU, ΔH, ΔG are obtained.

Quantum chemical calculations provide us with $E(\mathbf{R}) = U(0)_{\text{Electronic}}$, we now need to relate this to the internal energy at some (non-zero) temperature, T. To proceed, using the machinery of statistical mechanics, we assume that the different energetic degrees of freedom (electronic, vibration, rotation, translation) do not couple and we can write

$$U(T) = U(T)_{\text{Translation}} + U(T)_{\text{Rotation}} + U(T)_{\text{Vibration}} + U(T)_{\text{Electronic}} \qquad (1.44)$$

We now consider each term independently. In molar units we find:

$$U(T)_{\text{Translation}} = \frac{3}{2}RT \qquad (1.45)$$

$$U(T)_{\text{Rotation}} = \frac{3}{2}RT \text{ (for a non-linear molecule)} \qquad (1.46a)$$

$$U(T)_{\text{Rotation}} = RT \text{ (for a linear molecule)} \qquad (1.46b)$$

$$U(T)_{\text{Vibration}} = U(0)_{\text{Vibration}} + N_A \sum_{i}^{3N_{\text{atoms}}-6} \frac{h\nu_i}{e^{h\nu_i/k_B T} - 1} \qquad (1.47)$$

The vibrational term at 0 K is the quantum mechanical zero-point energy and is given by:

$$U(0)_{\text{Vibration}} = \frac{1}{2}N_A \sum_{i}^{3N_{\text{atoms}}-6} h\nu_i \qquad (1.48)$$

In eqns (1.45) – (1.48), R is the gas constant, N_A is Avogadro's number, k_B is Boltzmann's constant, h is Planck's constant and ν_i is the ith vibrational wavenumber (frequency).

The final term to consider is $U(T)_{\text{Electronic}}$. This is usually taken as the computed potential energy, $U(T)_{\text{Electronic}} \approx U(0)_{\text{Electronic}} = E(\mathbf{R})$, implying that there is no thermal correction to the electronic component. This is a reasonable assumption provided that the electronic levels in a molecule are energetically well separated.

The thermodynamic relationship between internal energy, U, and enthalpy, H, is

$$H = U + pV \qquad (1.49)$$

Quantum chemical calculations usually refer to the gas phase and for an ideal gas we know, $pV = nRT$ and can write (for molar quantities)

$$H(T) = U(T) + RT \qquad (1.50)$$

So given a calculation of $U(0)_{Electronic}$, it is a straightforward matter to obtain $H(T)$. It is often useful to discuss chemical processes in terms of the Gibbs energy, $G(T)$. To extend our treatment to cover this state function we need to know how to calculate the entropy, $S(T)$. At this stage, the reader should be aware that there are many equivalent ways of writing the statistical mechanical expressions for the entropy. We proceed as for the internal energy and write

$$S(T) = S(T)_{\text{Translation}} + S(T)_{\text{Rotation}} + S(T)_{\text{Vibration}} + S(T)_{\text{Electronic}} \qquad (1.51)$$

Taking each term in turn and again working in molar units, we obtain:

$$S(T)_{\text{Translation}} = R\left\{ \frac{5}{2} + \ln\left[\left(\frac{RT}{p}\right)\left(\frac{2\pi m k_B T}{h^2}\right)^{3/2} \right] \right\} \qquad (1.52)$$

In eqn (1.52), m is the molecular mass and in the SI system of units should be expressed in kg. To evaluate the rotational entropy for a non-linear molecule we need to know the three principal moments of inertia, I_A, I_B, I_C, (obtained by diagonalising the inertia tensor, see Appendix 1D) and the rotational symmetry number, σ. With these quantities at hand we can evaluate $S(T)_{\text{Rotation}}$ for a non-linear molecule as

$$S(T)_{\text{Rotation}} = R\left\{ \frac{3}{2} + \ln\left[\left(\frac{\sqrt{\pi}}{\sigma}\right)\left(\frac{8\pi^2 k_B T}{h^2}\right)^{3/2} (I_A I_B I_C)^{1/2} \right] \right\} \qquad (1.53a)$$

For a linear molecule, the moment of inertia along the bond axis will be zero and the moments of inertia in the two perpendicular directions will be equivalent. Denoting the moment of inertia as I, the rotational entropy for a linear molecule takes the form

$$S(T)_{\text{Rotation}} = R\left\{ 1 + \ln\left[\left(\frac{8\pi^2 I k_B T}{\sigma h^2}\right)^{3/2} \right] \right\} \qquad (1.53b)$$

The final two contributions are given in eqns (1.54) and (1.55).

$$S(T)_{\text{Vibration}} = R \sum_{i}^{3N_{\text{atoms}}-6} \left[\frac{h\nu_i}{k_B T} \frac{1}{e^{h\nu_i/k_B T}-1} - \ln\left(1 - e^{-h\nu_i/k_B T}\right) \right] \qquad (1.54)$$

$$S(T)_{\text{Electronic}} = R\ln(g_0) \qquad (1.55)$$

Some care must be exercised in using eqns (1.47) and (1.54). These are obtained within the vibrational harmonic-oscillator approximation. In some situations normal modes can correspond to low-frequency torsional motions, for example the torsional motion shown in Figure 1.12. The harmonic-oscillator approximation is not reliable here and simple application of eqns (1.47) and (1.54) can produce significant errors. Figure 1.13 shows a shallow anharmonic potential energy well (bold line) that deviates significantly from the harmonic form shown (light line). The first five vibrational energy levels are indicated for both potentials and it can be seen that, apart from the zero-point energy levels, the two potentials produce very different vibrational energy spacings. When the thermal energy corrections are of similar magnitude to the well depth, the molecule resides in the upper region of the potential energy well and this is the region that is poorly described by the harmonic-oscillator approximation. The general rule of thumb for the harmonic-oscillator approximation to be accurate is that the energy well depth should be $>10RT$. To put this into perspective, at 298 K, $RT = 2.5$ kJ mol^{-1}, while the rotational barrier in ethane is only about 12 kJ mol^{-1}. In such circumstances, one should ideally obtain the potential energy as a function of the normal mode and then solve the Schrödinger equation numerically for the vibrational energy levels of this

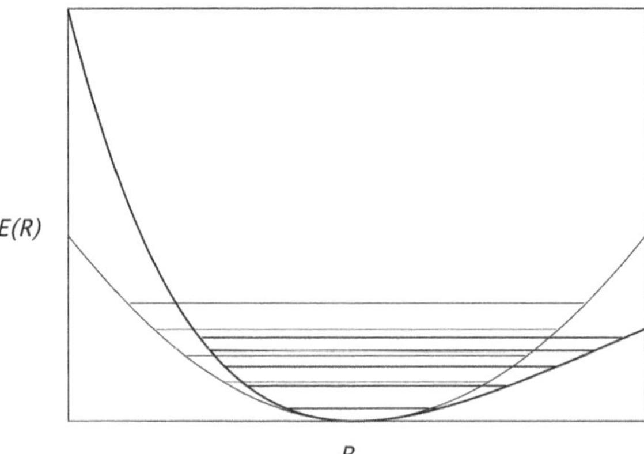

Figure 1.13   Harmonic (light line) and anharmonic (bold line) potential energy curves with vibrational energy levels superimposed.

potential. However, there are many other, simpler, approaches to this problem which provide good pragmatic accuracy.[3,4]

Knowing how to obtain H and S from our computations, it is an easy matter to evaluate the Gibbs energy, G, using the familiar thermodynamic relationship,

$$G = H - TS \qquad (1.56)$$

and hence many other chemically important quantities. For example, the equilibrium constant, K,

$$K = e^{-\Delta H/RT} e^{\Delta S/R} \qquad (1.57)$$

We now have a glimpse of how we can study geometry, vibrations and energetics using computational quantum chemistry. This already enables us to investigate a wealth of interesting chemical problems. For example, you may wish to calculate the geometric and electronic structure of a newly synthesised molecule, or predict the geometry and ground state structure of a yet to be synthesised molecule. Apart from the geometry and the electronic structure there are many other molecular properties that we shall want to be able to evaluate, for example the UV/Visible absorption spectrum, or the proton hyperfine coupling constant in a free radical. We shall discuss the calculation of molecular properties in more detail in chapter 3. Having touched on some of the molecular properties that may be obtained from quantum chemical computations we need to address the very large array of methods available for calculating $E(\mathbf{R})$. Understanding the strengths and weaknesses of commonly used quantum chemical procedures is key to successful scientific studies.

1.4   Standard Models of Electronic Structure

We have mentioned that computational quantum chemistry began in 1927 with the pioneering work of Heitler and London on the hydrogen molecule. In the intervening years a great number of techniques have emerged for dealing with the computation of electronic structure. The range of these techniques is vast and finding a suitable method for a given problem requires some careful consideration. In 1998, the Nobel prize in chemistry was awarded jointly to John Pople ("for his development of computational methods in quantum chemistry") and Walter Kohn ("for his development of the density functional theory"). Pople was one of the most important figures responsible for the widespread adoption of quantum chemical techniques. This he achieved through the development of widely available computer programs and also the development of theoretical models that can be successfully applied to chemical questions. In his Nobel lecture, Pople wrote:

"A theoretical model for any complex process is an approximate but well-defined mathematical procedure of simulation. When applied to chemistry, the task is to use input information about the number and character of component

particles (nuclei and electrons) to derive information and understanding of resultant molecular behaviour."

Here we shall discuss a number of these models by way of providing an overview. Specifically we shall look at the form of the wavefunction used to define these models and leave much of the detail to chapter 2.

1.4.1   The Hartree-Fock Model and Electron Correlation

At the simplest level we can start by taking a molecular wavefunction to be described by a single Slater determinant of the form given in eqn (1.26). The Slater determinant is composed of a set of molecular orbitals, and each molecular orbital is expanded in terms of a set of atomic basis functions, as shown in eqn (1.28). If we vary the expansion coefficients, $c_{\mu i}$, in eqn (1.28) so as to minimise the total molecular potential energy, we obtain a set of optimal molecular orbitals. This is the Hartree-Fock model and the orbital optimisation process is the SCF method alluded to in section 1.2.3. The resultant optimal molecular orbitals are one-electron wavefunctions, and the model treats the motion of the electrons as being largely independent of each other. Since electrons are negatively charged particles they repel each other. At any instant the repulsion between all the electrons in a molecule is dependent on their positions and a repulsive energy, $\dfrac{1}{r_{ij}}$, is experienced between all pairs of electrons, i and j. The Hartree-Fock model *cannot* describe this instantaneous repulsion between electrons of opposite spin. There are many ways to illustrate this and here we follow an elegant exposition due to Sinanoğlu.[5] For example, consider the two electrons of helium, denoted 1 and 2, which in the ground state reside in the $1s$ orbital. We familiarly write the electronic configuration as $1s^2$. The corresponding Slater determinant is

$$|\Psi(\mathbf{x}_1,\mathbf{x}_2)\rangle = \frac{1}{\sqrt{2}} \begin{vmatrix} \phi_{1s}(\mathbf{x}_1) & \bar{\phi}_{1s}(\mathbf{x}_1) \\ \phi_{1s}(\mathbf{x}_2) & \bar{\phi}_{1s}(\mathbf{x}_2) \end{vmatrix} \tag{1.58}$$

In this situation the two identically charged electrons have a coulombic repulsion inversely proportional to the distance between them. Consider the orbital, $\phi_{1s}(\mathbf{r})$, as centred at the nucleus and electron 1 located at its Bohr radius. Electron 2 now approaches electron 1 along a straight line and the inter-electronic coulombic repulsion, $\dfrac{1}{r_{12}}$, encloses electron 1 in a "coulomb hole" into which electron 2 cannot penetrate. The form of the coulomb hole is shown in Figure 1.14. Within the Hartree-Fock model, each electron experiences an averaged potential that arises from the electron density, $\phi_{1s}(\mathbf{r})^2$, of the other electron. This averaged potential gives rise to an inter-electronic repulsion of the form $\int \dfrac{\phi_{1s}(\mathbf{r}_2)^2}{r_{12}} d\mathbf{r}_2$. The orbitals and the electron

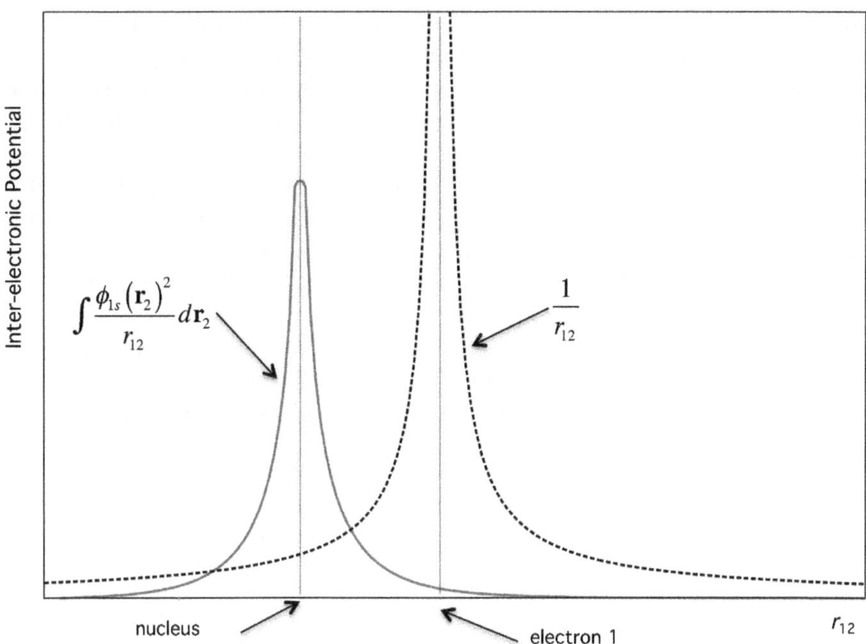

Figure 1.14   Inter-electronic coulombic potential (dashed line) around electron 1 and the averaged Hartree–Fock potential (solid line) experienced by an electron of opposite spin.

distribution are optimised in the SCF process under this averaged potential. In the averaged potential, the motion of electron 2 is not correctly correlated with that of electron 1 and the electrons are allowed to approach each other too closely. When the energy is evaluated, the repulsion energy is high and the overall energy is raised. The difference between the instantaneous and averaged forms of the inter-electronic repulsion is called the "electron correlation energy".

Expanding the form of the wavefunction in eqn (1.58), ignoring the normalisation factor, we can separate the space and spin components and write

$$\begin{aligned}
|\Psi(\mathbf{x}_1,\mathbf{x}_2)\rangle &= \phi_{1s}(\mathbf{x}_1)\bar{\phi}_{1s}(\mathbf{x}_2) - \bar{\phi}_{1s}(\mathbf{x}_1)\phi_{1s}(\mathbf{x}_2) \\
&= \phi_{1s}(\mathbf{r}_1)\phi_{1s}(\mathbf{r}_2)[\alpha(\mathbf{s}_1)\beta(\mathbf{s}_2) - \beta(\mathbf{s}_1)\alpha(\mathbf{s}_2)]
\end{aligned} \tag{1.59}$$

Considering the spatial component only, the two-electron density, $\rho(\mathbf{r}_1,\mathbf{r}_2) = ||\Psi(\mathbf{r}_1,\mathbf{r}_2)\rangle|^2$, gives the probability of finding an electron at $\mathbf{r}_2$ when one is known to be at $\mathbf{r}_1$. For the current case of helium

$$\rho(\mathbf{r}_1,\mathbf{r}_2) = ||\Psi(\mathbf{r}_1,\mathbf{r}_2)\rangle|^2 = \phi_{1s}(\mathbf{r}_1)^2\phi_{1s}(\mathbf{r}_2)^2 \tag{1.60}$$

The two-electron density is given by the product of the density of electron 1 situated at $\mathbf{r}_1$, and that of electron 2 situated at $\mathbf{r}_2$. There is nothing in this form to prevent the situation of having the two electrons at the same point, $\mathbf{r}_1 = \mathbf{r}_2$. Hence a Hartree-Fock wavefunction, for electrons of opposite spin, leads to an uncorrelated (independent electron) description of the two-electron density, $\rho(\mathbf{r}_1,\mathbf{r}_2)$. The importance of this type of electron correlation cannot be overstated, since its neglect leads to errors in calculations of a number of molecular properties: incorrect prediction of molecular geometries; incorrect electron distribution leading to incorrect electrostatic properties such as dipole moments; activation barriers which are too high, and overestimation of vibrational wavenumbers and bond dissociation energies.

For electrons of parallel spin, the Hartree–Fock model *does* provide a correlated description of the electron distribution. The lowest excited state of helium, in which the electrons have parallel spin, has the configuration $1s2s$. The corresponding Slater determinant is

$$|\Psi(\mathbf{x}_1,\mathbf{x}_2)\rangle = \frac{1}{\sqrt{2}}\begin{vmatrix} \phi_{1s}(\mathbf{x}_1) & \phi_{2s}(\mathbf{x}_1) \\ \phi_{1s}(\mathbf{x}_2) & \phi_{2s}(\mathbf{x}_2) \end{vmatrix} \tag{1.61}$$

Proceeding as before, ignoring normalisation, we can separate the space and spin components and write

$$\begin{aligned} |\Psi(\mathbf{x}_1,\mathbf{x}_2)\rangle &= \phi_{1s}(\mathbf{x}_1)\phi_{2s}(\mathbf{x}_2) - \phi_{2s}(\mathbf{x}_1)\phi_{1s}(\mathbf{x}_2) \\ &= [\phi_{1s}(\mathbf{r}_1)\phi_{2s}(\mathbf{r}_2) - \phi_{2s}(\mathbf{r}_1)\phi_{1s}(\mathbf{r}_2)]\alpha(\mathbf{s}_1)\alpha(\mathbf{s}_2) \end{aligned} \tag{1.62}$$

$\rho(\mathbf{r}_1,\mathbf{r}_2)$ now takes the form

$$\rho(\mathbf{r}_1,\mathbf{r}_2) = \phi_{1s}(\mathbf{r}_1)^2\phi_{2s}(\mathbf{r}_2)^2 - 2[\phi_{1s}(\mathbf{r}_1)\phi_{2s}(\mathbf{r}_2)\phi_{2s}(\mathbf{r}_1)\phi_{1s}(\mathbf{r}_2)] + \phi_{2s}(\mathbf{r}_1)^2\phi_{1s}(\mathbf{r}_2)^2 \tag{1.63}$$

Consider again the situation in which the two electrons coalesce at the same point, $\mathbf{r}_1 = \mathbf{r}_2 = \mathbf{r}$. In this case the first and third terms on the right hand side of eqn (1.63) are cancelled by the second term and $\rho(\mathbf{r},\mathbf{r}) = 0$. The two electrons cannot occupy the same point in space and so the electronic motion is correlated. This type of correlation, for electrons of the same spin, arises from the antisymmetric nature of the wavefunction and is termed "Fermi correlation". The $\rho(\mathbf{r},\mathbf{r}) = 0$ contour in the electron density defines the "Fermi hole".

For the reasons we have discussed, the Hartree-Fock energy, $E_{\mathrm{H-F}}$, is always higher than the exact energy obtainable from the Schrödinger equation, E_{Exact}, and this leads us to the most common definition of the correlation energy, E_{corr},

$$E_{\mathrm{corr}} = E_{\mathrm{Exact}} - E_{\mathrm{H-F}} \tag{1.64}$$

Since E_{Exact} is not generally within our reach, how do we proceed beyond the Hartree-Fock model?

We have introduced three principal assumptions: (i) the Born–Oppenheimer approximation (ii) the representation of the wavefunction by a single Slater determinant (iii) the expansion of the orbitals in a set of atomic functions. The orbitals, we have said, are one-electron wavefunctions arising from an approximate Schrödinger equation, namely the Hartree–Fock equation. The flexibility of the one-electron orbitals can be improved by enlarging the set of atomic functions from which they are built (eqn (1.28)). Eventually, any further expansion of the set of atomic functions will produce no useful improvement in the Hartree–Fock energy. At this point we have obtained the *Hartree-Fock limit*. Just as we can improve the one-electron orbitals by enlarging the number of functions used to expand them, we can also improve the many-electron wavefunction by mixing several Slater determinants together. The latter improves the description of the many-electron wavefunction. Hence the quality of our calculations will depend on the flexibility of the one-electron orbitals and also the form of the many-electron wavefunction. Using these two components, most computational models may be viewed in terms of a two–dimensional diagram, shown in Figure 1.15. The horizontal axis refers to the quality of the N-electron wavefunction. At the extreme left we have the case of a single Slater determinant, the Hartree-Fock model, and the accompanying lack of electron correlation. We can move to the right along the horizontal axis by combining more Slater determinants and so form a more accurate N-electron wavefunction. The vertical axis refers to the one-electron basis set used to expand the one-electron orbitals from which the Slater

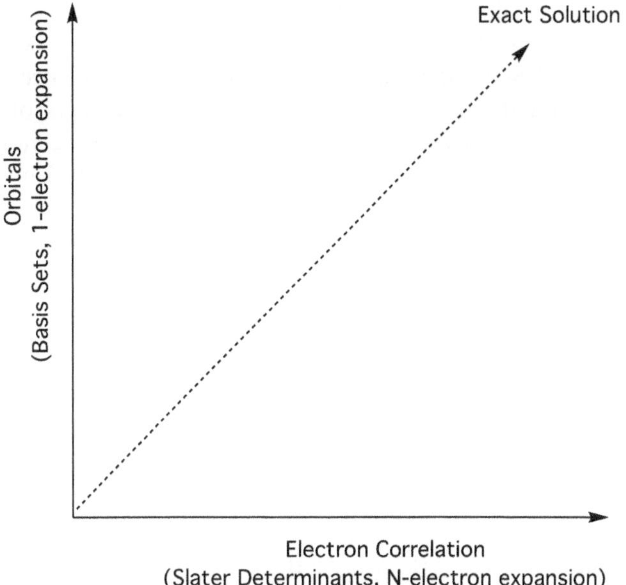

Figure 1.15   Systematic improvement of models (orbital sets and electron correlation) leads to the exact solution of the Schrödinger equation.

determinants are built. If we move along each axis until there is no useful change in the energy then we have effectively solved the electronic Schrödinger equation! Of course we are not able to do this for more than some very small molecules. However we are able to systematically improve the approximation to the Schrödinger equation, even if computational practicalities mean that we must terminate this process sooner than we would wish. In principle, any point on the diagram, specified by a coordinate along each axis, defines a computational "model". That is, a well-defined procedure by which the energy, and other properties of a molecule may be obtained.

The Hartree–Fock model produces a set of optimal occupied molecular orbitals, from which the Slater determinant describing the ground state is built. In addition a set of unoccupied, or virtual, orbitals is produced. The number of unoccupied orbitals is determined by the size of the one-electron atomic basis set used. If there are m atomic basis functions used in the LCAO expansion of the molecular orbitals, then there will m molecular orbitals produced. For example, a calculation on water might use a basis set consisting of a single $1s$ type basis function on each hydrogen atom and a $1s$, $2s$ and $2p$ ($2p_x$, $2p_y$, $2p_z$) set of basis functions on the oxygen atom. This gives a total of seven basis functions, and the Hartree–Fock procedure will mix these to produce seven molecular orbitals. Since water contains 10 electrons, only five of the orbitals will be doubly occupied and two will be unoccupied. Typically, the number of unoccupied orbitals is much larger than that of occupied orbitals if large flexible basis sets are used. The unoccupied orbitals do not contribute in any way to determining the Hartree–Fock energy, or any property within the Hartree-Fock model, but they are used for adding flexibility to the description of the N-electron wavefunction (moving along the horizontal axis in Figure 1.15). This is done by building additional Slater determinants in which the orbitals occupied in the Hartree–Fock ground state are now substituted by members of the set of unoccupied orbitals. For example, consider the lithium hydride molecule, with its four electrons in three molecular orbitals (ϕ_1, ϕ_2, ϕ_3) built from a basis set consisting of a $1s$ basis function on hydrogen and a $1s$ and $2s$ basis function on lithium. If we consider all possible spin states for N electrons distributed in m orbitals (including all M_S sub-levels) we can form a total of

$$\binom{2m}{N} = \frac{(2m)!}{(2m-N)!N!} \tag{1.65}$$

determinants. For the case here, $m = 3$ and $N = 4$ which gives 15 determinants. The orbital configurations corresponding to these 15 determinants, $|1\rangle - |15\rangle$, are shown in Figure 1.16. Note that determinants $|1\rangle - |9\rangle$ have the same number of α-spin electrons as β-spin electrons and consequently the total spin, $S = 0$. Determinants $|10\rangle - |12\rangle$, have three α-spin electrons and one β-spin electron, giving $S = 1$ and $M_S = 1$. Conversely, determinants $|13\rangle - |15\rangle$ have one α-spin electron and three β-spin electrons, giving $S = 1$ and $M_S = -1$. In the absence of a magnetic field, the sets of determinants $|10\rangle - |12\rangle$ and

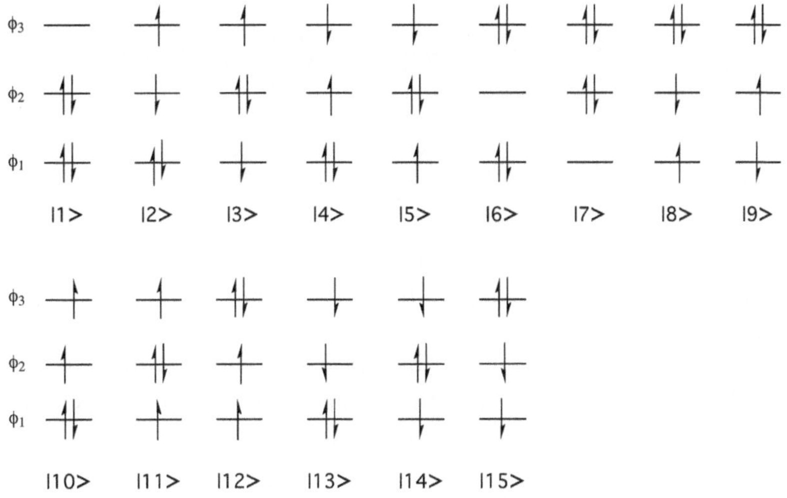

Figure 1.16 Four electrons distributed in all possible ways in three orbitals $\{\phi_1,\phi_2,\phi_3\}$ give rise to 15 determinants with the orbital occupancies shown.

$|13\rangle - |15\rangle$ will be degenerate and so for the $S = 1$ states, we need only consider three determinants instead of six. We are often interested in a particular spin state only, in the present case the ground state of lithium hydride will have $S = 0$ and so we can ignore the $S = 1$ states for many purposes. If we are only interested in a spin state characterised by a fixed number of α and β electrons, N_α and N_β respectively, we can form

$$\binom{m}{N_\alpha}\binom{m}{N_\beta} = \frac{m!}{(m-N_\alpha)!N_\alpha!}\frac{m!}{(m-N_\beta)!N_\beta!} \tag{1.66}$$

determinants. So for $S = 0$, we have $m = 3$, $N_\alpha = 2$ and $N_\beta = 2$ and eqn (1.66) yields 9 determinants in accord with Figure 1.16. Configuration 1 in Figure 1.16 corresponds to the Hartree–Fock, ground state, Slater determinant. All other determinants can be described in terms of "excitations" or, more correctly, substitutions from this determinant. For example, determinants $|2\rangle - |5\rangle$ correspond to substitution of a single spin-orbital while determinants $|6\rangle - |9\rangle$ are created by two substitutions or a "double excitation". The exact wavefunction for any state of a system can be written as a linear combination of all possible Slater determinants that can be formed within the basis set used. We shall denote the set of spin-orbitals occupied in the Hartree-Fock determinant by the letters $i, j, k, l \ldots$, and the spin-orbitals in the unoccupied set as $a, b, c, d \ldots$. The Hartree-Fock determinant will be denoted as $|\Psi_0\rangle$ and any determinant formed from $|\Psi_0\rangle$ by orbital substitutions, $i{\rightarrow}a, j{\rightarrow}b$ and so on as $|\Psi_{ij}^{ab}\rangle$. Using this notation the exact wavefunction (within the basis set used) can be written as

$$|\Psi\rangle = C_0|\Psi_0\rangle + \overset{\text{occupied}}{\underset{i}{\sum}}\;\overset{\text{unoccupied}}{\underset{a}{\sum}}\; C_i^a|\Psi_i^a\rangle + \overset{\text{occupied}}{\underset{i<j}{\sum}}\;\overset{\text{unoccupied}}{\underset{a<b}{\sum}}\; C_{ij}^{ab}|\Psi_{ij}^{ab}\rangle$$

$$+ \overset{\text{occupied}}{\underset{i<j<k}{\sum}}\;\overset{\text{unoccupied}}{\underset{a<b<c}{\sum}}\; C_{ijk}^{abc}|\Psi_{ijk}^{abc}\rangle + \overset{\text{occupied}}{\underset{i<j<k<l}{\sum}}\;\overset{\text{unoccupied}}{\underset{a<b<c<d}{\sum}}\; C_{ijkl}^{abcd}|\Psi_{ijkl}^{abcd}\rangle + \cdots \tag{1.67}$$

in which the Cs are coefficients to be determined by minimising the energy. For an N-electron system there must be, up to and including, N-electron substitutions.

A useful way to write eqn (1.67), which clearly exposes its connection to the Hartree–Fock wavefunction, is in terms of a general substitution operator, $\hat{T}$. For an N-electron system the operator, $\hat{T}$, takes the form

$$\hat{T} = \hat{T}_1 + \hat{T}_2 + \hat{T}_3 + \cdots + \hat{T}_N \tag{1.68}$$

For example $\hat{T}_1$ produces all single substitutions

$$\hat{T}_1 = \overset{\text{occupied}}{\underset{i}{\sum}}\;\overset{\text{unoccupied}}{\underset{a}{\sum}}\; t_i^a \hat{\tau}_i^a \tag{1.69}$$

where $\hat{\tau}_i^a$ is a specific substitution operator and t_i^a is a coefficient (related to the Cs in eqn (1.67)). Similarly $\hat{T}_2$ will produce all double substitutions

$$\hat{T}_2 = \overset{\text{occupied}}{\underset{i<j}{\sum}}\;\overset{\text{unoccupied}}{\underset{a<b}{\sum}}\; t_{ij}^{ab} \hat{\tau}_{ij}^{ab} \tag{1.70}$$

and so on, up to $\hat{T}_N$. To illustrate the function of the $\hat{\tau}$ operators, consider the Hartree–Fock determinant, $|1\rangle$, and the excited determinant, $|2\rangle$, in Figure 1.16. Determinants $|1\rangle$ and $|2\rangle$ are related by a single substitution and using the shorthand notation of eqn (1.27) we can write

$$\hat{\tau}_2^3|\phi_1\bar{\phi}_1\phi_2\bar{\phi}_2\rangle = |\phi_1\bar{\phi}_1\phi_3\bar{\phi}_2\rangle$$

$$\hat{\tau}_2^3|1\rangle = |2\rangle \tag{1.71}$$

$|1\rangle$ and $|8\rangle$ are related by a double substitution

$$\hat{\tau}_{12}^{\bar{3}3}|\phi_1\bar{\phi}_1\phi_2\bar{\phi}_2\rangle = |\phi_1\bar{\phi}_3\phi_3\bar{\phi}_2\rangle$$

$$\hat{\tau}_{\bar{1}2}^{\bar{3}3}|1\rangle = |8\rangle \tag{1.72}$$

and so on. This enables us to write the exact wavefunction in eqn (1.67) in a very compact notation, clearly illustrating the relationship between the Hartree–Fock determinant and the exact wavefunction, as

$$|\Psi\rangle = \left(1 + \hat{T}\right)|\Psi_0\rangle \tag{1.73}$$

It is important to note that the forms for the exact wavefunction given in eqn (1.67) and (1.73) are equivalent but that the C and t coefficients used in these two equations are subtly different. For example, the coefficient of $|\Psi_0\rangle$ in eqn (1.73) is taken as 1. We shall say more about these details in Chapter 2.

We now give a brief overview of a number of standard models which attempt to traverse the distance between the starting wavefunction, $|\Psi_0\rangle$, and the exact, $|\Psi\rangle$. These models achieve this in procedurally and theoretically distinct ways, but the aim is always to systematically improve an approximate starting wavefunction such that it approaches the exact wavefunction. This enables us to move along the horizontal axis in Figure 1.15 and, in principle, arrive at the exact solution of the electronic Schrödinger equation within a chosen basis set. The methods to be discussed are chosen because they are widely used and well understood. Additionally, each type of model has a substantial literature describing its strengths and weaknesses.

1.4.2 Configuration Interaction Methods

The first family of techniques is known as configuration interaction (CI). These methods are based closely on the ideas we introduced in the preceding section. The limiting case is known as full configuration interaction (FCI) and involves building all possible determinants, their number given by eqn (1.66), and determining the C coefficients in eqn (1.67) by minimising the energy. We can determine the C coefficients in this manner because certain types of method obey the variation principle. Such methods are termed "variational". This property that tells us the energy obtained by the model is an upper bound to the exact energy. This means that, whatever we do to the C coefficients, the resulting energy will never descend below the exact energy. So the lower we can drive the energy, by varying the C coefficients, the closer to the exact energy we shall come. The variational property has been shown to be of slight importance in terms of ensuring the accuracy of methods, but it can provide significant computational advantages in the computation of molecular properties.

The FCI method is simply stated and provides an exact solution of the electronic Schrödinger equation (within the chosen basis set). In practice it provides many computational challenges such that FCI can only be applied to systems with very few electrons: about $10 - 12$ electrons if a good basis set is used. The reason for this is that the number of determinants that can be formed grows factorially with the number of electrons and orbitals. For the very simple case of lithium hydride, all the determinants of the, $S = 0$, ground state are given by $|1\rangle - |9\rangle$ of Figure 1.16. However, this example deals with an unrealistically small basis set. If we were to use a slightly more realistic set in which we doubled all the $1s$ and $2s$ orbitals and included a set of $2p$ orbitals on each atom we would have a total of 12 basis functions. If $N = 4$ and $m = 12$, eqn (1.66) tells us that the number of determinants in the FCI wavefunction, for the $S = 0$ states, will be 4356. Some of these determinants can be eliminated by consideration of molecular symmetry. However, the dramatic increase in the number of determinants is clear. As a further example, consider the case of benzene (42 electrons) in a minimal basis set (36 molecular orbitals) which will have a FCI expansion of $>3.1 \times 10^{19}$ determinants. Despite the inadequacy of the basis set, this calculation is not currently tractable. The FCI method is still very useful in providing benchmark calculations, on very small molecules, against which all other more approximate methods can be tested.

To develop practical CI methods it is essential to restrict the level of substitutions. Returning to the benzene example above and restricting the level of substitutions to, $\hat{T}_1 + \hat{T}_2$, results in $<144 \times 10^3$ determinants which is a very manageable problem. This is known as the "CI singles and doubles" method, usually denoted CISD. It can be applied to small to medium sized molecules. CI in the space of single substitutions alone (CIS) does not produce any lowering of the energy relative to the Hartree–Fock method. This is a consequence of Brillouin's theorem, see section 2.6.1, which equates the orbital optimisation of the Hartree–Fock (SCF) method with the space of single subtitutions. CIS can be used to study excited states. CI in the space of double substitutions (CID) is the smallest type of CI expansion that produces an improvement over the Hartree–Fock energy. The CI method is general and higher levels of substitution can be included, such as triple or quadruple substitutions, but the number of determinants again increases very rapidly with the size of the molecule. Regardless of the level of truncation, all CI methods obey the variation principle and provide upper bounds to the exact energy.

The main drawback of any form of truncated CI, with the exception of CIS, is that it is not extensive, that is the calculated energy of a system does not scale linearly with the size of the system. In quantum chemistry we often refer to a closely related, but distinct, requirement of size-consistency. The definition originates from the work of John Pople, who defined a method as *size consistent* if the calculated energy of two systems, say two helium atoms, and that of the combined atoms when separated, beyond the range of any interaction, are equal.

Taking CID as an example of a truncated CI method we can illustrate the lack of size consistency by considering the case of two helium atoms described by two s type basis functions. Helium contains only two electrons and so CID is exact (single excitations do not contribute because of Brillouin's theorem). A CID calculation on an individual helium atom will contain two configurations as shown in Figure 1.17a. Now if the CID method is applied to two helium atoms, infinitely separated, the configurations shown in Figure 1.17(b) will be generated. Note that because the two atoms are infinitely separated there will be no transfer of electrons between atoms. The CID calculations on the

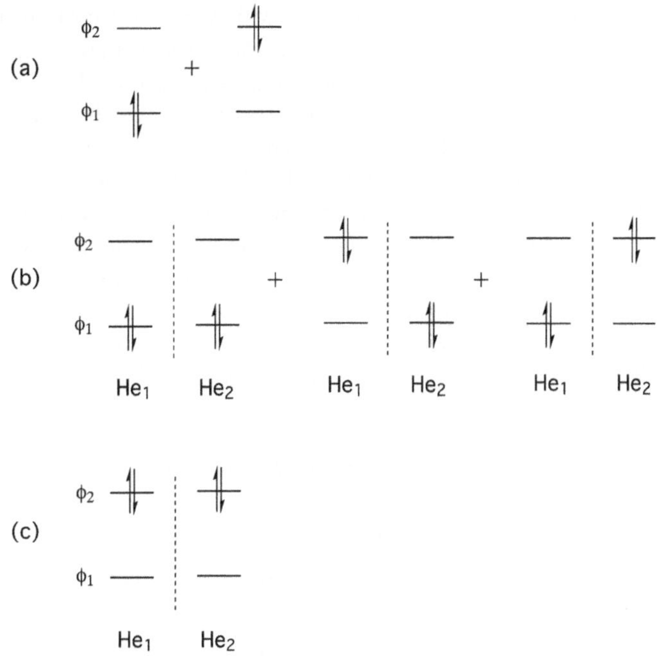

Figure 1.17 (a) CID configurations for helium in a basis set of two orbitals. (b) CID configurations for two helium atoms infinitely separated. (c) Configuration included in FCI on (helium)$_2$ but missing in CID.

individual helium atoms are exact, whereas for the two atoms infinitely separated, the exact wavefunction would have to include up to four-electron excitations. The difference between the FCI wavefunction and CID is the quadruple substitution shown in Figure 1.17(c). This is the size-consistency error for two helium atoms. If more atoms are considered then the difference between the FCI wavefunction and CID will increase. This means that the size-consistency error will increase with the number of electrons.

The size consistency of a computational method is extremely important when calculating relative energies, such as activation barriers or bond dissociation energies. When comparing the energies of systems containing different numbers of electrons, if a method is not size-consistent, the relative errors in the calculated energies will be different and give unreliable energy differences.

1.4.3 Perturbation Theory Methods

We can solve the configuration mixing problem in a different way, which retains the size-consistency property, using Rayleigh–Schrödinger perturbation theory. What defines a particular form of Rayleigh–Schrödinger perturbation theory is the choice of the unperturbed or model problem. In the context of computational quantum chemistry, the most widely used form of perturbation theory is due to Møller and Plesset. In the Møller–Plesset theory the unperturbed problem is chosen to be the Hartree–Fock model.

We can view the correlation energy as a perturbation which, when added to the model Hartree–Fock problem, should enable us to solve the correlation problem in a systematic fashion. Numerically the perturbation should yield a relatively small correction to the model problem, the latter having been solved exactly. The magnitude of the correlation energy is generally very much smaller than the Hartree–Fock energy and so the correction is relatively small. Additionally it is necessary that the Hartree–Fock wavefunction is a good approximation to the exact wavefunction, that is C_0 in eqn (1.67) should be close to 1. If these conditions are met, then a perturbed hamiltonian can be written as

$$\hat{H}(\lambda) = \hat{H}_0 + \lambda(\hat{H} - \hat{H}_0) = \hat{H}_0 + \lambda\hat{V} \tag{1.74}$$

where $\hat{H}_0$ is the Hartree-Fock hamiltonian and $\hat{V}$ is the perturbation, which is the difference between the exact hamiltonian $\hat{H}$ and the Hartree–Fock hamiltonian $\hat{H}_0$. λ is a dimensionless parameter that ranges between 0 and 1. When $\lambda = 0$, we have the Hartree–Fock reference or model problem

$$\hat{H}(\lambda) = \hat{H}_0 \quad (\lambda = 0) \tag{1.75}$$

and when $\lambda = 1$ we have the exact hamiltonian

$$\hat{H}(\lambda) = \hat{H}_0 + \hat{V} \quad (\lambda = 1) \tag{1.76}$$

The exact energy and wavefunction are expanded in powers of the perturbation parameter, λ,

$$
\begin{aligned}
E &= E^{(0)} + \lambda E^{(1)} + \lambda^2 E^{(2)} + \lambda^3 E^{(3)} + \cdots \\
\Psi &= \Psi^{(0)} + \lambda\Psi^{(1)} + \lambda^2\Psi^{(2)} + \lambda^3\Psi^{(3)} + \cdots
\end{aligned}
\tag{1.77}
$$

These expansions are then substituted back into the Schrödinger equation

$$
\begin{aligned}
(\hat{H}_0 + \lambda\hat{V})&\left(\Psi^{(0)} + \lambda\Psi^{(1)} + \lambda^2\Psi^{(2)} + \cdots\right) = \\
&\left(E^{(0)} + \lambda E^{(1)} + \lambda^2 E^{(2)} + \cdots\right)\left(\Psi^{(0)} + \lambda\Psi^{(1)} + \lambda^2\Psi^{(2)} + \cdots\right)
\end{aligned}
\tag{1.78}
$$

Terms are collected for a given order in λ yielding a series of equations that can be solved for the energy and wavefunction to any order in λ,

$$\hat{H}_0\Psi^{(0)} = E^{(0)}\Psi^{(0)}$$

$$\hat{H}_0\Psi^{(1)} + \hat{V}\Psi^{(0)} = E^{(0)}\Psi^{(1)} + E^{(1)}\Psi^{(0)}$$

$$\hat{H}_0\Psi^{(2)} + \hat{V}\Psi^{(1)} = E^{(0)}\Psi^{(2)} + E^{(1)}\Psi^{(1)} + E^{(2)}\Psi^{(0)} \tag{1.79}$$

$$\cdots$$

We shall discuss some more details of the Møller–Plesset perturbation theory in Chapter 2. The important feature being that all Møller–Plesset methods yield energies which are size-consistent, but lack the variational bound. The theory at second-order, denoted "MP2", is very widely used. Successive higher orders of the method are available, but are very much more complicated from both a theoretical and a computational perspective. Their formulation is best discussed in terms of diagrammatic techniques, which are outside the scope of this book.

1.4.4 Coupled-Cluster Methods

In formulating the configuration mixing problem, we previously defined an excitation operator, $\hat{T}$, in eqn (1.68). For application to general systems, this operator is only practicable when truncated at some low order. For example, $\hat{T}_2$, which leads to the CID method and its attendant lack of size-consistency. We saw for the case of two separated helium atoms, in Figure 1.17, that the size-consistency error comes about from a missing higher excitation, specifically a quadruple substitution. How might we correct for this missing term, without resorting to the inclusion of $\hat{T}_4$? In the coupled-cluster (CC) approach the exact wavefunction is generated by applying the exponential of the substitution operator to the reference Hartree–Fock wavefunction:

$$|\Psi\rangle = e^{\hat{T}}|\Psi_0\rangle \tag{1.80}$$

The expansion of the exponential function is

$$e^X = \sum_{p=0}^{\infty}\frac{X^p}{p!} = 1 + X + \frac{X^2}{2} + \frac{X^3}{6} + \frac{X^4}{24} + \cdots \tag{1.81}$$

Returning to the example of two separated helium atoms, let us apply this expansion to eqn (1.80) with the substitution operator restricted to $\hat{T}_2$

$$e^{\hat{T}_2}|\Psi_0\rangle = \left(1 + \hat{T}_2 + \frac{1}{2}\hat{T}_2\hat{T}_2\right)|\Psi_0\rangle \tag{1.82}$$

Since there are only four electrons in our example, we can truncate the expansion at the third term. In analogy with our previous notation, the wavefunction in eqn (1.82) is denoted CCD, coupled cluster in the space of double substitutions. The first two terms on the right hand side of eqn (1.82) produce substitutions equivalent to the CID wavefunction. The third term introduces a four-electron substitution as a product of two-electron substitutions. From our discussion of Figure 1.17, this last term will remove the size-consistency error. In fact all coupled-cluster methods, regardless of the level of truncation of $\hat{T}$, produce size-consistent energies.

If the complete $\hat{T}$ operator is included, then eqn (1.80) yields the exact wavefunction and is equivalent to FCI. Any truncation of $\hat{T}$, while including the effects of higher substitutions, will not be equivalent to FCI. This is because in the FCI method the contribution of the quadruple substitution has its own variationally determined coefficient, see eqn (1.67). Conversely, in eqn (1.82) the form of the $\hat{T}_2$ operator, eqn (1.70), means that the coefficient of the quadruple substitution is given as the product of the coefficients of two double substitutions. The double substitutions, which occur linearly, are termed "connected" clusters. Higher substitutions which appear as products of lower substitution operators, $\frac{1}{2}\hat{T}_2\hat{T}_2$, are termed "disconnected" clusters.

The coefficients t_{ij}^{ab}, eqn (1.70), are obtained by projection and the resultant energy, while size-consistent, is not variationally bound. Approximating the effects of higher substitutions as products of lower substitutions is often very reliable and CC techniques provide some of the most accurate results available for general chemical systems. The most widely used level is CCSD. Including triple substitutions, CCSDT, incurs a significant rise in computational cost and consequently a model termed CCSD(T) has found very widespread application. In CCSD(T) the effects of triple substitutions are included in an approximate fashion using ideas from the Møller–Plesset perturbation theory.

We have said something about the influence of computational cost on the choice of a method. There must always be a balance between cost and required accuracy. For the types of method we have discussed so far, is it possible to establish a hierarchy of accuracy? The vast literature associated with these methods suggests that the following is a plausible hierarchy:

$$HF \ll MP2 < CISD < CCSD < CCSD(T) < FCI$$

The extremes of 'best', FCI, and 'worst', HF, are irrefutable, but the intermediate methods are less clear and depend on the type of chemical problem being addressed. What about computational cost? This is not an easy thing to quantify because there are many details of implementation and approximation to be taken into account before meaningful comparisons can be made. For example, there have been significant advances in developing "linear-scaling techniques" that can be applied to large molecular systems. Linear-scaling techniques usually involve quite elaborate approximations, these may be valid for large systems but less appropriate for small to medium sized systems. However, we can look at the

Table 1.2 Formal scaling of computational cost with the size of the orbital space, *m*. *iter-m*^{x} indicates there are many iterative steps, each of computational cost m^x.

Method:	*HF*	*MP2*	*CISD*	*CCSD*	*CCSD(T)*	*FCI*
Formal Scaling	*iter-m*^{4}	m^5	*iter-m*^{6}	*iter-m*^{6}	*iter-m*^{6}+m^7	$m!$
Variational?	Yes	No	Yes	No	No	Yes
Size-consistent?	Yes	Yes	No	Yes	Yes	Yes

formal cost of the methods we have discussed, assuming no approximations, and this does provide a useful comparison, see Table 1.2.

1.4.5 Multiconfigurational and Multireference Methods

The families of correlated methods we have described so far all depend on the Hartree-Fock determinant dominating the final wavefunction, that is $C_0 \approx 1$ in eqn (1.67), and all the many other determinants providing a relatively small, but essential, component of the total wavefunction. In certain systems the reference wavefunction cannot be represented adequately by a single Slater determinant. Such situations arise, for example, when chemical bonds are stretched or when spin-paired electrons occupy energetically degenerate but spatially distinct orbitals. Consider the familiar molecular orbital picture of an electron pair bond between two identical atoms, in which atomic orbitals on the two atoms mix to form bonding and antibonding molecular orbitals, as shown in Figure 1.18a. If the bond between atom A and atom B is stretched, the energy separation between the bonding and antibonding molecular orbitals is reduced, as in Figure 1.18b. If the atoms are moved to very large separation, as might be required when calculating a diatomic potential energy curve, the bonding and antibonding combinations will become degenerate. In the first case, Figure 1.18a, it is obvious that the two electrons should be placed in the lower lying bonding orbital. In the final case, Figure 1.18c, there is no reason to place the electrons in the bonding orbital, since now it is degenerate with the antibonding orbital. In such situations we must include two determinants: one with the bonding orbital doubly occupied and the other with the antibonding orbital doubly occupied. This is the extreme case of exact degeneracy and we would expect each Slater determinant to contribute equally to the wavefunction. This is easily illustrated with the simple case of the hydrogen molecule described with a minimal basis set consisting of a single 1s type function on each atom. The problem arises in the Hartree-Fock method because, even at dissociation, both electrons are assigned to a single symmetric doubly occupied molecular orbital of the form

Molecular Orbitals

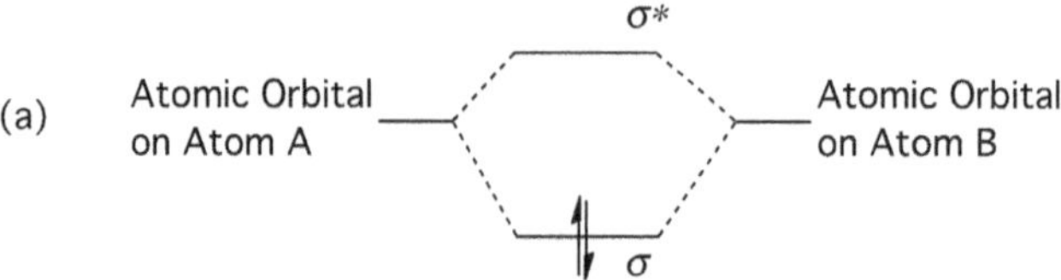

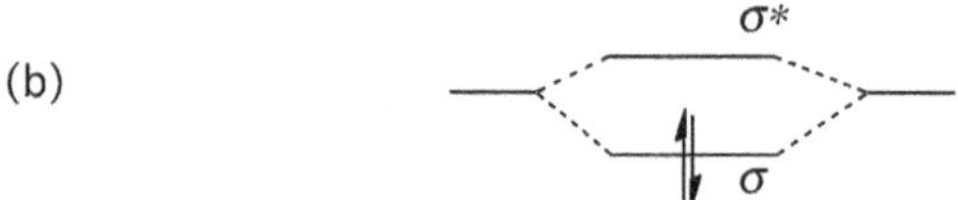

Figure 1.18   (a) Homonuclear diatomic bonding and antibonding orbital energy levels at equilibrium bond distance. (b) Bonding and antibonding energy levels at stretched bond distance. (c) Bonding and antibonding energy levels for completely separated atoms.

$$\sigma = \frac{1}{\sqrt{2}}[\chi_A + \chi_B] \quad H_A \cdots\cdots\cdots H_B \tag{1.83}$$

where χ_A is the $1s$ type function on atom A, and χ_B that on atom B. The corresponding Slater determinant is

$$|\Psi_0\rangle = |\sigma\bar{\sigma}\rangle = \frac{1}{\sqrt{2}}\sigma(\mathbf{r}_1)\sigma(\mathbf{r}_2)[\alpha(\mathbf{s}_1)\beta(\mathbf{s}_2) - \beta(\mathbf{s}_1)\alpha(\mathbf{s}_2)] \tag{1.84}$$

Regardless of the internuclear distance, the form of the bonding orbital remains that given in eqn (1.83). The spatial part of the wavefunction can be expanded to give

$$\sigma(\mathbf{r}_1)\sigma(\mathbf{r}_2) = \frac{1}{2}[\chi_A(\mathbf{r}_1)\chi_A(\mathbf{r}_2) + \chi_A(\mathbf{r}_1)\chi_B(\mathbf{r}_2) + \chi_B(\mathbf{r}_1)\chi_A(\mathbf{r}_2) + \chi_B(\mathbf{r}_1)\chi_B(\mathbf{r}_2)] \tag{1.85}$$

According to this the two electrons spend half of their time on different atoms, one on atom A and one on atom B, and the other half of their time on one atom, either atom A or atom B. In terms of structures, even at dissociation, the wavefunction contains 50% $H_A\cdot\cdot H_B$, 25% $H_A^-\,^+H_B$ and 25% $H_A^+\,^-H_B$. This is clearly incorrect, since the molecule should dissociate into 2H•. The correct form required for the wavefunction at dissociation is

$$\frac{1}{2}[\chi_A(\mathbf{r}_1)\chi_B(\mathbf{r}_2)+\chi_B(\mathbf{r}_1)\chi_A(\mathbf{r}_2)][\alpha(\mathbf{s}_1)\beta(\mathbf{s}_2)-\beta(\mathbf{s}_1)\alpha(\mathbf{s}_2)] \qquad (1.86)$$

This form of the wavefunction cannot be expressed as a single determinant. However if we use the antibonding orbital

$$\sigma^* = \frac{1}{\sqrt{2}}[\chi_A - \chi_B] \qquad (1.87)$$

and expand the spatial component of the corresponding determinant

$$\sigma^*(\mathbf{r}_1)\sigma^*(\mathbf{r}_2)=\frac{1}{2}[\chi_A(\mathbf{r}_1)\chi_A(\mathbf{r}_2)-\chi_A(\mathbf{r}_1)\chi_B(\mathbf{r}_2)-\chi_B(\mathbf{r}_1)\chi_A(\mathbf{r}_2)+\chi_B(\mathbf{r}_1)\chi_B(\mathbf{r}_2)] \quad (1.88)$$

We can see that the H_A• •H_B type terms occur with the opposite sign to that in eqn (1.85). If we subtract the determinants we obtain the required form for dissociated hydrogen

$$|\sigma\bar{\sigma}\rangle - |\sigma^*\bar{\sigma}^*\rangle = \frac{1}{2}[\chi_A(\mathbf{r}_1)\chi_B(\mathbf{r}_2)+\chi_B(\mathbf{r}_1)\chi_A(\mathbf{r}_2)][\alpha(\mathbf{s}_1)\beta(\mathbf{s}_2)-\beta(\mathbf{s}_1)\alpha(\mathbf{s}_2)] \quad (1.89)$$

In the intermediate situation of stretched bond lengths, we will not have exact degeneracy, but can speak of near-degeneracy of the bonding and antibonding orbitals. The contribution of the doubly substituted determinant, eqn (1.88), will not be 50% of the wavefunction but it will be significant. We can summarise the three situations depicted in Figure 1.18 for the case of the hydrogen molecule by writing a CI type wavefunction

$$|\Psi_{CI}\rangle = C_1|\sigma\bar{\sigma}\rangle + C_2|\sigma^*\bar{\sigma}^*\rangle \qquad (1.90)$$

At equilibrium, Figure 1.18a, $C_1 >> C_2 (C_1 \approx 1, C_2 \approx 0)$. At stretched bond distances, Figure 1.18b, $C_1 > C_2$ and at dissociation, Figure 1.18c, $C_1 = C_2$. In the latter two circumstances the qualitatively correct wavefunction is not given by eqn (1.84), but rather by eqn (1.90). It consists of more than one Slater determinant and is said to be multiconfigurational in character. Figure 1.19 shows the potential energy curve for molecular hydrogen calculated using eqns (1.84) and (1.90). The difference between the two curves in Figure 1.19 is a type of electron correlation energy, it is highly structure dependent and is termed non-dynamic electron correlation. The non-dynamic electron correlation comes from the inability of a single determinant to provide a qualitatively correct reference wavefunction. In Figure 1.19, if we were to adopt the definition of eqn (1.64) for the correlation energy, we can see that as the bond distance increases the correlation energy also increases, reaching a maximum

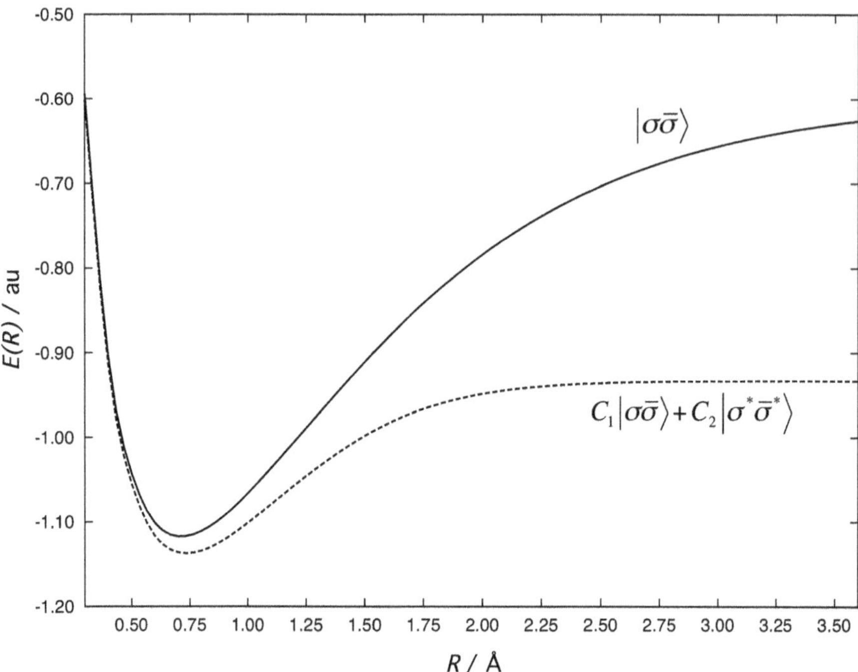

Figure 1.19   Potential energy curve for H_2 using the wavefunction forms given in eqns (1.84) (solid line) and (1.90) (dashed line).

value at dissociation! This is clearly nonsensical, since at dissociation we have two separated hydrogen atoms and there must be no electron correlation. In this case the correct reference wavefunction is given by eqn (1.90), since at dissociation eqn (1.84) describes a physically incorrect wavefunction.

The orbitals where near-degeneracy effects operate may be a very small subset of the full orbital space. The majority of the orbitals will be adequately described as doubly occupied or empty. To define an appropriate reference wavefunction we must optimise the orbitals, as in the Hartree-Fock method, but also allow a mixing of determinants, as in eqn (1.90), so that optimal mixing coefficients may be determined through minimisation of the energy. This is the realm of the multiconfigurational SCF (MCSCF) method. To proceed we partition the orbital space into three subspaces. The first subspace, termed "inactive", consists of orbitals that remain doubly occupied in all determinants. The second subspace, termed "active", consists of orbitals that are allowed to have variable occupancy through the mixing of determinants. The third subspace is the "virtual" space and contains orbitals that are unoccupied in all determinants. Returning to the lithium hydride example, with its four electrons in three molecular orbitals (ϕ_1, ϕ_2, ϕ_3) built from a basis set consisting of a $1s$ function on hydrogen and a $1s$ and $2s$ function on lithium, we could choose the inactive subspace to consist of the ϕ_1 orbital (essentially the $1s$ orbital on lithium) and the active subspace to be the ϕ_2 and ϕ_3 bonding and antibonding orbitals. In this limited basis there will be no virtual orbital subspace. The active space consists of two electrons and two orbitals, if we allow all possible distributions of the active electrons in the active orbitals for an overall $S = 0$ then we shall have a complete active space (CAS). The CAS will comprise the four determinants shown in Figure 1.20.

Keeping ϕ_1 as doubly occupied in all configurations has reduced the number of determinants from nine (Figure 1.16) to four (Figure 1.20). However if we were to consider the same approach for, say the nitrogen molecule, the number of active orbitals would have to include, at least, each bonding and antibonding orbital of the triple bond. The active space would now include six electrons and six orbitals and the CAS would involve 400 determinants. This small extension of the active space results in a dramatic increase in the number of determinants involved.

Multi-configurational methods are conceptually and computationally much more difficult than the single determinant Hartree-Fock method. For diatomic molecules the choice of active space is simple. However for general molecular systems, especially in chemical reactions, great care must be exercised in the choice of active space to keep it small enough to ensure that the calculation remains computationally tractable but flexible enough to adequately describe the chemical situation being explored. Provided that a suitable active space can be chosen, the orbitals are optimised in a similar fashion to the Hartree-Fock method. This defines a suitable reference wavefunction that can then be subjected to further elaboration using CI, Møller-Plesset or CC methods to account for the dynamic correlation energy. The techniques we have discussed for correlated calculations can all be extended to the case of multiconfigurational reference wavefunctions. Such multireference techniques (MR-CI, MR-MP2, MR-CC) are very much more computationally demanding but their use is sometimes essential in order to arrive at meaningful results.

Multi-configurational and multi-reference methods are not as widely used as their single-determinant-based counterparts. The calculations do require greater computational resources but perhaps more importantly they require a high degree of chemical understanding on the part of the user if they are to produce useful studies. The choice of active space provides a great degree of flexibility in the application of multi-reference techniques. This means that they are, in principle,

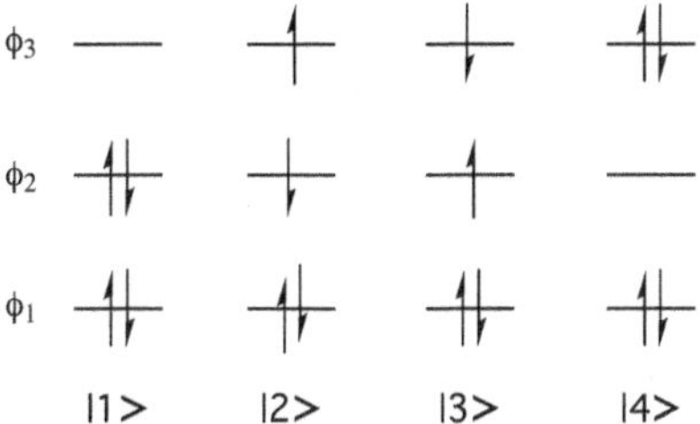

Figure 1.20 Complete active space (CAS) determinants for two electrons in two orbitals and $S = 0$.

applicable to any type of chemical structure but in practice the pragmatic choice of active space provides the opportunity for erroneous modeling. These are not automatic 'black-box' techniques and should be used with due care and caution.

1.4.6 Density Functional Methods

The final type of standard model we shall discuss is that based on density functional theory (DFT). These are currently the most widely used techniques for the study of general chemical problems. The ultimate aim is to deal directly with the one-electron density and circumvent consideration of the many-electron wavefunction. An N-electron wavefunction, $|\Psi(\mathbf{x}_1,\mathbf{x}_2,\ldots\mathbf{x}_N)\rangle\rangle$, contains all information on the system and depends on $4N$ coordinates, that is three spatial and one spin coordinate per electron. A remarkable theorem due to Hohenberg and Kohn establishes that the ground state energy and all ground state properties can be obtained from functionals of the one-electron density alone. Formally, the one-electron density, $\rho(\mathbf{r})$, can be obtained from the N-electron wavefunction, $|\Psi(\mathbf{x}_1,\mathbf{x}_2,\ldots\mathbf{x}_N)\rangle\rangle$, by integrating over all electronic spin coordinates and over all spatial coordinates, except those of electron 1.

$$\rho(\mathbf{r}_1) = N \int |\psi(\mathbf{x}_1,\mathbf{x}_2, \ldots\mathbf{x}_N)|^2 d\mathbf{s}_1 d\mathbf{x}_2 \ldots d\mathbf{x}_N \tag{1.91}$$

Consequently $\rho(\mathbf{r}_1)$ depends only on the three spatial coordinates of electron 1, regardless of the number of electrons in the system. If we can work with $\rho(\mathbf{r}_1)$ as a fundamental quantity, instead of $|\Psi(\mathbf{x}_1,\mathbf{x}_2,\ldots\mathbf{x}_N)\rangle\rangle$, it should lead to less demanding computations. Additionally it provides us with a simple quantity defined in three-dimensional physical space to deal with, rather than the many-dimensional space of the N-electron wavefunction.

Unfortunately the Hohenberg-Kohn theorem does not tell us how we might find suitable functionals of $\rho(\mathbf{r}_1)$. To proceed we can make use of the known exact solution to an idealised electronic problem, that of the uniform electron gas. The "uniform" refers to the distribution of the electron density throughout the gas. The Coulomb and Fermi correlation properties of the uniform electron gas can be extracted in the form of exchange (Fermi) – correlation (Coulomb) functionals of the one-electron density. These exchange-correlation functionals can be added to a self-consistent field type formalism providing a computationally efficient technique. This approach, with suitable modifications, is capable of wide application and good accuracy for chemical problems. Modifications and refinements are necessary since the uniform electron gas is generally not a good approximation to the electron distribution in chemical systems. For example, the electron distribution around two bonded atoms of very different electronegativity will produce a very non-uniform distribution of the electron density. Many formal results in DFT are known and formal developments in the theory continue. If the exact form of the exchange and correlation terms was known for a general electronic system, rather than the uniform electron gas, then DFT would provide an exact solution to the Schrödinger equation.

Practical forms of DFT have developed with appeal to empirical input. This can be a contentious issue, but it is clear that many of the 'parameters' that enter practical DFT methods have been obtained by comparison to some form of external data, for example atomic correlation energies or thermochemical databases. DFT methods have received huge efforts in their development and implementation and provide a cost-effective approach to many chemical questions. Comparing with the methods in Table 1.2, pure DFT methods have a formal scaling, with basis set size, of m^3. At present there is no systematic way to refine the density functional approach to the exact limit in analogy with Figure 1.15. A similar construct to Figure 1.15 has been suggested for DFT methods,[6] but it refers to an elaboration of the formal techniques used and provides no guarantee of approaching the limiting case of exactly solving the Schrödinger equation.

1.5 How Do You Select a Theoretical Model?

From the preceding section it should be clear that the choice of theoretical model on which to base the study of a chemical question is vast. A key consideration in narrowing down the method to be used is the balance of computational cost against accuracy. Computational costs include: the amount of processor time used; the quantity of disk space needed; the amount of core (random access) memory necessary to efficiently complete the task. It is invariably the case that there must be a compromise between cost, accuracy and ease of carrying out a chemical study. A study will usually involve many calculations of different types and it is not enough to be able to perform a single demanding calculation and stop. Certain methods are only feasible for small molecules and may provide very accurate results. For larger molecules, less accurate methods must be employed. Even if large computational resources are available, the skill and experience of the user in applying more complex techniques must be a consideration. As in any computational endeavour the maxim "garbage in, garbage out", holds unassailable sway.

We have already mentioned that it is a difficult task to quantify the precise cost of a computational method. There are many considerations of computational implementation that can influence the cost and scaling of a given technique. For example, we have already stated in Table 1.2 that the Hartree-Fock technique formally scales with the basis set size as m^4. This scaling comes from the $\sim m^4/8$ unique two-electron integrals that must be evaluated in a Hartree-Fock calculation. Various integral approximations may be introduced to reduce this dependence to between m^2 and m^3. These approximations are typically integral screening techniques, which circumvent the evaluation of batches of integrals, thereby reducing the computational effort. For very large molecules other types of distant-dependent approximations may be employed to further reduce the scaling to m.

Many of the standard methods we have discussed in the previous section have well-known strengths and weaknesses. This knowledge has been gained through wide application of the techniques over many years to a wide range of chemical problems. For example, the Hartree-Fock method is arguably the simplest approach. Its weakness stems from the lack of correlation between the motion of electrons of opposite spin. Despite this severe approximation, the Hartree-Fock method is both variational and size-consistent and able to provide reasonable estimates of equilibrium bond lengths and conformational energy differences. Its ability to provide accurate estimates of reaction energies and activation barriers is rather poor.

To assess the accuracy of a method for a given task we can look at the performance of the method in the calculation of reaction energies; equilibrium geometries; harmonic vibrational wavenumbers; dipole moments and any other molecular properties of relevance. Many molecular databases exist which provide this type of information for a variety of methods. If the methods of interest have not been assessed for a given property then it is important that a validation is carried out by the user on a selection of molecules for which reliable data is available for comparison. For some chemical questions, for example the geometry of a transition structure for a reaction, experimental data will not be available for comparison. In such cases comparison may be made with higher-level computational techniques. Table 1.3 lists a number of databases that can be used for validation against experimental data or high-level computational results.

1.6   The Apparatus: Hardware and Software

Since the 1980s, computer hardware has developed at an extraordinary pace. In real terms this has meant that fast computing platforms have become available, at a decreasing cost, to a wide range of users. There have been a number of key developments in the structure of central processing units (CPUs) that have enabled this remarkable progress. It would take us too far from our purpose here to attempt any detailed discussion of these. However, it is to be hoped that some of these remarkable historical developments in computer design are of sufficient interest to merit the briefest mention here.

There has been an impressive increase in the operational clock speed at which modern CPUs work, amounting to some three orders of magnitude. There have also been four key changes in the way instructions are processed.

(a)   A CPU design, termed "RISC" (Reduced Instruction Set Computing) exploited the idea that it was possible to carry out a small set of simple instructions in a highly optimised manner and so achieve much faster effective speeds than when using more complex instructions. For example, a complex instruction might involve fetching from memory, executing an arithmetic operation and reloading memory with the result. In the RISC CPU this complex instruction is decomposed into constituent lower level instructions, each being performed very efficiently.

Table 1.3 Selected databases for the assessment of computational methods.

1	**Atomic Reference Data for Electronic Structure Calculations**
	Location: http://www.nist.gov/physlab/data/dftdata/index.cfm/
	References: S. Kotochigova, Z. H. Levine, E. L. Shirley, M. D. Stiles and C. W. Clark, *Phys. Rev. A*, 1997, **55**, 191 and S. Kotochigova, Z. H. Levine, E. L. Shirley, M. D. Stiles and C. W. Clark, *Phys. Rev. A*, 1997, **56**, 5191.
2	**Benchmark Energy and Geometry Database**
	Location: http://www.begdb.com
	Reference: J. Řezáč, P. Jurečka, K. E. Riley, J. Černý, H. Valdes, K. Pluháčková, K. Berka, T. Řezáč, M. Pitoňák, J. Vondrášek and P. Hobza, *Collect. Czech. Chem. Commun.*, 2008, **73**, 1261.
3	**Barrier Heights of Diverse Reaction Types**
	References: J. Zheng, Y. Zhao and D. G. Truhlar, *J. Chem. Theory Comput.*, 2007, **3**, 569 and J. Zheng, Y. Zhao and D. G. Truhlar, *J. Chem. Theory Comput.*, 2009, **5**, 808.
4	**NIST: Computational Chemistry Comparison and Benchmark Database**
	Location: http://cccbdb.nist.gov
5	**The Computational Results Database**
	Location: http://tyr3.chem.wsu.edu/~feller/Site/Database.html/
	Reference: D. Feller, *J. Comput. Chem.*, 1996, **17**, 1571.

(b) The RISC structure lends itself to pipelining, in which a CPU module is dedicated to performing each of the basic tasks of fetching from memory, decoding the instruction, executing the instruction and writing back the result. The advantage of pipelining is that, as the first instruction exits the first operational module, a second instruction can begin processing. As in a pipeline, several instructions can begin processing before the outcome of the first instruction is completed, see Figure 1.21. Pipelining may not be faster for the execution of a single instruction but will be faster when several instructions must be performed.

(c) Multiple-issue techniques are complex procedures which effectively allow more than one instruction to be performed per clock cycle, for example by fetching a word in which several instructions are packed.

(d) More efficiency can be gained by enabling several instructions to begin execution simultaneously. Multi-core processors are now commonplace in everyday desktop machines. Multi-core processors are single components (chips) with two (dual-core) or more (*e.g.* quad-core) independent processors, which read and execute instructions. Since the cores are independent of each other, parallel execution of instructions may be performed.

More recently, graphical processing units (GPUs) have been the focus of much interest in the computational chemistry community, since they allow high levels of parallelism at relatively low cost. We now also have access to chips with very large numbers of core processing units. The question of how these new hardware structures are to be exploited efficiently for computational quantum chemistry has led to new avenues of research and endeavour.

Clock Cycle:	1	2	3	4	5	6	7	8	9	10	11	12
Non-Pipelined CPU												
Instruction 1	F	D	X	W								
Instruction 2					F	D	X	W				
Instruction 3									F	D	X	W
Pipelined CPU												
Instruction 1	F	D	X	W								
Instruction 2		F	D	X	W							
Instruction 3			F	D	X	W						

F=Fetch; D=Decode; X=Execute; W=Write.

Figure 1.21 The execution of three consecutive instructions on a non-pipelining CPU compared to a pipelining CPU.

Developments in computer hardware continue unabated. The speed of these developments means that almost anything one writes about hardware will be out of date by the time it is read! For the interested reader, a recent review of high performance computing in chemistry may be found in ref. 7.

The cost of computer hardware has decreased and its efficiency has increased. Consequently, computational quantum chemistry can be carried out on a wide range of systems, from simple desktop workstations to massive machines with hundreds of thousands of CPUs. In monetary currency, this range spans some eight orders of magnitude as indicated in Table 1.4. The top supercomputers in the world are constantly changing and what was impressive in the recent past is quickly overtaken. At the time of writing, the top supercomputer in the world is deemed to be the Sequoia, which is installed at the U.S. Department of Energy's Lawrence Livermore National Laboratory in California. The Sequoia contains more than 1.5×10^6 processing cores!

The other essential component of computational quantum chemistry is the software: the sets of instructions, to be carried out by a computer to produce energies, geometries, molecular orbitals and anything else of interest. Computational quantum chemistry software is complicated because it contains many, many, different components. For example, any program must start with a procedure to read in the specification of a molecule in terms of coordinates and atom types, which is a relatively simple task. Having read in a molecule, the program must assign basis functions to the atoms, and then begin efficiently calculating the many necessary integrals involving the basis functions, which is a relatively complicated task.

Efficient quantum chemistry programs, see Table 1.5 for a partial list, have often undergone many tens, and in some cases hundreds, of man-years of development. It is possible to put together a simple program to perform quantum chemical calculations, but it should be borne in mind that a working program is not necessarily an efficient program. To perform some of the elaborate studies currently being carried out by researchers does require validated programs of extreme efficiency. The list in Table 1.5 is by no means exhaustive, but illustrates that there are many programs available offering a

Table 1.4 Relative cost of some typical computing platforms used for computational quantum chemistry.

Platform	Relative cost
Multi-core desktop PC	1
Multi-CPU, multi-core workstation	10
Workstation cluster	100
Supercomputer	10^6–10^8

wide range of functionality. The specialist abilities of these programs varies considerably and the interested reader should explore which programs might be most suitable for the work they wish to carry out. This should include considerations of the methods available, interfaces to other software (for example for graphical display of output), availability of source code to enable addition of functionality and the nature and limitations of the software Licence.

Table 1.5 An incomplete list of widely used computational quantum chemistry programs, listed alphabetically.

1	**ACES**
	Website: http://www.qtp.ufl.edu/ACES
2	**ADF**
	Website: http://www.scm.com/Products/ADF.html
3	**Columbus**
	Website: http://www.univie.ac.at/columbus
4	**Dalton**
	Website: http://dirac.chem.sdu.dk/daltonprogram.org
5	**DeMon**
	Website: http://www.demon-software.com/public_html/program.html
6	**GAMESS**
	Website: http://www.msg.ameslab.gov/gamess/index.html
7	**GAMESS-UK**
	Website: http://www.cfs.dl.ac.uk
8	**Gaussian**
	Website: http://www.gaussian.com
9	**Jaguar**
	Website: http://www.schrodinger.com/products/14/7
10	**MOLCAS**
	Website: http://www.molcas.org/introduction.html
11	**Molpro**
	Website: http://www.molpro.net
12	**NWChem**
	Website: http://www.nwchem-sw.org/index.php/Main_Page
13	**ORCA**
	Website: http://www.mpibac.mpg.de/bac/logins/neese/description.php
14	**Q-Chem**
	Website: http://www.q-chem.com/features.htm
15	**TURBOMOLE**
	Website: http://www.cosmologic.de/index.php?cosId=3010&crId=3

Appendix 1A   Physical Constants, Atomic Units and Conversion Factors

Quantity	Symbol or Formula	Atomic Units	SI Units
Planck's constant	h	2π	6.626069×10^{-34} J s
Planck's constant/2π	$\hbar$	1	1.054571×10^{-34} J s
Charge	e	1	1.602176×10^{-19} C
	$4\pi\varepsilon_0$	1	1.112650×10^{-10} J^{-1} C^2 m^{-1}
Unified atomic mass unit	u	1822.888	1.660538×10^{-27} kg
Electron rest mass	m_e	1	9.109382×10^{-31} kg
Proton mass	m_p	1836.152	1.672621×10^{-27} kg
Neutron mass	m_n	1838.683	1.674927×10^{-27} kg
Fine structure constant	α		7.297352×10^{-3}
Speed of light in vacuum	$c \;\left(=\dfrac{1}{\alpha}\right)$	137.035999	299792458 m s^{-1}
Energy	$E_h = \dfrac{e^2}{4\pi\varepsilon_0 a_0}$	1	4.359744×10^{-18} J (27.21138 eV)
Time	$\dfrac{\hbar}{E_h}$	1	2.418884×10^{-17} s
Length	a_0	1	$0.5291772 \times 10^{-10}$ m
Charge density	$\dfrac{e}{a_0^3}$	1	1.081202×10^{12} C m^{-3}
Electric Field	$\dfrac{E_h}{ea_0}$	1	5.142206×10^{11} V m^{-1}
Dipole Moment	ea_0	1	8.478353×10^{-30} C m
Quadrupole moment	ea_0^2	1	4.486551×10^{-40} C m^2
Electric polarisability	$\dfrac{e^2 a_0^2}{E_h}$	1	1.648777×10^{-41} C^2 m^2 J^{-1}
Magnetic dipole moment	$\dfrac{\hbar e}{m_e}$	1	1.854801×10^{-23} J T^{-1}
Bohr Magneton	μ_B	$\dfrac{1}{2}$	9.274009×10^{-24} J T^{-1}
Nuclear magneton	μ_N	2.723085×10^{-4}	5.050783×10^{-27} J T^{-1}

It is often convenient to express energy in terms of energy equivalents according to the equations: $E = h\nu = \dfrac{hc}{\lambda} = mc^2 = kT.$

$$1 \text{ au of energy} \equiv \frac{E}{h} = 6.579683 \times 10^{15}\text{Hz}$$

$$\equiv \frac{E}{hc} = 2.194746 \times 10^{7}\text{m}^{-1}$$

$$\equiv \frac{E}{c^2} = 4.850869 \times 10^{-35}\text{kg} = 2.921262 \times 10^{-8}u$$

$$\equiv \frac{E}{k_{\mathrm{B}}} = 3.157750 \times 10^{5}\text{K}$$

A complete listing of currently accepted values of physical constants is available in P. J. Mohr, B. N. Taylor, and D. B. Newell, *Rev. Mod. Phys.*, 2012, **84**, 1527.

Appendix 1B   Elementary Properties of Determinants

Consider the square 2 $\times$ 2 matrix

$$\mathbf{A} = \begin{pmatrix} a & b \\ c & d \end{pmatrix} \tag{1B.1}$$

The determinant of **A**, det **A**, which we shall denote as **D** is defined as

$$\mathbf{D} = \begin{vmatrix} a & b \\ c & d \end{vmatrix} = ad - bc \tag{1B.2}$$

For determinants of larger dimension we can compute the value of the determinant by a Laplace expansion along any row or column. The idea of the Laplace expansion is to reduce the larger determinant into a combination of 2 $\times$ 2 determinants that can be evaluated using eqn (1B.2). To illustrate this consider the 3 $\times$ 3 determinant

$$\mathbf{D} = \begin{vmatrix} a & b & c \\ d & e & f \\ g & h & i \end{vmatrix} \tag{1B.3}$$

We shall expand this determinant along the top row. To begin we take the first element in row 1 and column 1, which is a. Now strike out the elements in the row and column containing a

$$\mathbf{D} = \begin{vmatrix} \cancel{a} & \cancel{b} & \cancel{c} \\ \cancel{d} & e & f \\ \cancel{g} & h & i \end{vmatrix} \tag{1B.4}$$

This leaves the 2 $\times$ 2 determinant

$$\begin{vmatrix} e & f \\ h & i \end{vmatrix} \tag{1B.5}$$

The first term in the expansion is

$$a\begin{vmatrix} e & f \\ h & i \end{vmatrix} \tag{1B.6}$$

We now proceed to the second element in row 1 column 2, b, and strike out the row and column containing b

$$\begin{vmatrix} \cancel{a} & \cancel{b} & \cancel{c} \\ d & \cancel{e} & f \\ g & \cancel{h} & i \end{vmatrix} \tag{1B.7}$$

leaving the determinant

$$\begin{vmatrix} d & f \\ g & i \end{vmatrix} \tag{1B.8}$$

The second term in the expansion becomes

$$b\begin{vmatrix} d & f \\ g & i \end{vmatrix} \tag{1B.9}$$

Repeating the same process for the element in row 1 column 3 gives

$$\begin{vmatrix} \cancel{a} & \cancel{b} & \cancel{c} \\ d & e & \cancel{f} \\ g & h & \cancel{i} \end{vmatrix} \tag{1B.10}$$

leaving as the third term in the expansion

$$c\begin{vmatrix} d & e \\ g & h \end{vmatrix} \tag{1B.11}$$

Each of the three terms in eqn (1B.6), (1B.9) and (1B.11) are called the *minors* of the 3×3 determinant $\mathbf{D}$. To complete the expansion we must include a phase factor, $(-1)^{i+j}$ where i is the row and j the column which has been struck out. The phase factor takes the form

$$\begin{vmatrix} + & - & + \\ - & + & - \\ + & - & + \end{vmatrix} \qquad (1B.12)$$

Now we can write **D** in terms of the three *cofactors* (being the minors multiplied by the phase factor) as

$$\mathbf{D} = \begin{vmatrix} a & b & c \\ d & e & f \\ g & h & i \end{vmatrix} = (-1)^{1+1}a\begin{vmatrix} e & f \\ h & i \end{vmatrix} + (-1)^{1+2}b\begin{vmatrix} d & f \\ g & i \end{vmatrix}(-1)^{1+3}c\begin{vmatrix} d & e \\ g & h \end{vmatrix} \qquad (1B.13)$$

$$= a(ei - fh) - b(di - fg) + c(dh - eg)$$

The expansion of an $N \times N$ determinant contains $N!$ terms. For **D**, $N = 3$ and $N! = 6$. Moving to $N = 4$, will yield $N! = 24$ terms. The procedure remains the same, the $N \times N$ determinant is reduced to an expansion of cofactors containing $(N-1) \times (N-1)$ determinants. These in turn are reduced until all the cofactors contain 2×2 determinants that can be evaluated using eqn (1B.2).

For our purposes we need note only three important properties of determinants that we will illustrate with **D** given in eqn (1B.3).

1. A determinant changes sign if two rows are interchanged

$$\mathbf{D} = a(ei - fh) - b(di - fg) + c(dh - eg)$$

swap the top two rows

$$\mathbf{D}' = \begin{vmatrix} d & e & f \\ a & b & c \\ g & h & i \end{vmatrix} = dbi - dch - eai + ecg + fah - fbg \qquad (1B.14)$$

Comparing term by term we find that $\mathbf{D}' = -\mathbf{D}$.

2. A determinant changes sign if two columns are interchanged
    Swapping the first two columns of **D** gives

$$\mathbf{D}'' = \begin{vmatrix} b & a & c \\ e & d & f \\ h & g & i \end{vmatrix} = bdi - bfg - aei + afh + ceg - cdh \qquad (1B.15)$$

Again comparing term by term we find that $\mathbf{D}'' = -\mathbf{D}$.

3. A determinant is zero if two rows or columns are the same
   Taking the first two columns as the same

$$\mathbf{D'''} = \begin{vmatrix} a & a & c \\ d & d & f \\ g & g & i \end{vmatrix} = adi - afg - adi + afg + cdg - cdg = 0 \tag{1B.16}$$

In the context of Slater determinants, consider the LiH^+ molecule with orbital occupancy $1\sigma^2 2\sigma$

$$
\begin{aligned}
\Psi^{LiH^+} &= \frac{1}{\sqrt{6}} \begin{vmatrix} \phi_{1\sigma}(1) & \bar{\phi}_{1\sigma}(1) & \phi_{2\sigma}(1) \\ \phi_{1\sigma}(2) & \bar{\phi}_{1\sigma}(2) & \phi_{2\sigma}(2) \\ \phi_{1\sigma}(3) & \bar{\phi}_{1\sigma}(3) & \phi_{2\sigma}(3) \end{vmatrix} \\
&= \frac{1}{\sqrt{6}} \left\{ \begin{array}{l} \phi_{1\sigma}(1)\left[\bar{\phi}_{1\sigma}(2)\phi_{2\sigma}(3) - \phi_{2\sigma}(2)\bar{\phi}_{1\sigma}(3)\right] \\ -\bar{\phi}_{1\sigma}(1)\left[\phi_{1\sigma}(2)\phi_{2\sigma}(3) - \phi_{2\sigma}(2)\phi_{1\sigma}(3)\right] \\ +\phi_{2\sigma}(1)\left[\phi_{1\sigma}(2)\bar{\phi}_{1\sigma}(3) - \bar{\phi}_{1\sigma}(2)\phi_{1\sigma}(3)\right] \end{array} \right\}
\end{aligned}
\tag{1B.17}
$$

If the $1\sigma^2$ electrons were of parallel spin, then we would substitute $\bar{\phi}_{1\sigma} \to \phi_{1\sigma}$. In which case the first two terms in eqn (1B.17) would cancel each other and the part of the third term in square brackets would be zero. Consequently $\Psi^{LiH^+} = 0$, in accordance with the Pauli exclusion principle which states that such a wavefunction cannot describe the electronic structure of LiH^+.

Appendix 1C   Diagonalisation of Matrices

For symmetric matrices, such as the ones we shall deal with here, there are many techniques for obtaining the eigenvalues and eigenvectors. We shall briefly describe only one technique, known as the *Jacobi method*. It is not the most efficient method but is easy to explain and fairly robust in numerical application.

A symmetric $n \times n$ matrix, $\mathbf{A}$, can be reduced to diagonal form, $\mathbf{a}$, through a unitary transformation, $\mathbf{U}$. A unitary transformation is one which preserves orthonormality of vectors. Hence, $\mathbf{A}$ obeys the eigenvalue equation

$$\mathbf{AU} = \mathbf{Ua} \tag{1C.1}$$

where

$$\mathbf{A} = \begin{pmatrix} A_{11} & A_{12} & \cdots & A_{1n} \\ A_{21} & A_{22} & \cdots & A_{2n} \\ \vdots & \vdots & & \vdots \\ A_{n1} & A_{n2} & \cdots & A_{nn} \end{pmatrix} \qquad \mathbf{U} = \begin{pmatrix} U_{11} & U_{12} & \cdots & U_{1n} \\ U_{21} & U_{22} & \cdots & U_{2n} \\ \vdots & \vdots & & \vdots \\ U_{n1} & U_{n2} & \cdots & U_{nn} \end{pmatrix}$$

$$\mathbf{a} = \begin{pmatrix} a_{11} & 0 & \cdots & 0 \\ 0 & a_{11} & \cdots & 0 \\ \vdots & \vdots & \ddots & 0 \\ 0 & 0 & \cdots & a_{nn} \end{pmatrix}$$

The diagonal elements of the matrix $\mathbf{a}$ contain the eigenvalues of the matrix $\mathbf{A}$, and the *columns* of the transformation matrix $\mathbf{U}$ contain the eigenvectors of the matrix $\mathbf{A}$.

A unitary matrix such as $\mathbf{U}$ has the special property that its inverse, $\mathbf{U}^{-1}$, is equal to its *conjugate transpose*, $\mathbf{U}^*$. The conjugate transpose of a matrix is obtained by taking the complex conjugate of all elements, U_{ij}^*, and then swapping all elements U_{ij}^* with U_{ji}. In most of this book we are concerned with real matrices, and so $\mathbf{U}$ is orthogonal rather than unitary and the inverse matrix is simply the transpose of $\mathbf{U}$: $\mathbf{U}^{-1} = \mathbf{U}^{\mathrm{T}}$. Thus to obtain $\mathbf{a}$, we pre-multiply $\mathbf{A}$ by $\mathbf{U}^{\mathrm{T}}$ and post-multiply by $\mathbf{U}$

$$\mathbf{U}^{\mathrm{T}}\mathbf{A}\mathbf{U} = \mathbf{a} \tag{1C.2}$$

All that remains is to specify the form of $\mathbf{U}$. We can specify an elementary unitary matrix, $\mathbf{u}$, as $\mathbf{u}(i,j,\theta)$, for example

$$\mathbf{u}(2,4,\theta) = \begin{pmatrix} 1 & 0 & 0 & 0 \\ 0 & \cos\theta & 0 & \sin\theta \\ 0 & 0 & 1 & 0 \\ 0 & -\sin\theta & 0 & \cos\theta \end{pmatrix} \tag{1C.3}$$

This matrix has 1 along the diagonal and 0 everywhere else except for 4 entries in the "*ij-plane*". In the example above the *ij* refers to 2,4. The entries in the 2,2 and 4,4 positions are replaced by $\cos\theta$ and those in the 2,4 and 4,2 positions by $\sin\theta$ and $-\sin\theta$, respectively. A matrix such as $\mathbf{u}$ describes a rotation in the *ij* (2,4) plane

$$\mathbf{u}^{\mathrm{T}}\mathbf{A}\mathbf{u} = \mathbf{A}' \tag{1C.4}$$

in which we have omitted the dependence on the plane ij and the rotation angle θ, for notational convenience. The angle of rotation, θ, may be chosen so as to eliminate the ij and ji elements of the matrix $\mathbf{A}'$. This is most easily illustrated by a 2 x 2 example. Consider the symmetric matrix,

$$\mathbf{A} = \begin{pmatrix} 5 & -2 \\ -2 & 2 \end{pmatrix}$$

there is only one off-diagonal element to eliminate and

$$\mathbf{u} = \begin{pmatrix} \cos\theta & \sin\theta \\ -\sin\theta & \cos\theta \end{pmatrix} \text{ and } \tan 2\theta = \frac{2A_{12}}{A_{22} - A_{11}}$$

$$\Rightarrow \theta = 0.463647 \text{ radians}; \cos\theta = 0.894427; \sin\theta = 0.447213$$

$$\mathbf{u}^\mathrm{T}\mathbf{A}\mathbf{u} = \begin{pmatrix} 0.894427 & -0.447213 \\ 0.447213 & 0.894427 \end{pmatrix} \begin{pmatrix} 5 & -2 \\ -2 & 2 \end{pmatrix} \begin{pmatrix} 0.894427 & 0.447213 \\ -0.447213 & 0.894427 \end{pmatrix}$$

$$= \begin{pmatrix} 6 & 0 \\ 0 & 1 \end{pmatrix}$$

So the matrix $\mathbf{A}$ has eigenvalues of 6 and 1 with corresponding eigenvectors of $(0.894427, -0.447213)$ and $(0.447213, 0.894427)$.

For a matrix of dimension >2, we proceed by searching the lower triangle (since the matrix is symmetric) for the largest off-diagonal matrix element, which we denote as A_{ij}. The angle of rotation required to eliminate this off-diagonal element is evaluated as

$$\tan 2\theta = \frac{2A_{ij}}{A_{ii} - A_{jj}} \left\{ \text{if } A_{ii} = A_{jj} \text{ then } \theta = \frac{\pi}{4} \right\} \tag{1C.5}$$

and we form $\mathbf{u}_1^\mathrm{T}\mathbf{A}\mathbf{u}_1 = \mathbf{A}'$. We next search the lower triangle of the matrix $\mathbf{A}'$ for the largest of-diagonal element, A'_{ij}, and repeat our procedure until no off-diagonal elements remain

$$\mathbf{u}_2^\mathrm{T}\mathbf{A}'\mathbf{u}_2 = \mathbf{A}''$$

$$\mathbf{u}_3^\mathrm{T}\mathbf{A}''\mathbf{u}_3 = \mathbf{A}'''$$

$$\vdots$$

$$\mathbf{u}_\mathrm{m}^\mathrm{T}\mathbf{A}'''\mathbf{u}_\mathrm{m} = \mathbf{a} \text{ (diagonal)}$$

The total rotation matrix $\mathbf{U}$ is then given by

$$U = u_1 u_2 u_3 \cdots u_m \tag{1C.6}$$

By such a procedure we can find the eigenvalues and eigenvectors of a symmetric matrix of essentially any size (provided it fits into the memory of a computer). This also provides a general scheme for computing *powers of a matrix, e.g.* $\mathbf{A}^3, \mathbf{A}^{\frac{1}{2}}, \mathbf{A}^{-\frac{3}{2}}$. In general

$$\mathbf{A}^x = \mathbf{U} \mathbf{a}^x \mathbf{U}^{-1}$$

First the eigenvalues and eigenvectors must be found. Then the eigenvalues (diagonal elements of $\mathbf{a}$) are raised to the appropriate power to produce $\mathbf{a}^x$, finally this matrix is back transformed by the eigenvector matrix to yield $\mathbf{A}^x$.

To implement this diagonalisation procedure as a computer program we can exploit the fact that each plane rotation matrix $\mathbf{u}(i,j,\theta)$ given by eqn (1C.3), only affects the rows and columns of the matrix $\mathbf{A}$ that contain the indices i and j. The updated elements of $\mathbf{A}' = \mathbf{u}^{\mathrm{T}}(i,j,\theta)\,\mathbf{A}\,\mathbf{u}(i,j,\theta)$ are:

$$A'_{ij} = A'_{ji} = 0$$

$$A'_{ii} = \frac{1}{2}\left(A_{ii} + A_{jj}\right) + \frac{1}{2}\left(A_{ii} - A_{jj}\right)\cos(2\theta) - A_{ij}\sin(2\theta)$$

$$A'_{jj} = \frac{1}{2}\left(A_{ii} + A_{jj}\right) - \frac{1}{2}\left(A_{ii} - A_{jj}\right)\cos(2\theta) + A_{ij}\sin(2\theta) \tag{1C.7}$$

$$A'_{ik} = A'_{ki} = A_{ik}\cos(\theta) - A_{jk}\sin(\theta) \quad [k \neq i,j]$$

$$A'_{jk} = A'_{kj} = A_{ik}\sin(\theta) + A_{jk}\cos(\theta) \quad [k \neq i,j]$$

To evaluate these quantities we need to know the cosine and sine of θ and 2θ. These can be obtained directly in terms of the elements of $\mathbf{A}$ as:

$$\cos(2\theta) = \frac{\left|A_{ii} - A_{jj}\right|}{\sqrt{\left(A_{ii} - A_{jj}\right)^2 + 4A_{ij}^2}}$$

$$\sin(2\theta) = sign \frac{2A_{ij}}{\sqrt{\left(A_{ii} - A_{jj}\right)^2 + 4A_{ij}^2}} \tag{1C.8}$$

$$\cos(\theta) = \sqrt{\frac{1}{2}(\cos(2\theta) + 1)}$$

$$\sin(\theta) = \frac{\sin(2\theta)}{2\cos(\theta)}$$

In the second equation above, the *sign* factor is taken depending on whether the quantity $\left(A_{ii} - A_{jj}\right)$ is positive or negative. Similarly, the elements of

$U' = U\,\mathbf{u}(i,j,\theta)$ can be accumulated as:

$$U'_{ki} = U_{ki}\cos(\theta) - U_{kj}\sin(\theta)$$
$$U'_{kj} = U_{ki}\sin(\theta) + U_{kj}\cos(\theta)$$

$$(1C.9)$$

This process will produce the eigenvalues along the diagonal of $\mathbf{A}$ in random order and it will be necessary to order them into ascending (or descending) order and, in doing so, remembering to swap the columns of the accumulated transformation matrix $\mathbf{U}$.

Appendix 1D   Moments of Inertia and the Inertia Tensor

Let us assume that we have the cartesian coordinates of a general molecule relative to some origin, $X' = (x', y', z')$. The centre of mass in this system has coordinates:

$$x^{\text{CoM}} = \frac{\sum\limits_{A}^{\text{atoms}} m_A x'_A}{\sum\limits_{A}^{\text{atoms}} m_A} \quad : \quad y^{\text{CoM}} = \frac{\sum\limits_{A}^{\text{atoms}} m_A y'_A}{\sum\limits_{A}^{\text{atoms}} m_A} \quad : \quad z^{\text{CoM}} = \frac{\sum\limits_{A}^{\text{atoms}} m_A z'_A}{\sum\limits_{A}^{\text{atoms}} m_A} \quad (1D.1)$$

The coordinates must now be shifted so that the origin is at the centre of mass:

$$x_A = x'_A - x^{\text{CoM}} \quad : \quad y_A = y'_A - y^{\text{CoM}} \quad : \quad z_A = z'_A - z^{\text{CoM}} \quad (1D.2)$$

Relative to the centre of mass coordinate system, the inertia tensor, I, has elements

$$I = \begin{pmatrix} I_{xx} & I_{xy} & I_{xz} \\ I_{yx} & I_{yy} & I_{yz} \\ I_{zx} & I_{zy} & I_{zz} \end{pmatrix} \quad (1D.3)$$

The tensor is symmetric: $I_{yx} = I_{xy}$; $I_{zx} = I_{xz}$; $I_{zy} = I_{yz}$. The elements of I are given by:

$$I_{xx} = \sum\limits_{A}^{\text{atoms}} m_A\left(y_A^2 + z_A^2\right) \quad : \quad I_{yy} = \sum\limits_{A}^{\text{atoms}} m_A\left(x_A^2 + z_A^2\right) \quad : \quad I_{zz} = \sum\limits_{A}^{\text{atoms}} m_A\left(x_A^2 + y_A^2\right)$$

$$(1D.4)$$

$$I_{xy} = \sum\limits_{A}^{\text{atoms}} m_A x_A y_A \quad : \quad I_{xz} = \sum\limits_{A}^{\text{atoms}} m_A x_A z_A \quad : \quad I_{yz} = \sum\limits_{A}^{\text{atoms}} m_A y_A z_A$$

To obtain the principal moments of inertia, I_A, I_B, I_C, we must assemble and diagonalise I. The eigenvalues of I give the principal moments of inertia, and by convention we choose $I_A \leq I_B, \leq I_C$.

For example consider the hydrogen peroxide molecule with coordinates (au):

Coordinates (au) of HOOH in initial frame

Atom	x	y	z	Masses of atoms / u
O	-7.58339	-1.97542	0.49607	15.99491
H	-8.11665	-0.22102	0.49684	1.00783
O	-4.82324	-1.61428	0.49604	15.99491
H	-4.28999	-3.36869	0.49715	1.00783

Using eqns (1D.1) and (1D.2) we can change the origin of the coordinate system to the centre of mass to obtain

---

Coordinates (au) of HOOH with origin at centre of mass

Atom	x	y	z
O	-1.38007	-0.18057	-0.00004
H	-1.91333	1.57384	0.00073
O	1.38007	0.18057	-0.00007
H	1.91333	-1.57384	0.00103

We can now form the elements of the inertia tensor using eqn (1D.4)

Components of the inertia tensor

$I_{xx} = 6.03576$
$I_{yy} = 68.30694$
$I_{zz} = 74.34270$
$I_{xy} = 1.90217$
$I_{xz} = -0.00016$
$I_{yz} = -0.00058$

Assembling I as in eqn (1D.3) and diagonalising yields the eigenvalues and eigenvectors:

Eigenvalues:	(1) 5.97768	(2) 68.36489	(3) 74.34257
Eigenvectors:			
x	0.99566	-0.09302	-0.00001
y	0.09302	0.99566	0.00010
z	0.00000	-0.00010	1.00000

From which we determine that $I_A = 5.97768$, $I_B = 68.36489$, $I_C = 74.34257$, with the units being $u\,a_0^2$.

References

1. W. G. Richards, *Nature*, 1979, **278**, 507.
2. H. F. Schaefer III, *Science*, 1986, **231**, 1100.
3. P. Y. Ayala and H. B. Schlegel, *J. Chem. Phys.*, 1998, **108**, 2314.
4. D. G. Truhlar, *J. Comput. Chem.*, 1991, **12**, 266.
5. O. Sinanoğlu, *J. Chem. Phys.*, 1962, **36**, 706.
6. J. P. Perdew and K. Schmidt, *AIP Conf. Proc.*, 2001, **577**, 1.
7. W. A. de Jong, E. Bylaska, N. Govind, C. L. Janssen, K. Kowalski, T. Müller, D. I. M. B. Nielsen, H. J. J. van Dam, V. Veryazovb and R. Lindh, *Phys. Chem. Chem. Phys.*, 2010, **12**, 6896.

Computational Electronic Structure Theory

2.1 A Few Essential Notions and Requirements

This chapter explores some of the details associated with the evaluation of the molecular potential energy using the various methods that were introduced in Section 1.4. To begin we need to establish some ideas and notation.

We have already met the idea of a potential energy surface for a ground-state molecule. It is important to recall that the time-independent Schrödinger equation applies to all stationary states of a molecule and not just the ground state. For example, consider the diatomic potential energy curves shown in Figure 2.1. The lowest energy curve refers to the ground state, above this are the curves for two electronically excited states. The lower of the two excited state potential energy curves shows a shallow minimum and so corresponds to a weakly bound electronic state. The highest potential energy curve does not have a minimum and so corresponds to a repulsive, unbound, electronic state. If we denote these three electronic states as A, B and C, then the corresponding electronic Schrödinger equations are

$$\hat{H}|\Psi_A\rangle = E_A|\Psi_A\rangle$$

$$\hat{H}|\Psi_B\rangle = E_B|\Psi_B\rangle \qquad (2.1)$$

$$\hat{H}|\Psi_C\rangle = E_C|\Psi_C\rangle$$

RSC Theoretical and Computational Chemistry Series No. 5
Computational Quantum Chemistry: Molecular Structure and Properties *in Silico*
By Joseph J W McDouall
© Joseph J W McDouall 2013
Published by the Royal Society of Chemistry, www.rsc.org

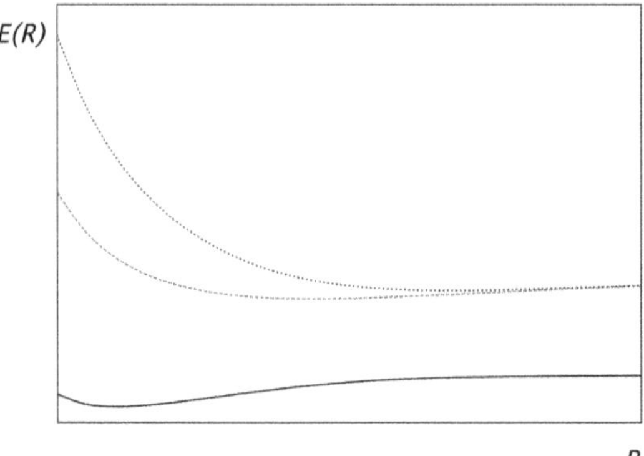

Figure 2.1   Potential energy curves of the lowest three electronic states of a diatomic molecule.

Each electronic state, within the Born–Oppenheimer approximation, has its own wavefunction and energy. An important idea to associate with eqn (2.1) is that the wavefunctions, Ψ_A, Ψ_B and Ψ_C have no overlap with each other. That is, they are orthogonal, implying that the integral over all space of their products is zero. For example,

$$\int_{-\infty}^{+\infty} \Psi_A^* \Psi_B d\tau = 0 \tag{2.2}$$

where we have used τ to indicate all the variables on which Ψ depends. Additionally, we shall take all wavefunctions to be normalised, for example

$$\int_{-\infty}^{+\infty} \Psi_C^* \Psi_C d\tau = 1 \tag{2.3}$$

The conditions of orthogonality in eqn (2.2) and that of normalisation in eqn (2.3) can be succinctly stated by introducing the Kronecker delta function, δ_{XY}. The property of δ_{XY} is that when $X = Y$, $\delta_{XY} = 1$ and if $X \neq Y$, $\delta_{XY} = 0$. Eqns (2.2) and (2.3) can be combined as a statement of the orthonormality of the electronic states which, using Dirac notation, may be written as

$$\langle \Psi_X | \Psi_Y \rangle = \delta_{XY} \tag{2.4}$$

The normalisation of the electronic wavefunctions simplifies the evaluation of the energy since now eqn (1.5) reduces to

$$E = \langle \Psi | H | \Psi \rangle \tag{2.5}$$

2.1.1 Matrix Elements and Integrals

To evaluate the energy requires that we pre- and post-multiply the hamiltonian by the wavefunction and integrate appropriately. Let us look at this statement in a little more detail. Using the form of the electronic hamiltonian given in eqn (1.14), consider the ground state of the lithium hydride molecule with its four electrons and two doubly occupied molecular orbitals, $\phi_{1\sigma}$ and $\phi_{2\sigma}$. The hamiltonian contains two one-electron type operators, corresponding to the kinetic energy of each electron and the attractive potential energy between each electron and the positively charged nuclei. Let us combine these two operators into a single one-electron operator, which we shall denote as $h(\mathbf{r})$. In general for electron i

$$\hat{h}(\mathbf{r}_i) = -\frac{1}{2}\nabla_i^2 - \sum_A^{\text{nuclei}} \frac{Z_A}{r_{iA}} \tag{2.6}$$

For the lithium hydride example each electron will have associated with it an operator as given in eqn (2.6), that is

$$h(\mathbf{r}_1) + h(\mathbf{r}_2) + h(\mathbf{r}_3) + h(\mathbf{r}_4) \tag{2.7}$$

The operator $h(\mathbf{r})$ contains no reference to spin coordinates. The two-electron operator given in eqn (1.14) yields for the current example

$$\frac{1}{r_{12}} + \frac{1}{r_{13}} + \frac{1}{r_{14}} + \frac{1}{r_{23}} + \frac{1}{r_{24}} + \frac{1}{r_{34}} \tag{2.8}$$

To obtain the energy we must substitute these operators into eqn (2.5). Next, we take the wavefunction to be the single Slater determinant

$$|\Psi_0\rangle = |\phi_{1\sigma}\bar{\phi}_{1\sigma}\phi_{2\sigma}\bar{\phi}_{2\sigma}\rangle \tag{2.9}$$

The expansion of $|\Psi_0\rangle$ will produce $4! = 24$ orbital products, an example of which is $\phi_{1\sigma}(\mathbf{x}_1)\bar{\phi}_{1\sigma}(\mathbf{x}_2)\phi_{2\sigma}(\mathbf{x}_3)\bar{\phi}_{2\sigma}(\mathbf{x}_4)$. We must now consider the $(4!)^2$ orbital products that result when we substitute the expansion of eqn (2.9) into eqn (2.5). Considering, for example, the term

$$\int \phi_{1\sigma}^*(\mathbf{x}_1)\bar{\phi}_{1\sigma}^*(\mathbf{x}_2)\phi_{2\sigma}^*(\mathbf{x}_3)\bar{\phi}_{2\sigma}^*(\mathbf{x}_4)h(\mathbf{r}_1)\phi_{1\sigma}(\mathbf{x}_1)\bar{\phi}_{1\sigma}(\mathbf{x}_2)\phi_{2\sigma}(\mathbf{x}_3)\bar{\phi}_{2\sigma}(\mathbf{x}_4)d\mathbf{r}_1 \tag{2.10}$$

$h(\mathbf{r}_1)$ can only act on the spatial coordinate of electron 1, which gives

$$\int \phi_{1\sigma}^*(\mathbf{r}_1)h(\mathbf{r}_1)\phi_{1\sigma}(\mathbf{r}_1)d\mathbf{r}_1 \times \int \alpha^*(\mathbf{s}_1)\alpha(\mathbf{s}_1)d\mathbf{s}_1$$

$$\times \int \phi_{1\sigma}^*(\mathbf{r}_2)\phi_{1\sigma}(\mathbf{r}_2)d\mathbf{r}_2 \times \int \beta^*(\mathbf{s}_2)\beta(\mathbf{s}_2)d\mathbf{s}_2$$

$$\times \int \phi_{2\sigma}^*(\mathbf{r}_3)\phi_{2\sigma}(\mathbf{r}_3)d\mathbf{r}_3 \times \int \alpha^*(\mathbf{s}_3)\alpha(\mathbf{s}_3)d\mathbf{s}_3 \tag{2.11}$$

$$\times \int \phi_{2\sigma}^*(\mathbf{r}_4)\phi_{2\sigma}(\mathbf{r}_4)d\mathbf{r}_4 \times \int \beta^*(\mathbf{s}_4)\beta(\mathbf{s}_4)d\mathbf{s}_4$$

This messy expression can be simplified by noting two further orthonormality conditions, and using Dirac notation. The first condition is between the spin functions, $\alpha(\mathbf{s})$ and $\beta(\mathbf{s})$

$$\int \alpha^*(\mathbf{s})\alpha(\mathbf{s})d\mathbf{s} = 1 \quad \equiv \langle \alpha \mid \alpha \rangle = 1$$

$$\int \beta^*(\mathbf{s})\beta(\mathbf{s})d\mathbf{s} = 1 \quad \equiv \langle \beta \mid \beta \rangle = 1$$

$$\int \alpha^*(\mathbf{s})\beta(\mathbf{s})d\mathbf{s} = 0 \quad \equiv \langle \alpha \mid \beta \rangle = 0 \tag{2.12}$$

$$\int \beta^*(\mathbf{s})\alpha(\mathbf{s})d\mathbf{s} = 0 \quad \equiv \langle \beta \mid \alpha \rangle = 0$$

and the second between the molecular orbitals

$$\int \phi_i^*(\mathbf{r})\,\phi_j(\mathbf{r})d\mathbf{r} = \delta_{ij} \quad \equiv \langle \phi_i \mid \phi_j \rangle = \delta_{ij} \tag{2.13}$$

Taking into account eqns (2.12) and (2.13), we find that the complex expression in eqn (2.11) is reduced to

$$\int \phi_{1\sigma}^*(\mathbf{r}_1)h(\mathbf{r}_1)\phi_{1\sigma}(\mathbf{r}_1)d\mathbf{r}_1 \equiv \langle \phi_{1\sigma}|h|\phi_{1\sigma} \rangle \tag{2.14}$$

By a similar process it is possible to show that the two-electron operator, $1/\mathbf{r}_{12}$, for the same orbital products as in eqn (2.10), yields a single integral

$$\iint \phi_{1\sigma}^*(\mathbf{r}_1)\phi_{1\sigma}^*(\mathbf{r}_2)\frac{1}{r_{12}}\phi_{1\sigma}(\mathbf{r}_1)\phi_{1\sigma}(\mathbf{r}_2)d\mathbf{r}_1 d\mathbf{r}_2 \equiv \langle \phi_{1\sigma}\phi_{1\sigma}|\frac{1}{r_{12}}|\phi_{1\sigma}\phi_{1\sigma} \rangle \tag{2.15}$$

These are the basic one- and two-electron integrals that we shall meet throughout the rest of this book. Being so ubiquitous, it is useful to introduce a shorthand notation for them. The one-electron integrals, between molecular orbitals ϕ_i and ϕ_j, will be denoted as

$$h_{ij} = \langle \phi_i | h | \phi_j \rangle \tag{2.16}$$

The two-electron integrals are usually represented by two shorthand notations. The first is mostly found in the physics literature and involves the angled brackets, $< >$. The second notation dominates in the chemistry literature and involves parentheses, $(\ )$. A general two-electron integral between molecular orbitals ϕ_i, ϕ_j, ϕ_k and ϕ_l is written as

$$\langle \phi_i \phi_k | \frac{1}{r_{12}} | \phi_j \phi_l \rangle = \langle \phi_i \phi_k \mid \phi_j \phi_l \rangle = (\phi_i \phi_j | \phi_k \phi_l) = (ij|kl) \tag{2.17}$$

The difference between the $< >$ and the $(\ )$ notation is that in the $(\ )$ notation the functions associated with electron 1 are on the left of the vertical bar and those of electron 2 on the right. The $(\ )$ notation is also known as "Mulliken notation" or "charge cloud" notation, an allusion to the charge cloud of the electron on the left of the vertical bar interacting with the charge cloud of the electron on the right.

The evaluation of the energy of a Slater determinant can be carried out using general rules for a system of arbitrary complexity. Consider a system of N electrons described by a Slater determinant of N spin–orbitals, $|\Psi\rangle = |\phi_i \phi_j \phi_k \phi_l \cdots\rangle$. The matrix element of this Slater determinant over the electronic hamiltonian, $\langle \Psi | H | \Psi \rangle$, can be written compactly using the notations we have introduced as

$$\langle \Psi | H | \Psi \rangle = \langle \phi_i \phi_j \phi_k \phi_l \cdots | H | \phi_i \phi_j \phi_k \phi_l \cdots \rangle$$
$$= \sum_i^N h_{ii} + \sum_{i<j}^N [(ii|jj) - (ij|ji)] \tag{2.18}$$

This expression is written over spin–orbitals. Comparing the two-electron terms we note that in the final integral the electronic coordinates are exchanged between the orbitals ϕ_i and ϕ_j. Consequently unless the spin–orbitals ϕ_i and ϕ_j are of the same spin, the final integral will be zero by the spin-orthogonality of eqn (2.12).

Before leaving our discussion on evaluating matrix elements of the type $\langle \Psi | H | \Psi \rangle$, we must consider what happens when the Slater determinants on the left and right of the hamiltonian are different. Consider two Slater determinants, $|\Psi\rangle$ and $|\Psi'\rangle$

$$|\Psi\rangle = |\phi_i \phi_j \phi_k \phi_l \cdots\rangle$$
$$|\Psi'\rangle = |\phi_{i'} \phi_j \phi_k \phi_l \cdots\rangle \tag{2.19}$$

The two determinants differ only in the substitution of ϕ_i in $|\Psi\rangle$ with $\phi_{i'}$ in $|\Psi'\rangle$. The matrix element can be evaluated as

$$\langle\Psi|H|\Psi'\rangle = \langle\phi_i\phi_j\phi_k\phi_l\cdots|H|\phi_{i'}\phi_j\phi_k\phi_l\cdots\rangle$$
$$= h_{ii'} + \sum_{j\neq i,i'}^{N}\left[(ii'|jj) - (ij|ji')\right] \tag{2.20}$$

The only other non-vanishing matrix element we need consider will be when there are two spin–orbital differences. For example,

$$|\Psi\rangle = |\phi_i\phi_j\phi_k\phi_l\cdots\rangle$$
$$|\Psi''\rangle = |\phi_{i'}\phi_{j'}\phi_k\phi_l\cdots\rangle \tag{2.21}$$

The resulting matrix element is

$$\langle\Psi|H|\Psi''\rangle = \langle\phi_i\phi_j\phi_k\phi_l\cdots|H|\phi_{i'}\phi_{j'}\phi_k\phi_l\cdots\rangle$$
$$= (ii'|jj') - (ij'|ji') \tag{2.22}$$

If two determinants differ by more than two spin–orbitals then the matrix element over the hamiltonian in eqn (1.14) is zero. Before applying the rules in eqns (2.18), (2.20) and (2.22) the two determinants must be brought into maximum coincidence by permuting the orbital indices and noting the change in phase that accompanies each permutation. For example, consider the determinants

$$|\Psi_I\rangle = |\phi_1\bar{\phi}_1\phi_2\bar{\phi}_2\rangle$$
$$|\Psi_{II}\rangle = |\phi_1\bar{\phi}_1\phi_2\bar{\phi}_3\rangle \tag{2.23}$$
$$|\Psi_{III}\rangle = |\phi_1\bar{\phi}_1\bar{\phi}_3\phi_4\rangle$$

Applying eqn (2.18), remembering the spin orthogonality will eliminate the last term unless the spins match, the energy of $|\Psi_I\rangle$ is given over spatial integrals by

$$\langle\Psi_I|H|\Psi_I\rangle = \langle\phi_1\bar{\phi}_1\phi_2\bar{\phi}_2|H|\phi_1\bar{\phi}_1\phi_2\bar{\phi}_2\rangle$$
$$= 2h_{11} + 2h_{22} + (11|11) + 4(11|22) + (22|22) - 2(12|21) \tag{2.24}$$

The matrix element $\langle\Psi_I|H|\Psi_{II}\rangle$ corresponds to a single spin–orbital difference and is given by

$$\langle\Psi_I|H|\Psi_{II}\rangle = \langle\phi_1\bar{\phi}_1\phi_2\bar{\phi}_2|H|\phi_1\bar{\phi}_1\phi_2\bar{\phi}_3\rangle$$
$$= h_{23} + 2(23|11) + (23|22) - (21|13) \tag{2.25}$$

The matrix element $\langle \Psi_{II} | H | \Psi_{III} \rangle$ also corresponds to a single spin–orbital difference. Before evaluating $\langle \Psi_{II} | H | \Psi_{III} \rangle$ we must bring the spin–orbitals into maximum coincidence

$$
\begin{aligned}
\langle \Psi_{II} | H | \Psi_{III} \rangle &= \langle \phi_1 \bar{\phi}_1 \phi_2 \bar{\phi}_3 | H | \phi_1 \bar{\phi}_1 \bar{\phi}_3 \phi_4 \rangle \\
&= -\langle \phi_1 \bar{\phi}_1 \phi_2 \bar{\phi}_3 | H | \phi_1 \bar{\phi}_1 \phi_4 \bar{\phi}_3 \rangle
\end{aligned}
\tag{2.26}
$$

The evaluation now proceeds as for $\langle \Psi_I | H | \Psi_{II} \rangle$. As a final example, $\langle \Psi_I | H | \Psi_{III} \rangle$ can be evaluated as

$$
\begin{aligned}
\langle \Psi_I | H | \Psi_{III} \rangle &= \langle \phi_1 \bar{\phi}_1 \phi_2 \bar{\phi}_2 | H | \phi_1 \bar{\phi}_1 \bar{\phi}_3 \phi_4 \rangle \\
&= -\langle \phi_1 \bar{\phi}_1 \phi_2 \bar{\phi}_2 | H | \phi_1 \bar{\phi}_1 \phi_4 \bar{\phi}_3 \rangle \\
&= -(24|23)
\end{aligned}
\tag{2.27}
$$

2.1.2 Permutational Symmetry of One- and Two-Electron Integrals

We introduced one- and two-electron integrals in the preceding section. In all the calculations that we shall discuss, these integrals will be evaluated over real orbitals, which means that $\phi^* = \phi$. This gives a permutational equivalence to the integrals, such that for the one-electron terms there is a two-fold symmetry

$$
h_{ji} = h_{ij}
\tag{2.28}
$$

and for the two-electron integrals we have an eight-fold symmetry

$$
(ij|kl) = (ji|kl) = (ij|lk) = (ji|lk) = (kl|ij) = (lk|ij) = (kl|ji) = (lk|ji)
\tag{2.29}
$$

These symmetries can be used to reduce the number of integrals that must be evaluated. For the one-electron integrals we can adopt the convention that $i \geq j$. The number of integrals to be evaluated, for m orbitals, is reduced from m^2 to $m(m+1)/2$. For efficiency, these can be stored in a linear array and the address of element ij can be obtained as

$$
ij = i(i-1)/2 + j
\tag{2.30}
$$

The two-electron integrals are reduced from m^4 to $\frac{1}{2}[m(m+1)/2][m(m+1)/2+1] \approx m^4/8$. We now need four indices to address a two-electron integral, $(ij|kl)$. We similarly adopt the restrictions: $i \geq j; k \geq l; ij \geq kl$ and evaluate the address of element $ijkl$ as

$$ijkl = ij(ij-1)/2 + kl \tag{2.31}$$

where ij and kl are given by eqn (2.30).

2.1.3 Spin Symmetry

The hamiltonian in eqn (1.14) makes no reference to spin coordinates, it is "spin-free". This means that the many-electron spin operators, $\hat{S}_z$ and $\hat{S}^2$ commute with this hamiltonian, that is

$$\hat{H}\hat{S}_z - \hat{S}_z\hat{H} = \left[\hat{H},\hat{S}_z\right] = 0$$
$$\hat{H}\hat{S}^2 - \hat{S}^2\hat{H} = \left[\hat{H},\hat{S}^2\right] = 0 \tag{2.32}$$

The square bracket is a compact notation for the commutator of two operators, $\left[\hat{A},\hat{B}\right] = \hat{A}\hat{B} - \hat{B}\hat{A}$. We shall denote the spin quantum numbers of a single electron as s and m_s, and that of a many-electron system as S and M_S.

It is a property of quantum mechanical operators that if they commute, they may have simultaneous solutions (eigenfunctions). This suggests that in forming wavefunctions, corresponding to the electronic hamiltonian, we should require that they be solutions of the electronic Schrödinger equation and also the operators $\hat{S}_z$ and $\hat{S}^2$.

The $\hat{S}_z$ operator gives the z-axis projection of the spin, z is taken as the axis of spin quantisation. $\hat{S}^2$ gives the magnitude squared of the spin angular momentum. Quantum mechanical rules tell us that we can simultaneously measure the magnitude squared of the spin angular momentum and at most one cartesian component. Conventionally that component is taken as the z-axis, and so

$$\left[\hat{S}^2,\hat{S}_z\right] = 0 \tag{2.33}$$

Consequently our electronic wavefunction must satisfy eqn (1.16) and also

$$\hat{S}_z|\Psi\rangle = M_s|\Psi\rangle \tag{2.34}$$

$$\hat{S}^2|\Psi\rangle = S(S+1)|\Psi\rangle \tag{2.35}$$

It is possible to obtain simple expressions for the action of $\hat{S}_z$ and $\hat{S}^2$ on a Slater determinant. The case of $\hat{S}_z$ is particularly simple since the spin–orbitals from which the Slater determinant is built are individually solutions to eqn

(1.23), and hence

$$\hat{S}_z|\Psi\rangle = \frac{1}{2}\left(N_\alpha - N_\beta\right)|\Psi\rangle$$

$$= M_S|\Psi\rangle \tag{2.36}$$

where N_α and N_β are the number of α and β electrons, respectively. The action of $\hat{S}^2$ is a little more complicated and may be written as

$$\hat{S}^2|\Psi\rangle = \left\{ \sum_{\substack{\alpha,\beta \\ permutations}} \hat{P}_{\alpha\beta} + \frac{1}{4}\left[(N_\alpha - N_\beta)^2 + 2N_\alpha + 2N_\beta\right]\right\}|\Psi\rangle \tag{2.37}$$

$\hat{P}_{\alpha\beta}$ is an operator that swaps α and β spin functions between the orbitals of the determinant. For example, the ground state of lithium hydride has total spin $S = 0$ and $S\,(S+1) = 0$. Applying eqn (2.37) to the Slater determinant in eqn (2.9) gives

$$\frac{1}{4}\left[(N_\alpha - N_\beta)^2 + 2N_\alpha + 2N_\beta\right]|\phi_{1\sigma}\bar{\phi}_{1\sigma}\phi_{2\sigma}\bar{\phi}_{2\sigma}\rangle = 2|\phi_{1\sigma}\bar{\phi}_{1\sigma}\phi_{2\sigma}\bar{\phi}_{2\sigma}\rangle$$

$$\sum_{\substack{\alpha,\beta \\ permutations}} \hat{P}_{\alpha\beta}|\phi_{1\sigma}\bar{\phi}_{1\sigma}\phi_{2\sigma}\bar{\phi}_{2\sigma}\rangle = |\bar{\phi}_{1\sigma}\phi_{1\sigma}\phi_{2\sigma}\bar{\phi}_{2\sigma}\rangle$$

$$+ |\phi_{1\sigma}\phi_{1\sigma}\bar{\phi}_{2\sigma}\bar{\phi}_{2\sigma}\rangle \tag{2.38}$$

$$+ |\bar{\phi}_{1\sigma}\bar{\phi}_{1\sigma}\phi_{2\sigma}\phi_{2\sigma}\rangle$$

$$+ |\phi_{1\sigma}\bar{\phi}_{1\sigma}\bar{\phi}_{2\sigma}\phi_{2\sigma}\rangle$$

$$= -2|\phi_{1\sigma}\bar{\phi}_{1\sigma}\phi_{2\sigma}\bar{\phi}_{2\sigma}\rangle$$

so $\hat{S}^2|\phi_{1\sigma}\bar{\phi}_{1\sigma}\phi_{2\sigma}\bar{\phi}_{2\sigma}\rangle = 0$ in accord with eqn (2.35). If we applied eqn (2.37) to the cation of lithium hydride, $|\phi_{1\sigma}\bar{\phi}_{1\sigma}\phi_{2\sigma}\rangle$, we would find

$$\frac{1}{4}\left[(N_\alpha - N_\beta)^2 + 2N_\alpha + 2N_\beta\right]|\phi_{1\sigma}\bar{\phi}_{1\sigma}\phi_{2\sigma}\rangle = \frac{7}{4}|\phi_{1\sigma}\bar{\phi}_{1\sigma}\phi_{2\sigma}\rangle$$

$$\sum_{\substack{\alpha,\beta \\ permutations}} \hat{P}_{\alpha\beta}|\phi_{1\sigma}\bar{\phi}_{1\sigma}\phi_{2\sigma}\rangle = |\bar{\phi}_{1\sigma}\phi_{1\sigma}\phi_{2\sigma}\rangle + |\phi_{1\sigma}\phi_{1\sigma}\bar{\phi}_{2\sigma}\rangle \tag{2.39}$$

$$= -|\phi_{1\sigma}\bar{\phi}_{1\sigma}\phi_{2\sigma}\rangle$$

and $\hat{S}^2|\phi_{1\sigma}\bar{\phi}_{1\sigma}\phi_{2\sigma}\rangle = \frac{3}{4}|\phi_{1\sigma}\bar{\phi}_{1\sigma}\phi_{2\sigma}\rangle$. The wavefunctions we have used for lithium hydride and its cation both obey eqns (2.34) *and* (2.35) and are said to be spin eigenfunctions.

Electronic determinants that contain electrons in different spatial orbitals that are spin-paired, for example $|\phi_1\bar{\phi}_1\phi_2\bar{\phi}_3\rangle$, will not satisfy eqn (2.35)

$$\frac{1}{4}\left[(N_\alpha - N_\beta)^2 + 2N_\alpha + 2N_\beta\right]|\phi_1\bar{\phi}_1\phi_2\bar{\phi}_3\rangle = 2|\phi_1\bar{\phi}_1\phi_2\bar{\phi}_3\rangle$$

$$\sum_{\substack{\alpha,\beta \\ permutations}} \hat{P}_{\alpha\beta}|\phi_1\bar{\phi}_1\phi_2\bar{\phi}_3\rangle = |\bar{\phi}_1\phi_1\phi_2\bar{\phi}_3\rangle + |\phi_1\phi_1\bar{\phi}_2\bar{\phi}_3\rangle$$

$$+ |\bar{\phi}_1\bar{\phi}_1\phi_2\phi_3\rangle + |\phi_1\bar{\phi}_1\bar{\phi}_2\phi_3\rangle$$

$$= -|\phi_1\bar{\phi}_1\phi_2\bar{\phi}_3\rangle + |\phi_1\bar{\phi}_1\bar{\phi}_2\phi_3\rangle$$

$$\hat{S}^2|\phi_1\bar{\phi}_1\phi_2\bar{\phi}_3\rangle = |\phi_1\bar{\phi}_1\phi_2\bar{\phi}_3\rangle + |\phi_1\bar{\phi}_1\bar{\phi}_2\phi_3\rangle \tag{2.40}$$

Clearly the action of $\hat{S}^2$ is not returning a number multiplied by the original Slater determinant. Instead, a combination of determinants is produced. $|\phi_1\bar{\phi}_1\phi_2\bar{\phi}_3\rangle$ is not a spin eigenfunction, since it only obeys eqn (2.34) but not eqn (2.35). To obtain a spin eigenfunction for such an electronic configuration requires that we take a linear combination of Slater determinants, for example

$$|\Psi_I\rangle = |\phi_1\bar{\phi}_1\phi_2\bar{\phi}_3\rangle + |\phi_1\bar{\phi}_1\bar{\phi}_2\phi_3\rangle$$

$$\hat{S}_z|\Psi_I\rangle = M_S|\Psi_I\rangle = 0|\Psi_I\rangle \tag{2.41}$$

$$\hat{S}^2|\Psi_I\rangle = S(S+1)|\Psi_I\rangle = 2|\Psi_I\rangle \quad \Rightarrow S = 1$$

or

$$|\Psi_I\rangle = |\phi_1\bar{\phi}_1\phi_2\bar{\phi}_3\rangle - |\phi_1\bar{\phi}_1\bar{\phi}_2\phi_3\rangle$$

$$\hat{S}_z|\Psi_I\rangle = M_S|\Psi_I\rangle = 0|\Psi_I\rangle \tag{2.42}$$

$$\hat{S}^2|\Psi_I\rangle = S(S+1)|\Psi_I\rangle = 0|\Psi_I\rangle \quad \Rightarrow S = 0$$

Spin eigenfunctions can always be formed as linear combinations of Slater determinants.

2.1.4 The Variation Theorem

We mentioned the variational optimisation of certain types of wavefunction in our survey of methods in Section 1.4. The variation theorem establishes this as a theoretically sound procedure.

We know the form of the hamiltonian, $\hat{H}$, but we must find the wavefunction as a solution to the Schrödinger equation. Let us assume that we don't know the exact solution $|\Psi\rangle$, but have "guessed" a normalised approximation to $|\Psi\rangle$, which we shall denote as $|\Phi\rangle$. The variation theorem states that the expectation value, that is the energy, for the guessed wavefunction will always be greater than or equal to the exact ground state energy, E_{Exact},

$$E_{\text{Exact}} \leq \langle\Phi|H|\Phi\rangle \tag{2.43}$$

This property will hold for any choice of $|\Phi\rangle$. If $|\Phi\rangle$ contains parameters which can be varied to obtain the lowest energy, that set of parameters will correspond to a "best" wavefunction and "best" estimate of the energy obtainable with the chosen form of $|\Phi\rangle$.

Suppose that the exact solutions of the Schrödinger equation exist in the set $\{\Psi\}$. We can write $|\Phi\rangle$ in terms of the set $\{\Psi\}$ as

$$|\Phi\rangle = \sum_M C_M |\Psi_M\rangle \tag{2.44}$$

Normalisation of $|\Phi\rangle$ and the orthonormality of the set $\{\Psi\}$ gives two useful properties,

$$\sum_M C_M^2 = 1$$
$$\langle\Psi_M|\Psi_N\rangle = \delta_{MN} \tag{2.45}$$

We now proceed to evaluate the energy using $|\Phi\rangle$

$$
\begin{aligned}
E[\Phi] &= \langle\Phi|H|\Phi\rangle \\[1em]
&= \left\langle \sum_M C_M\Psi_M \middle| H \middle| \sum_N C_N\Psi_N \right\rangle \\[1em]
&= \sum_M \sum_N \langle C_M\Psi_M | H | C_N\Psi_N \rangle \\[1em]
&= \sum_M \sum_N \langle C_M\Psi_M | E_N | C_N\Psi_N \rangle \qquad (\hat{H}\Psi_N = E_N\Psi_N) \\[1em]
&= \sum_M \sum_N C_M C_N E_N \langle\Psi_M|\Psi_N\rangle \\[1em]
&= \sum_N C_N^2 E_N
\end{aligned}
\tag{2.46}
$$

Since $C_N^2 \geq 0$ and $E_N \geq E_{\text{Exact}}$ then $\langle \Phi | H \Phi \rangle \geq E_{\text{Exact}}$. The variational principle provides the basis for the optimisation of molecular orbitals in the Hartree–Fock theory and the optimisation of the Kohn–Sham orbitals in density functional methods. As we shall see in later chapters, the use of variational methods that obey eqn (2.43) can also provide significant simplifications in the evaluation of molecular properties.

2.2   Hartree–Fock Theory

We met the Hartree–Fock method in Section 1.4. Here we return to it in order to address some matters of detail. The Hartree–Fock method was once the most widely used quantum chemical technique, and it is still the basis for a number of "post Hartree–Fock" electron correlation methods.

In the Hartree–Fock approach the N-electron wavefunction is taken to be a single Slater determinant, $|\Psi_0\rangle$. In Section 2.1.1 we noted that the energy associated with a single Slater determinant of orthonormal spin–orbitals is given by

$$
E_0 = \langle \Psi_0 | H | \Psi_0 \rangle
$$

$$
= \sum_i^N h_{ii} + \sum_{i<j}^N [(ii|jj) - (ij|ji)] \tag{2.47}
$$

in which we have used the compact notations of eqns (2.16) and (2.17) to represent the integrals. For the development here let us restrict consideration to the case where $|\Psi_0\rangle$ represents a closed-shell molecule consisting of N electrons in $N/2$ doubly occupied orbitals, as in Figure 2.2, with $|\Psi_0\rangle = |\phi_i \bar{\phi}_i \phi_j \bar{\phi}_j \phi_k \bar{\phi}_k\rangle$. Consider the one-electron integrals in eqn (2.47). The spin–orbitals ϕ_i and $\bar{\phi}_i$ differ only in their spin functions. Since the operator $\hat{h}$ only works on spatial coordinates we have, $h_{\bar{i}\bar{i}} = h_{ii}$. Separating the summation over the N spin–orbitals

$$
\sum_i^N = \sum_i^{N_\alpha} + \sum_{\bar{i}}^{N_\beta} \tag{2.48}
$$

allows us to write

$$
\sum_i^N h_{ii} = \sum_i^{N_\alpha} h_{ii} + \sum_{\bar{i}}^{N_\beta} h_{\bar{i}\bar{i}} = 2 \sum_i^{N/2} h_{ii} \tag{2.49}
$$

For the two-electron integrals we can proceed similarly. Noting that when $i = j$, the two-electron contribution in eqn (2.47) vanishes, we can write

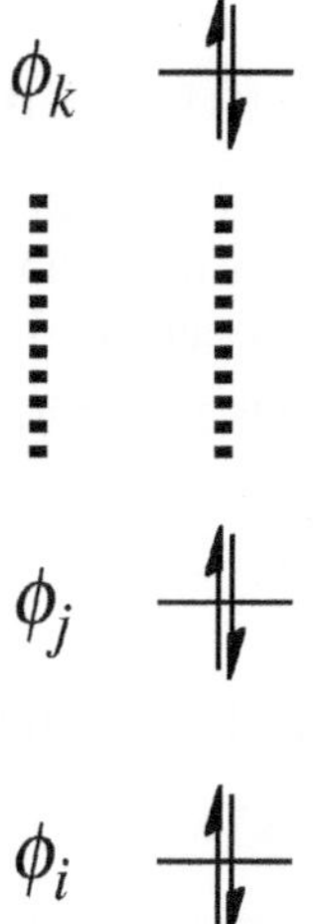

Figure 2.2 Electronic occupancy corresponding to a closed-shell molecule.

$$\sum_{i<j}^{N} = \frac{1}{2}\sum_{i}^{N}\sum_{j}^{N} = \frac{1}{2}\left(\sum_{i}^{N_\alpha} + \sum_{\bar{i}}^{N_\beta}\right)\left(\sum_{j}^{N_\alpha} + \sum_{\bar{j}}^{N_\beta}\right) \tag{2.50}$$

The unrestricted summations can now be applied to the two-electron integrals

$$\frac{1}{2}\left\{\begin{array}{l} \sum_{i}^{N_\alpha}\sum_{j}^{N_\alpha}[(ii|jj)-(ij|ji)] + \sum_{i}^{N_\alpha}\sum_{\bar{j}}^{N_\beta}(ii|\bar{j}\bar{j}) \\[1em] + \sum_{\bar{i}}^{N_\beta}\sum_{j}^{N_\alpha}(\bar{i}\bar{i}|jj) + \sum_{\bar{i}}^{N_\beta}\sum_{\bar{j}}^{N_\beta}[(\bar{i}\bar{i}|\bar{j}\bar{j})-(\bar{i}\bar{j}|\bar{j}\bar{i})] \end{array}\right\} = \sum_{ij}^{N/2}[2(ii|jj)-(ij|ji)] \tag{2.51}$$

where we have used the fact that the integrals are evaluated over spatial functions and so $(ii|jj) = (ii|\bar{j}\bar{j}) = (\bar{i}\bar{i}|jj) = (\bar{i}\bar{i}|\bar{j}\bar{j})$, $(ij|ji) = (\bar{i}\bar{j}|\bar{j}\bar{i})$ and any integral for which the spin functions of an electron do not match is zero by spin orthogonality, eqn (2.12). Using eqns (2.49) and (2.51) enables us to write the energy for a closed-shell molecule as a sum over doubly occupied orbitals

$$E_0 = 2\sum_{i}^{N/2} h_{ii} + \sum_{ij}^{N/2}[2(ii|jj)-(ij|ji)] \tag{2.52}$$

Of the two-electron integrals the first is called a "Coulomb integral" and has the form

$$(ii|jj) = \langle \phi_i(\mathbf{r}_1)\phi_i(\mathbf{r}_1)|\frac{1}{r_{12}}|\phi_j(\mathbf{r}_2)\phi_j(\mathbf{r}_2)\rangle = J_{ij} \tag{2.53}$$

The factor of two that pre-multiplies this term in eqn (2.52) arises because each electron in ϕ_i is repelled by two electrons in ϕ_j. We can also think of this integral as arising from a Coulomb operator, $\hat{J}_j$

$$\hat{J}_j = \int \phi_j(\mathbf{r}_2)\frac{1}{r_{12}}\phi_j(\mathbf{r}_2)d\mathbf{r}_2 \tag{2.54}$$

pre- and post-multiplying by ϕ_i and integrating gives

$$\langle \phi_i|J_j|\phi_i\rangle = J_{ij} \tag{2.55}$$

The second type of two-electron integral that appears in eqn (2.52) arises from the antisymmetry of the wavefunction and has the form

$$(ij|ji) = \langle \phi_i(\mathbf{r}_1)\phi_j(\mathbf{r}_1)|\frac{1}{r_{12}}|\phi_j(\mathbf{r}_2)\phi_i(\mathbf{r}_2)\rangle = K_{ij} \tag{2.56}$$

Notice that, in contrast to J_{ij}, the coordinates of electrons 1 and 2 have been swapped between orbitals ϕ_i and ϕ_j. Accordingly, K_{ij} is usually called an "exchange integral" and can be associated with an exchange operator, $\hat{K}_j$, which acting on an orbital ϕ_i has the property

$$\hat{K}_j \phi_i(\mathbf{r}_1) = \int \phi_j(\mathbf{r}_2)\frac{1}{r_{12}}\phi_i(\mathbf{r}_2)d\mathbf{r}_2\,\phi_j(\mathbf{r}_1) \tag{2.57}$$

We can write

$$\langle \phi_i|K_j|\phi_i\rangle = K_{ij} \tag{2.58}$$

in analogy with the Coulomb operator.

In terms of the Coulomb and exchange integrals we have introduced, the Hartree–Fock energy given in eqn (2.52) can be written as

$$E_0 = 2\sum_i^{N/2} h_{ii} + \sum_{ij}^{N/2} \left[2\langle \phi_i|J_j|\phi_i\rangle - \langle \phi_i|K_j|\phi_i\rangle\right]$$

$$\tag{2.59}$$

$$= 2\sum_i^{N/2} h_{ii} + \sum_{ij}^{N/2} \left[2J_{ij} - K_{ij}\right]$$

2.2.1 Minimisation of the Hartree–Fock Energy

The expressions for the Hartree–Fock energy we obtained in the previous section are entirely equivalent to eqn (2.18) when specialised to a Slater determinant composed of a set of doubly occupied molecular orbitals. The purpose of the Hartree–Fock method is to vary the forms of $\{\phi\}$ such that the energy is minimised in accordance with the variation theorem. Once the orbitals have been optimised the electronic energy can be evaluated using these molecular orbitals. This section develops the Hartree–Fock equations that define the orbital optimisation problem.

Let us return to the energy expression for a closed-shell N-electron system, we shall use a slightly less compact notation to make what follows clear,

$$E_0 = 2\sum_i^{N/2} \langle \phi_i|h|\phi_i \rangle + \sum_{ij}^{N/2} \left[2(\phi_i\phi_i|\phi_j\phi_j) - (\phi_i\phi_j|\phi_j\phi_i) \right] \qquad (2.60)$$

Now consider the variation of an orbital, ϕ_i, such that

$$\phi_i \rightarrow \phi_i + \delta\phi_i \qquad (2.61)$$

The variation is $\delta\phi_i$ and if we substitute this into the energy expression, the one-electron terms become

$$\langle \phi_i + \delta\phi_i|h|\phi_i + \delta\phi_i \rangle = \langle \phi_i|h|\phi_i \rangle + \langle \delta\phi_i|h|\phi_i \rangle + \langle \phi_i|h|\delta\phi_i \rangle + \langle \delta\phi_i|h|\delta\phi_i \rangle \qquad (2.62)$$

We can identify the order of the variation by the frequency of occurrence of $\delta\phi_i$. The first term corresponds to a zero-order change and is simply the one-electron integral in the original orbital set. The second and third terms correspond to first-order variations and the final term to a second-order variation. To optimise the energy we must make the first-order variation stationary, that is $\delta E_0 = 0$. Applying the orbital variation to the full energy expression we obtain the first-order variation as

$$\delta E_0 = 2\sum_i^{N/2} \langle \delta\phi_i|h|\phi_i \rangle + 2\sum_i^{N/2} \langle \phi_i|h|\delta\phi_i \rangle + \sum_{ij}^{N/2} 2(\delta\phi_i\phi_i|\phi_j\phi_j)$$

$$+ \sum_{ij}^{N/2} 2(\phi_i\delta\phi_i|\phi_j\phi_j) + \sum_{ij}^{N/2} 2(\phi_i\phi_i|\delta\phi_j\phi_j) + \sum_{ij}^{N/2} 2(\phi_i\phi_i|\phi_j\delta\phi_j) \qquad (2.63)$$

$$- \sum_{ij}^{N/2} (\delta\phi_i\phi_j|\phi_j\phi_i) - \sum_{ij}^{N/2} (\phi_i\delta\phi_j|\phi_j\phi_i) - \sum_{ij}^{N/2} (\phi_i\phi_j|\delta\phi_j\phi_i) - \sum_{ij}^{N/2} (\phi_i\phi_j|\phi_j\delta\phi_i)$$

Using the definitions of the Coulomb and exchange operators in eqns (2.55) and (2.58) we can rewrite δE_0 as

$$\delta E_0 = 2\sum_i^{N/2} \langle \delta\phi_i|h|\phi_i\rangle + 2\sum_i^{N/2}\langle\phi_i|h|\delta\phi_i\rangle + \sum_{ij}^{N/2} 2\langle\delta\phi_i|J_j|\phi_i\rangle$$

$$+ \sum_{ij}^{N/2} 2\langle\phi_i|J_j|\delta\phi_i\rangle + \sum_{ij}^{N/2} 2\langle\delta\phi_j|J_i|\phi_j\rangle + \sum_{ij}^{N/2} 2\langle\phi_j|J_i|\delta\phi_j\rangle \qquad (2.64)$$

$$- \sum_{ij}^{N/2} \langle\delta\phi_i|K_j|\phi_i\rangle - \sum_{ij}^{N/2}\langle\phi_i|K_j|\delta\phi_i\rangle - \sum_{ij}^{N/2}\langle\delta\phi_j|K_i|\phi_j\rangle - \sum_{ij}^{N/2}\langle\phi_j|K_i|\delta\phi_j\rangle$$

Collecting terms gives

$$\delta E_0 = \sum_i^{N/2} \langle\delta\varphi_i|2h+\sum_j^{N/2}2J_j-K_j|\varphi_i\rangle + \sum_i^{N/2}\langle\varphi_i|2h+\sum_j^{N/2}2J_j-K_j|\delta\varphi_i\rangle$$

$$+ \sum_{ij}^{N/2}\langle\delta\varphi_j|2J_i-K_i|\varphi_j\rangle + \sum_{ij}^{N/2}\langle\varphi_j|2J_i-K_i|\delta\varphi_j\rangle \qquad (2.65)$$

Note that each summation over the two-electron integrals is equivalent in both indices, i and j. This allows us to rearrange the summations, for example

$$\sum_{ij}^{N/2}\langle\delta\phi_j|2J_i-K_i|\phi_j\rangle = \sum_{ij}^{N/2}\langle\delta\phi_i|2J_j-K_j|\phi_i\rangle \qquad (2.66)$$

This enables δE_0 to be written more compactly as

$$\delta E_0 = 2\sum_i^{N/2}\langle\delta\phi_i|h+\sum_j^{N/2}2J_j-K_j|\phi_i\rangle + 2\sum_i^{N/2}\langle\phi_i|h+\sum_j^{N/2}2J_j-K_j|\delta\phi_i\rangle \quad (2.67)$$

Collecting the operators into a single term, $\hat{F}$,

$$\hat{F} = \hat{h} + \sum_j^{N/2} 2\hat{J}_j - \hat{K}_j \qquad (2.68)$$

gives

$$\delta E_0 = 2\sum_i^{N/2}\{\langle\delta\phi_i|F|\phi_i\rangle + \langle\phi_i|F|\delta\phi_i\rangle\} \qquad (2.69)$$

We now have an expression for the variation of the energy with respect to the molecular orbitals.

At this stage we must recall that the variation of the orbitals is not a free variation but is constrained to satisfy the orbital orthonormality condition expressed in eqn (2.13). Constrained variations are conveniently dealt with using the method of Lagrange multipliers, see Appendix 2A for a very brief introduction. A Lagrange function, L, is formed from the function to be minimised, here the energy E_0, and the constraints that must be satisfied. The constraints of orthonormality between molecular orbitals are written in the form

$$\langle \phi_i | \phi_j \rangle - \delta_{ij} = 0 \tag{2.70}$$

The Lagrange function is

$$L = E_0 - 2 \sum_{ij} \lambda_{ij} \left(\langle \phi_i | \phi_j \rangle - \delta_{ij} \right) \tag{2.71}$$

The λ_{ij} are elements of a matrix and are known as "Lagrange multipliers". The factor of two is an expedience, as will be seen, but is not essential since the satisfaction of the constraint depends on the term in parentheses. The variation of the energy under the constraint of orbital orthonormality is given by the Lagrange function and so we must find the condition for which $\delta L = 0$.

$$\delta L = \delta E_0 - 2 \sum_{ij} \lambda_{ij} \left(\langle \delta \phi_i | \phi_j \rangle + \langle \phi_i | \delta \phi_j \rangle \right)$$

$$= 2 \sum_i^{N/2} \langle \delta \phi_i | F | \phi_i \rangle + 2 \sum_i^{N/2} \langle \phi_i | F | \delta \phi_i \rangle - 2 \sum_{ij} \lambda_{ij} \langle \delta \phi_i | \phi_j \rangle \tag{2.72}$$

$$- 2 \sum_{ij} \lambda_{ij} \langle \phi_i | \delta \phi_j \rangle = 0$$

$|\phi_i\rangle$ and $\langle \phi_i|$ are related by complex conjugation and their variation is independent of each other. Hence we can separate eqn (2.72) into two conditions

$$\sum_i^{N/2} \langle \delta \phi_i | F | \phi_i \rangle - \sum_{ij} \lambda_{ij} \langle \delta \phi_i | \phi_j \rangle = 0$$

$$\tag{2.73}$$

$$\sum_i^{N/2} \langle \phi_i | F | \delta \phi_i \rangle - \sum_{ij} \lambda_{ij} \langle \phi_i | \delta \phi_j \rangle = 0$$

The Hermitian conjugate, denoted by †, of an integral such as $\langle \phi_i | \delta \phi_j \rangle$ is the complex conjugate transpose, or for real functions simply the transpose

$$\left[\lambda_{ij} \langle \phi_i | \delta \phi_j \rangle \right]^\dagger = \lambda_{ji} \langle \delta \phi_i | \phi_j \rangle \tag{2.74}$$

Taking the Hermitian conjugate of the second line of eqn (2.73) gives

$$\sum_i^{N/2} \langle \delta\phi_i | F | \phi_i \rangle - \sum_{ij} \lambda_{ji} \langle \delta\phi_i | \phi_j \rangle = 0 \tag{2.75}$$

and subtracting from the first line in eqn (2.73) gives

$$\left(\lambda_{ji} - \lambda_{ij} \right) \langle \delta\phi_i | \phi_j \rangle = 0 \tag{2.76}$$

Since $\langle \delta\phi_i |$ is arbitrary, this implies $\lambda_{ji} = \lambda_{ij}$, so the matrix of Lagrange multipliers is symmetric (or in general Hermitian).

Returning to eqn (2.73) we can write the equation for the optimal form of $|\phi_i\rangle$ as

$$\hat{F} | \phi_i \rangle = \sum_j \lambda_{ij} | \phi_j \rangle \tag{2.77}$$

This is the Hartree–Fock equation. The operator, $\hat{F}$, is called the "Fock operator" and is a pseudo one-electron operator. Pseudo because the interactions of all other electrons are included in an averaged way through the Coulomb and exchange operators, $\sum_j 2\hat{J}_j - \hat{K}_j$. Since $\hat{F}$ depends on $\{\phi\}$ through $\hat{J}_j$ and $\hat{K}_j$ it can be seen that the solution to eqn (2.77) is $|\phi_i\rangle$, but the solution depends on itself through the $\hat{J}_j$ and $\hat{K}_j$ operators. We shall return to this point in Section 2.4.1.

2.2.2 The Canonical Hartree–Fock Equations

The Hartree–Fock equations for all $\{\phi\}$ can be written in matrix form as

$$\mathbf{F\Phi} = \mathbf{\Phi\Lambda} \tag{2.78}$$

Note that here the matrix of Lagrange multipliers, $\mathbf{\Lambda}$, is not diagonal. We can find a transformation, $\mathbf{U}$, which is unitary in general and orthogonal for real matrices ($\mathbf{U}^\dagger\mathbf{U} = \mathbf{I} = \mathbf{UU}^\dagger$), such that

$$\mathbf{U\Lambda U}^\dagger = \mathbf{\varepsilon} \tag{2.79}$$

The matrix $\mathbf{\varepsilon}$ is diagonal. This corresponds to a transformation of the orbitals $\mathbf{\Phi}' = \mathbf{\Phi U}$. Substituting into eqn (2.78) gives the Hartree–Fock equations in terms of the transformed orbitals

$$\mathbf{F\Phi U} = \mathbf{\Phi U\Lambda} \tag{2.80}$$

Post-multiplying by $\mathbf{U}^{\dagger}$ we obtain

$$\mathbf{F\Phi UU}^{\dagger} = \mathbf{\Phi U\Lambda U}^{\dagger}$$
$$\mathbf{F\Phi} = \mathbf{\Phi\varepsilon}$$

(2.81)

Choosing to write the Hartree–Fock equations in terms of the diagonal ε rather than the non-diagonal $\mathbf{\Lambda}$ amounts to solving the same equation and we are at liberty to use either form. The equations in eqn (2.81) are known as the "canonical Hartree–Fock equations". They are the favoured form for implementation as they are amenable to solution by matrix diagonalisation, see Appendix 1C. Additionally, it is possible to associate the elements of ε with orbital energies.

2.2.3 Understanding Solutions to the Hartree–Fock Equations: Canonical Molecular Orbitals

Solution of the canonical Hartree–Fock equations yields the set of occupied molecular orbitals, specifically the set of canonical molecular orbitals. It happens that the Fock operator commutes with the symmetry operators of the molecular point group, and as we noted in Section 2.1.3, this means that eigenfunctions of the Fock operator can simultaneously be eigenfunctions of the molecular symmetry operators. The molecular orbitals arising from the canonical Hartree–Fock equation will have symmetry properties of the molecular point group. Each orbital will transform according to an irreducible representation of the molecular point group.

The Fock operator is a pseudo one-electron operator (with two-electron interactions being included in an averaged way) and its eigenfunctions are one-electron molecular orbitals. It is important to note that these one-electron wavefunctions are used to build the N-electron wavefunction, which in the Hartree–Fock model is a single Slater determinant. The quantity that may be probed in experiments is the N-electron state, *not* the one-electron molecular orbitals. The one-electron orbitals, nevertheless, provide a great deal of chemical insight since they are the components from which the N-electron wavefunction is built. The molecular orbitals allow us to concentrate on key features of the electron distribution and to explain and predict many chemical properties of molecules.

As an illustration consider the water molecule in its bent and linear geometries (Figure 1.8). Water contains 10 electrons and we obtain five doubly occupied molecular orbitals as solutions to the canonical Hartree–Fock equations. The corresponding Slater determinant is

$$\left| \Psi_0^{H_2O} \right\rangle = \left| \phi_1 \bar{\phi}_1 \phi_2 \bar{\phi}_2 \phi_3 \bar{\phi}_3 \phi_4 \bar{\phi}_4 \phi_5 \bar{\phi}_5 \right\rangle$$

(2.82)

Both bent and linear geometries will have a wavefunction of this type, but the shapes of the five doubly occupied orbitals will differ according to the geometry. The form of the four valence shell molecular orbitals, $\phi_2 - \phi_5$, is shown in Figure 2.3 (we shall look in much more detail at orbitals and their graphical representation in Chapter 4). The lowest energy orbital, ϕ_1, is essentially a $1s$ orbital located on oxygen and is too low in energy to mix with the hydrogen orbitals. In both geometries, ϕ_2 is a bonding molecular orbital distributed over the whole molecular frame. ϕ_3 is also a bonding molecular orbital, but contains a nodal plane which passes through the oxygen atom, bisecting the molecule. ϕ_4 and ϕ_5 are both lone-pair orbitals. In the bent structure the two lone pairs are quite distinct and well separated in energy but in the linear structure they are energetically equivalent, being distributed solely on the oxygen atom and perpendicular to each other.

2.2.4 Understanding Solutions to the Hartree–Fock Equations: Orbital Energies

We have mentioned the idea of orbital energies being associated with the eigenvalues of the canonical Hartree–Fock equations. To understand this we must look at the specific interactions included in the Fock operator

$$\hat{F}(\mathbf{x}_1) = -\frac{1}{2}\nabla_1^2 - \sum_A^{nuclei} \frac{Z_A}{r_{1A}} + \sum_j^{N/2} \left(2\hat{J}_j - \hat{K}_j\right) \tag{2.83}$$

The canonical Hartree–Fock equation for orbital $|\phi_i\rangle$ is

$$\hat{F}|\phi_i\rangle = \varepsilon_i|\phi_i\rangle \tag{2.84}$$

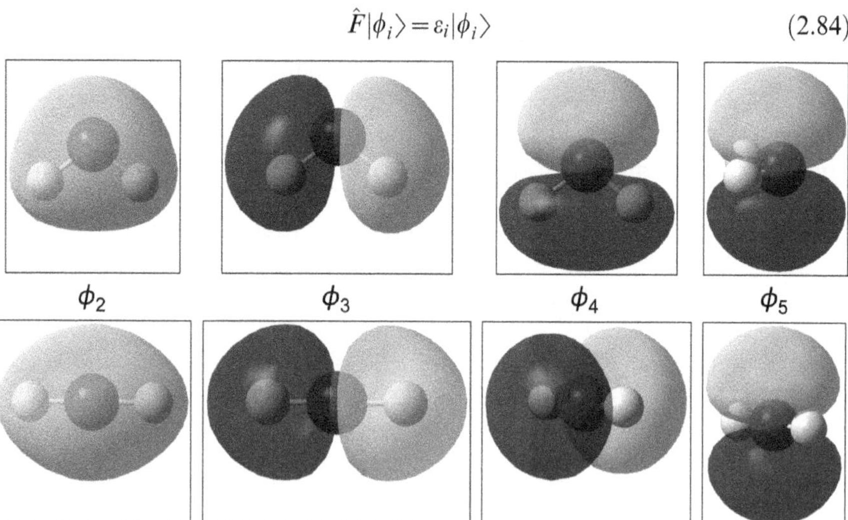

Figure 2.3 Surface plots of the four occupied valence molecular orbitals of water in bent and linear geometries.

To obtain the orbital energy, ε_i, we pre-multiply by $\langle\phi_i|$ and integrate

$$\langle\phi_i|\hat{F}|\phi_i\rangle = \varepsilon_i \tag{2.85}$$

Considering each term in the Fock operator

$$\langle\phi_i|-\frac{1}{2}\nabla_1^2|\phi_i\rangle - \langle\phi_i|\sum_A^{nuclei}\frac{Z_A}{r_{1A}}|\phi_i\rangle + \langle\phi_i|\sum_j^{N/2}2\hat{J}_j|\phi_i\rangle - \langle\phi_i|\sum_j^{N/2}\hat{K}_j|\phi_i\rangle \tag{2.86}$$

The first integral gives the kinetic energy of electron 1 (the choice of electron is arbitrary) moving in orbital $|\phi_i\rangle$. The second integral gives the potential energy of attraction between the electron in orbital $|\phi_i\rangle$ and all the positively charged nuclei. The third integral is the coulombic potential energy of repulsion between the electron in orbital $|\phi_i\rangle$ and two electrons in each of the other doubly occupied orbitals (the index j sums over these). The final integral is the exchange integral arising from the antisymmetry of the wavefunction, between the electron in $|\phi_i\rangle$ and the single electron of the same spin in each of the other doubly occupied orbitals. Hence ε_i gives the average kinetic energy of an electron in $|\phi_i\rangle$, plus the averaged electron-nuclear attraction. The instant-aneous electron–electron repulsion, $1/r_{12}$, is replaced by the averaged Coulomb and exchange interactions, $\sum_{ij}2J_{ij}-K_{ij}$. Note that when $i = j$, the Coulomb and exchange integrals are equivalent, $J_{ii}=K_{ii}$, so that $2J_{ii}-K_{ii}=J_{ii}$. This being the only interaction when both electrons are in the same orbital. Combining the two one-electron integrals into h_{ii} as in eqn (2.16), we can write the orbital energy ε_i as

$$\varepsilon_i = h_{ii} + \sum_{j=1}^{N/2}\left(2J_{ij}-K_{ij}\right) \tag{2.87}$$

Returning to the example of water in the bent and linear configurations, the lowest molecular orbital corresponds to an atomic core shell and is consequently of much lower energy (-20.55108 au for the bent structure) than the four valence shell molecular orbitals. Figure 2.4 shows how the energies of the valence shell molecular orbitals change when the geometry is altered from the optimal bent structure to the linear form. The lowest energy valence shell molecular orbital is raised in energy by this geometric perturbation but the second lowest is stabilised. The highest two molecular orbitals become degenerate in the linear configuration.

2.2.5 The Total Hartree–Fock Energy

The total electronic energy in the Hartree–Fock formalism refers to the N-electron system and can be *related* to the one-electron orbital energies, it is *not*

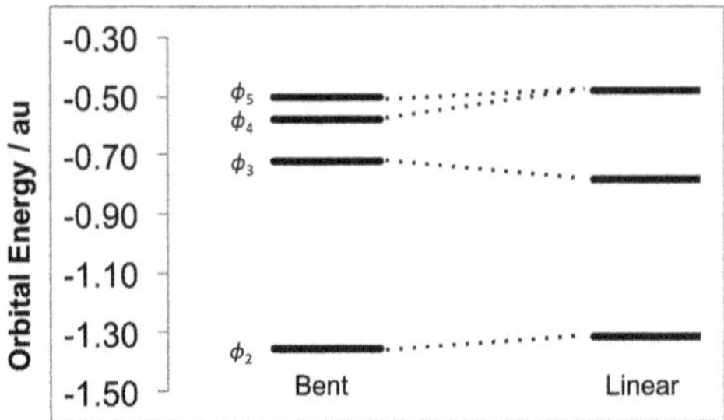

Figure 2.4 Orbital energies of the four occupied valence molecular orbitals of water in bent and linear geometries.

the sum of the orbital energies. To see why the total energy is not the sum of orbital energies consider the simple case of the ground state of helium. If we sum the corresponding orbital energies we obtain

$$2\varepsilon_i = 2h_{ii} + 2\sum_{ij}^{N/2}\left(2J_{ij} - K_{ij}\right)$$

$$= 2h_{11} + 2J_{11}$$

$$(2.88)$$

We obtained the second line of this equation by noting that helium will contain one doubly occupied orbital, $i = j = 1$, and that reduces the two-electron terms to $2J_{11} - K_{11} = J_{11}$. Since there are only two electrons there must be only one coulombic repulsion term, J_{11}. Yet summing of the orbital energies has resulted in double counting of this interaction. This observation is trivially extended to any number of electrons pairs and we conclude that simple summation of orbital energies leads to the inter-electronic terms being counted twice. Hence we must subtract the two-electron terms from the sum of orbital energies to obtain the correct electronic energy, that is

$$E_{\text{Electronic}} = 2\sum_i^{N/2}\varepsilon_i - \sum_{ij}^{N/2}\left(2J_{ij} - K_{ij}\right)$$

$$(2.89)$$

Substituting the first line of eqn (2.88) into this expression gives

$$E_{\text{Electronic}} = 2\sum_i^{N/2}h_{ii} + 2\sum_i^{N/2}\left(2J_{ij} - K_{ij}\right) - \sum_i^{N/2}\left(2J_{ij} - K_{ij}\right)$$

$$= 2\sum_i^{N/2}h_{ii} + \sum_i^{N/2}\left(2J_{ij} - K_{ij}\right)$$

$$(2.90)$$

$$= \sum_{ij}^{N/2}\left(h_{ii} + \varepsilon_i\right)$$

This yields a convenient expression for the electronic energy. To obtain the total energy we must add the nuclear repulsion energy

$$E = E_{\text{Electronic}} + V_{AB} = E_{\text{Electronic}} + \sum_{A<B}^{M} \frac{Z_A Z_B}{R_{AB}} \tag{2.91}$$

In all that follows we shall drop the "Electronic" subscript for notational convenience and assume, unless otherwise stated, that we refer to the electronic energy and that the nuclear repulsion energy must be included.

2.2.6　Ionisation Energies: Koopmans' Theorem

The one-electron orbital energies that we have discussed can be related to molecular ionisation energies and electron affinities. In general, if we wish to know the energetic requirement of a chemical or physical process we calculate the energies of the final and initial states and take the difference. This requires two calculations, one on each state. In the case of ionisation of a molecule, $\mathbf{m}$,

$$\mathbf{m} \to \mathbf{m}^+ + e^- \tag{2.92}$$

The energy of an unbound electron is zero, by definition, and so we can write for the ionisation energy

$$IE = E^{\mathbf{m}^+} - E^{\mathbf{m}} \tag{2.93}$$

From the preceding section we know that $E^{\mathbf{m}}$ is given by

$$E^{\mathbf{m}} = 2 \sum_{i}^{N/2} h_{ii} + \sum_{ij}^{N/2} \left[2J_{ij} - K_{ij} \right] \tag{2.94}$$

Suppose that the cation, $\mathbf{m}^+$, is formed by removal of an electron from orbital, ϕ_p. Applying Slater's rules (eqn (2.18)) and using the definition of the Coulomb and exchange integrals (eqns (2.55) and (2.58)) we can write the energy of $\mathbf{m}^+$ as

$$E_p^{\mathbf{m}+} = 2 \sum_{i \neq p}^{N/2} h_{ii} + \sum_{i \neq p}^{N/2} \sum_{j \neq p}^{N/2} \left[2J_{ij} - K_{ij} \right] + h_{pp} + \sum_{i \neq p}^{N/2} \left[2J_{ip} - K_{ip} \right] \tag{2.95}$$

The first two terms give the energy due to all electrons except that of the electron in the orbital being ionised. ϕ_p contains only one electron in the cation and the third term in eqn (2.95) gives the kinetic energy and nuclear attraction energy for this electron. The final term gives the Coulomb and exchange interaction of the electron in ϕ_p with the electrons in all other doubly occupied orbitals. Note that the final term is equal to the component omitted in the

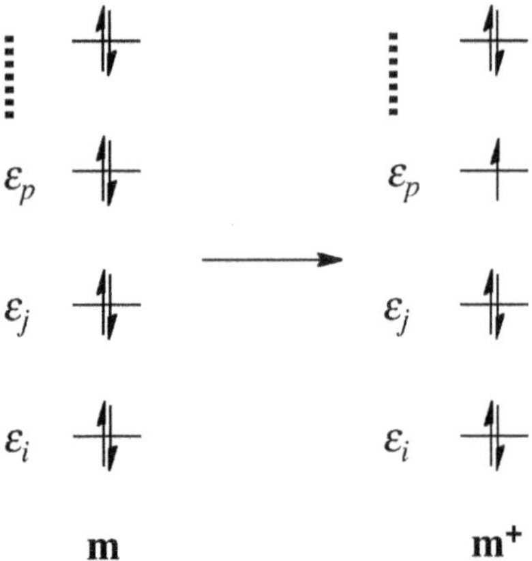

Figure 2.5 Cation (**m$^+$**) of molecule (**m**) formed by ionisation out of orbital ϕ_p.

summation of the second term, when restricted to $j \neq p$, so we can simplify eqn (2.95) as

$$E_p^{\mathbf{m}+} = 2 \sum_{\substack{i \neq p}}^{N/2} h_{ii} + \sum_{\substack{i \neq p}}^{N/2} \sum_{j}^{N/2} \left[2J_{ij} - K_{ij}\right] + h_{pp} \tag{2.96}$$

We can remove the last restriction if we include all indices in the summation and then subtract the energy corresponding to one electron in ϕ_p

$$E_p^{\mathbf{m}+} = 2 \sum_{i}^{N/2} h_{ii} + \sum_{ij}^{N/2} \left[2J_{ij} - K_{ij}\right] - h_{pp} - \sum_{j}^{N/2} \left[2J_{jp} - K_{jp}\right] \tag{2.97}$$

$$= E^{\mathbf{m}} - \varepsilon_p$$

Rearranging the second line in the equation above yields the ionisation energy

$$IE_p = E_p^{\mathbf{m}+} - E^{\mathbf{m}} = -\varepsilon_p \tag{2.98}$$

This is Koopmans' theorem, which gives the ionisation energy as the negative of the orbital energy associated with the ionisation. Using the orbital energies of water, we can obtain estimates of the ionisation energies. Table 2.1 compares the ionisation energies obtained using Koopmans' theorem with those obtained by experiment. The relationship is approximate and generally produces ionisation energies that are too large. Koopmans showed that the form of the orbital being ionised is the same in the wavefunction of the neutral molecule and that of the cation, however the other orbitals which remain doubly occupied in the cation, must change since the potential in the ion arises from $N - 1$ electrons as opposed to N electrons in the neutral molecule. This

Table 2.1 The first three ionisation energies (eV) of water obtained using Koopmans' theorem $(-\varepsilon)$ compared with experiment.

Orbital	$-\varepsilon$	Experiment	Error
ϕ_5	13.75	12.62	+9%
ϕ_4	15.73	14.74	+7%
ϕ_3	19.64	18.51	+6%

means that the energy of the cation is higher than its optimal value, when all orbitals are optimised. Hence Koopmans' theorem estimates of ionisation energies tend to be too large.

A similar relationship can be obtained between electron affinities and the orbital energies of unoccupied levels. In practice this correlation is quite poor as the Hartree–Fock method is prone to produce unoccupied orbitals that are excessively diffuse.

2.3   Open-Shell Systems in Hartree–Fock Theory

We have considered the case of closed-shell molecules and atoms, containing $N/2$ doubly occupied orbitals, in some detail. The spin–orbitals, $\{\phi(\mathbf{x})\}$, are taken to possess the same spatial form for each doubly occupied level, differing only in the spin function that they carry. This is the spin-restricted or, more commonly, restricted Hartree–Fock (RHF) approach. The RHF wavefunction for any closed-shell molecule can be written as

$$|\Psi_0\rangle = \left|\phi_1\bar{\phi}_1\phi_2\bar{\phi}_2\cdots\phi_{\frac{N}{2}}\bar{\phi}_{\frac{N}{2}}\right\rangle \tag{2.99}$$

2.3.1   The Restricted Open-Shell Hartree–Fock (ROHF) Method

We can extend the spin-restricted Hartree–Fock approach to open-shell molecules. Consider a system with an overall spin, $S > 0$. By convention each unpaired electron carries a spin of $+\frac{1}{2}$, such that all unpaired electrons are associated with α-spin. This implies that there must be $2S$ orbitals carrying a single unpaired spin. Of the N electrons in the system, $N - 2S$, will be paired in doubly occupied orbitals and $2S$ electrons will reside in singly occupied orbitals. The situation is depicted in Figure 2.6. Denoting the number of doubly occupied orbitals as $N_D = \frac{1}{2}(N - 2S)$ and the number of singly occupied orbitals as $N_S = 2S$, the Hartree–Fock determinant can be written as

$$|\Psi_0\rangle = \left|\phi_1\bar{\phi}\phi_2\bar{\phi}_2\cdots\phi_{N_D}\bar{\phi}_{N_D}\phi_{N_D+1}\phi_{N_D+2}\cdots\phi_{N_D+N_S}\right\rangle \tag{2.100}$$

Denoting doubly occupied orbitals by labels i, j, … and singly occupied orbitals by r, s, …, the electronic energy for this wavefunction is given by

$$E = 2\sum_{i=1}^{N_D} h_{ii} + \sum_{r=N_D+1}^{N_D+N_S} h_{rr} + \sum_{i,j=1}^{N_D} \left(2J_{ij} - K_{ij}\right)$$

$$+ \sum_{i=1}^{N_D} \sum_{r=N_D+1}^{N_D+N_S} \left(2J_{ir} - K_{ir}\right) + \frac{1}{2}\sum_{r,s=N_D+1}^{N_D+N_S} \left(J_{rs} - K_{rs}\right) \tag{2.101}$$

The advantage of using a determinant such as that given in eqn (2.100) is that it is an eigenfunction of $\hat{S}_z$ and $\hat{S}^2$. However it turns out that the form of the Fock operator, which is required in the orbital optimisation process, is not unique and may be formulated in a number of ways. This makes the ROHF method less well defined than its closed-shell counterpart. A further disadvantage with the ROHF scheme is the restriction of the spatial form of the doubly occupied orbitals. The α spin–orbitals of the doubly occupied set include an exchange interaction with the high-spin singly occupied orbitals. The β spin–orbitals do not include these exchange interactions, yet the ROHF scheme requires the spatial form for α- and β-spins to be the same. This has a particular significance for the evaluation of spin dependent properties, for example spin densities, electronic **g** matrices and hyperfine coupling constants. For molecular properties such as these the ROHF is generally avoided.

2.3.2 The Unrestricted Hartree–Fock (UHF) Method

For the description of open shells the ROHF approach is too constrained to allow a proper description of spin-dependent properties. A partial remedy to this limitation is provided if we allow the spatial forms for α- and β- spin–orbitals to differ. This is the idea of the spin-unrestricted Hartree–Fock (UHF) method. Rather than the single set of orthonormal orbitals $\{\phi\}$ to which we append either α or β spin functions, the UHF method adopts two sets of orbitals, $\left\{\phi^{\alpha}\right\}$ and $\left\{\phi^{\beta}\right\}$. Each set of spatial orbitals is orthonormal within itself

$$\langle \phi_i^{\alpha} | \phi_j^{\alpha} \rangle = \delta_{ij}$$

$$\langle \phi_i^{\beta} | \phi_j^{\beta} \rangle = \delta_{ij} \tag{2.102}$$

but the two sets are not orthonormal to each other

$$\langle \phi_i^{\alpha} | \phi_j^{\beta} \rangle = S_{ij}^{\alpha\beta} \tag{2.103}$$

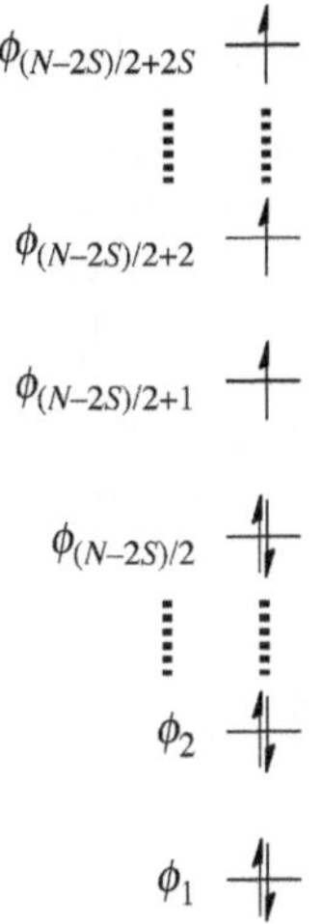

Figure 2.6 Structure of an ROHF determinant with $(N-2S)/2$ doubly occupied orbitals and $2S$ singly occupied orbitals.

The integrals $S_{ij}^{\alpha\beta}$ constitute the elements of the $\alpha\beta$ overlap matrix. Each member of these two sets can accommodate only one electron, in accordance with the Pauli exclusion principle, since each set is specific to either α-spin or β-spin electrons. When the spin coordinates are included the two sets do not overlap because of the orthogonality of the spin functions.

A single Slater determinant built of unrestricted spin–orbitals, for a system of N_α electrons of α-spin and N_β electrons of β-spin, can be written as

$$|\Psi_0\rangle = \left|\phi_1^\alpha \phi_2^\alpha \cdots \phi_{N_\alpha}^\alpha \phi_1^\beta \phi_2^\beta \cdots \phi_{N_\beta}^\beta\right\rangle \tag{2.104}$$

where $N = N_\alpha + N_\beta$. Such unrestricted determinants are eigenfunctions of $\hat{S}_z$ but not of $\hat{S}^2$. It is possible to show that unrestricted determinants correspond to a mixture of spin states. This is termed "spin contamination", since each determinant of total spin $\left[(N_\alpha - N_\beta)/2\right]$ can contain a contribution of higher spin components corresponding to $\left[(N_\alpha - N_\beta)/2\right]+1, \left[(N_\alpha - N_\beta)/2\right]+2, \cdots$. The degree of spin contamination can vary from negligible to very high levels. It can be assessed quantitatively by evaluating the eigenvalue of $\hat{S}^2$. We know from Section 2.1.3 that for a total spin, S, the eigenvalue of the $\hat{S}^2$ operator is $S(S+1)$. For an unrestricted determinant the eigenvalue of $\hat{S}^2$ is given by

$$\langle \hat{S}^2 \rangle = \left(\frac{N_\alpha - N_\beta}{2}\right)\left(\frac{N_\alpha - N_\beta}{2}+1\right) + N_\beta - \sum_i^{N_\alpha}\sum_j^{N_\beta}\left|S_{ij}^{\alpha\beta}\right|^2 \tag{2.105}$$

If the two sets of orbitals $\{\phi^\alpha\}$ and $\{\phi^\beta\}$ become equivalent, then the last term in eqn (2.105) will sum to the value $-N_\beta$ and an exact eigenfunction of $\hat{S}^2$ will be obtained. This simply corresponds to the RHF/ROHF situation and the resultant wavefunction cannot be described as spin-unrestricted. However if the deviation of the spatial forms of $\{\phi^\alpha\}$ and $\{\phi^\beta\}$ is small, the spin contamination of the UHF wavefunction will also be small.

In practice we must now deal with two Fock operators, one for each spin type,

$$\hat{F}^\alpha = \hat{h} + \sum_i^{N_\alpha} \left(\hat{J}_i^\alpha - \hat{K}_i^\alpha \right) + \sum_i^{N_\beta} \hat{J}_i^\beta$$

$$\hat{F}^\beta = \hat{h} + \sum_i^{N_\beta} \left(\hat{J}_i^\beta - \hat{K}_i^\beta \right) + \sum_i^{N_\alpha} \hat{J}_i^\alpha$$

(2.106)

The Coulomb and exchange operators are now associated with a specific spin, for example

$$\hat{J}_i^\sigma = \int \phi_i^\sigma(\mathbf{r}_2) \frac{1}{\mathbf{r}_{12}} \phi_i^\sigma(\mathbf{r}_2) d\mathbf{r}_2$$

$$\hat{K}_i^\sigma \phi_j^\sigma = \left\{ \int \phi_i^\sigma(\mathbf{r}_2) \frac{1}{\mathbf{r}_{12}} \phi_j^\sigma(\mathbf{r}_2) d\mathbf{r}_2 \right\} \phi_i^\sigma(\mathbf{r}_1)$$

(2.107)

where $\sigma = \alpha, \beta$. The resultant integrals being

$$h_{ii}^\sigma = \langle \phi_i^\sigma | h | \sigma_i^\sigma \rangle$$

$$J_{ij}^{\sigma\sigma'} = \left\langle \phi_i^\sigma \phi_i^\sigma \left| \frac{1}{\mathbf{r}_{12}} \right| \phi_j^{\sigma'} \phi_j^{\sigma'} \right\rangle$$

$$K_{ij}^{\sigma\sigma} = \left\langle \phi_i^\sigma \phi_j^\sigma \left| \frac{1}{\mathbf{r}_{12}} \right| \phi_j^\sigma \phi_i^\sigma \right\rangle$$

(2.108)

The electronic energy can be evaluated in terms of these integrals as

$$E = \sum_i^{N_\alpha} h_{ii}^\alpha + \sum_i^{N_\beta} h_{ii}^\beta + \frac{1}{2} \sum_{ij}^{N_\alpha} \left(J_{ij}^{\alpha\alpha} - K_{ij}^{\alpha\alpha} \right) + \frac{1}{2} \sum_{ij}^{N_\beta} \left(J_{ij}^{\beta\beta} - K_{ij}^{\beta\beta} \right) + \sum_i^{N_\alpha} \sum_j^{N_\beta} J_{ij}^{\alpha\beta}$$

(2.109)

and the orbital energies can be expressed, in analogy to eqn (2.87), as

$$\varepsilon_i^\sigma = h_{ii}^\sigma + \sum_i^{N_\sigma} \left(J_{ij}^{\sigma\sigma} - K_{ij}^{\sigma\sigma} \right) + \sum_j^{N_{\sigma'}} J_{ij}^{\sigma\sigma'} \quad (\sigma' \neq \sigma) \qquad (2.110)$$

2.3.3 UHF Method for $S = 0$ and the Dissociation Problem

In the UHF method the two sets of orbitals $\left\{\phi^\alpha\right\}$ and $\left\{\phi^\beta\right\}$ are spatially distinct. A useful consequence of this is that it provides a route to treating the dissociation problem encountered in the RHF method, as discussed in Section 1.4.5. In Figure 1.19 we noted the incorrect behavior of the $|\sigma\bar{\sigma}\rangle$ wavefunction of molecular hydrogen at long inter-nuclear distances. The origin of the problem was that the $|\sigma\bar{\sigma}\rangle$ wavefunction retains a large fraction of ionic components, which at dissociation erroneously leads to the products 2H• + 2H$^-$ + 2H$^+$. Since the α and β spin–orbitals of the UHF approach are distinct it should be possible to obtain the lower energy solution corresponding to 2H• at dissociation. Figure 2.7 shows the potential energy curve for hydrogen obtained using the RHF determinant $|\sigma\bar{\sigma}\rangle$ and the UHF determinant $|\sigma^\alpha\sigma^\beta\rangle$. Note that the energy obtained by the RHF and UHF methods are the same until about $R = 1.2$ Å, at which point the UHF energy falls below the RHF energy. At shorter distances the additional flexibility of the UHF wavefunction does not provide any improvement in the energy and the UHF calculation is producing an RHF solution! At $R = 1.3$ Å and beyond, the two methods produce different energies. The point at which this happens is known as the "singlet/triplet instability". Figure 2.8 shows the eigenvalue of $\hat{S}^2$ obtained using the UHF method along the potential energy curve. At the singlet/triplet instability the eigenvalue of $\hat{S}^2$ starts to deviate from the exact value of zero. At dissociation, the wavefunction is an equal mixture of $S = 0$ (singlet) and $S = 1$ (triplet) states. The eigenvalue of $\hat{S}^2$ reflects this with the value of 1 being the average of the eigenvalues for $S = 0$ and $S = 1$ states. So while the potential energy curve shows the correct dissociation behaviour, the underlying spin of the electronic state is wrong. Whether this situation poses a problem for studying chemical properties depends on the system under investigation. Spin-contaminated UHF states can often provide a reasonable zero-order description of the electronic structure of open-shell systems. Figure 2.9 shows the two occupied UHF spin–orbitals, ϕ_σ^α and ϕ_σ^β, and the corresponding doubly occupied ϕ_σ of the RHF method for the hydrogen molecule at $R = 0.7$, 1.30 and 3.4 Å. The UHF orbitals are able to break the delocalisation imposed by the RHF method and provide spatially separated orbitals, as required to describe dissociation. Unlike the RHF ϕ_σ, the UHF ϕ_σ^α and ϕ_σ^β do not carry the symmetry of the molecular frame, $D_{\infty h}$. Situations in which the electronic wavefunction has a lower degree of symmetry than the corresponding molecular geometry are referred to as "broken-symmetry solutions".

2.3.4 Spin Polarisation

The UHF method is much more successful than the ROHF method for the description of spin-dependent properties. As an illustration of this we shall discuss the classical example of the spin density in the methyl radical. Experimentally the spin density can be probed using electron paramagnetic resonance (EPR) spectroscopy. The spin density is the difference in the α and β spin densities. For the ROHF method

$$\rho^s(\mathbf{r}) = \sum_i^{N_\alpha} |\phi_i(\mathbf{r})|^2 - \sum_i^{N_\beta} |\bar{\phi}_i(\mathbf{r})|^2 \tag{2.111}$$

Since the α and β orbitals have equivalent spatial parts the sum may be restricted to the unpaired α-spin electrons. In the methyl radical there are eight electrons in four doubly occupied orbitals and one unpaired α-spin electron in ϕ_5, see Figure 2.10(a). ϕ_5 is a pure π orbital composed of $2p$-type atomic basis functions located on the carbon atom. The spin density in the ROHF description is simply

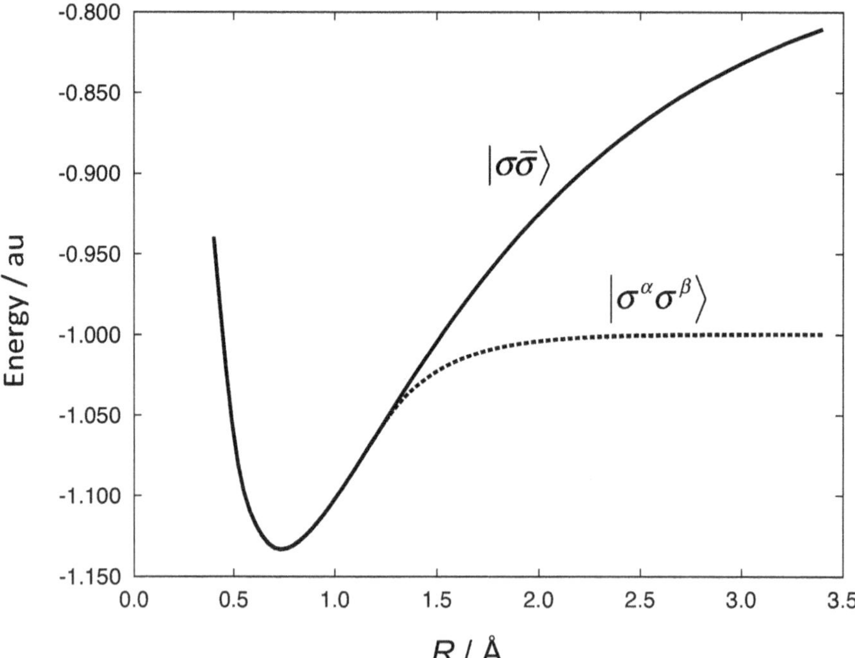

Figure 2.7 Potential energy curves of molecular hydrogen obtained with RHF (solid line) and UHF (dashed line) wavefunctions.

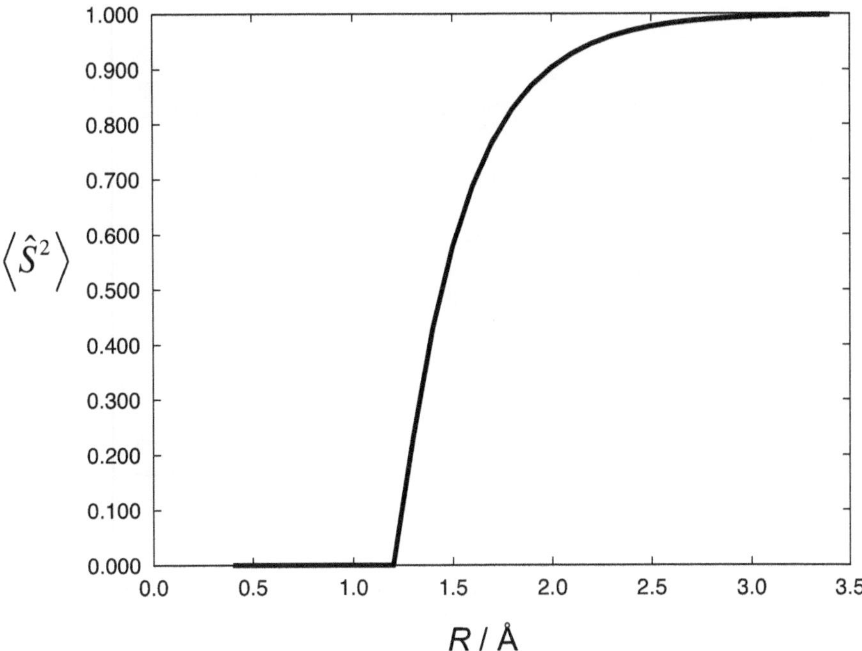

Figure 2.8 Eigenvalue of $\hat{S}^2$ in molecular hydrogen, obtained with a UHFwavefunction, as a function of the bond distance.

$$\rho^s(\mathbf{r}) = |\phi_5(\mathbf{r})|^2 \tag{2.112}$$

and is shown in Figure 2.10(b). In EPR spectroscopy the isotropic hyperfine coupling constant, A_X^{iso}, is proportional to the spin density at nucleus X. For the methyl radical $A_C^{iso} = +3.834$ mT and $A_H^{iso} = -2.304$ mT, implying non-zero spin densities at carbon and hydrogen with opposite signs. Since the spin density arising from ϕ_5, shown in Figure 2.10(b), contains no contribution of hydrogen, we must conclude that $A_H^{iso}(\text{ROHF}) = 0$! ϕ_5 contains a node at the carbon nucleus, consequently $A_C^{iso}(\text{ROHF}) = 0$, also.

If we perform a UHF calculation on the methyl radical and form the spin density as

$$\rho^s(\mathbf{r}) = \sum_i^{N_\alpha} |\phi_i^\alpha(\mathbf{r})|^2 - \sum_i^{N_\beta} |\phi_i^\beta(\mathbf{r})|^2 \tag{2.113}$$

we now obtain the orbital energies and spin density shown in Figure 2.11. The shading shows the spin density at carbon and hydrogen to be non-zero and

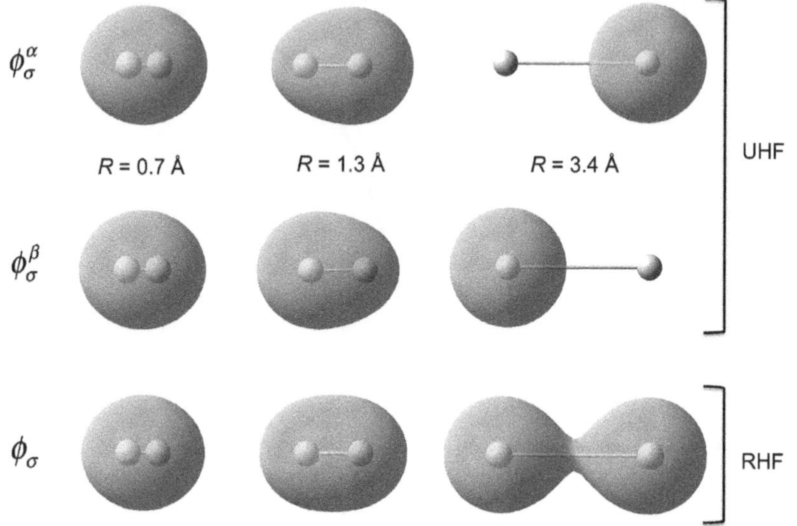

Figure 2.9 Occupied molecular orbitals of molecular hydrogen at different bond distances obtained with UHF $\phi_\sigma^\alpha, \phi_\sigma^\beta$ and RHF (ϕ_σ) wavefunctions.

also of opposite phase. Hence the UHF treatment has produced a qualitatively correct physical description of the spin distribution in the methyl radical.

2.4 Computational Realisation of the Hartree–Fock Theory

In our discussion of the Hartree–Fock method we have referred to the central importance of the set of molecular orbitals, $\{\phi\}$. All integrals have been written, so far, to be evaluated over this set of molecular orbitals. The canonical Hartree–Fock equation, eqn (2.84), is an integro–differential equation and it is possible to solve it using numerical integration techniques in certain limited cases. No, generally applicable, numerical technique is available for the direct solution of the Hartree–Fock equations. In 1951 Clemens C. J. Roothaan[1] and George G. Hall,[2] independently, showed that the introduction of a set of spatial basis functions enabled the Hartree–Fock equations to be written as a set of algebraic equations which lend themselves to easy solution by standard matrix techniques.

2.4.1 The Roothaan–Hall Equations and Basis Set Expansions

In the Roothaan–Hall approach each molecular orbital, ϕ_i, is expanded in a set of m atomic basis functions, $\{\chi\}$,

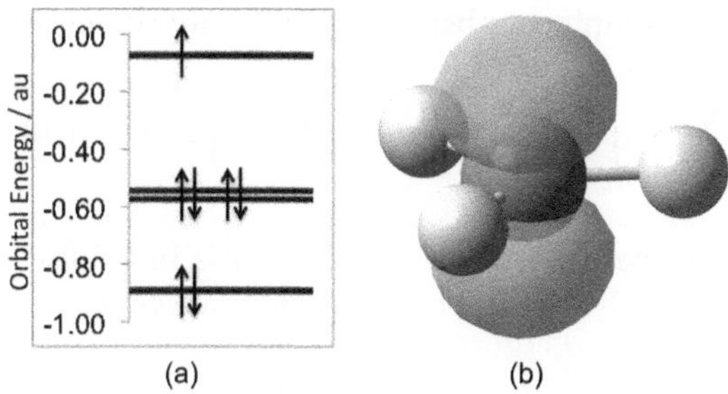

Figure 2.10   (a) ROHF orbital energies of the occupied valence molecular orbitals of the methyl radical. (b) Surface plot of the open-shell ROHF orbital of the methyl radical.

$$\phi_i(\mathbf{r}) = \sum_{\mu}^{m} c_{\mu i}\chi_{\mu}(\mathbf{r}) \tag{2.114}$$

We shall denote the members of $\{\chi\}$ using the Greek letters: μ,ν,λ,σ. Substituting eqn (2.114) into eqn (2.84) gives

$$\hat{F}\sum_{\nu}^{m} c_{\nu i}\chi_{\nu}(\mathbf{r}) = \varepsilon_i \sum_{\nu}^{m} c_{\nu i}\chi_{\nu}(\mathbf{r}) \tag{2.115}$$

We now multiply both sides of this equation by $\chi_{\mu}(\mathbf{r})$ and integrate

$$\int \chi_{\mu}(\mathbf{r})\hat{F}\sum_{\nu}^{m} c_{\nu i}\chi_{\nu}(\mathbf{r})d\mathbf{r} = \varepsilon_i \int \sum_{\nu}^{m} c_{\nu i}\chi_{\mu}(\mathbf{r})\chi_{\nu}(\mathbf{r})d\mathbf{r} \tag{2.116}$$

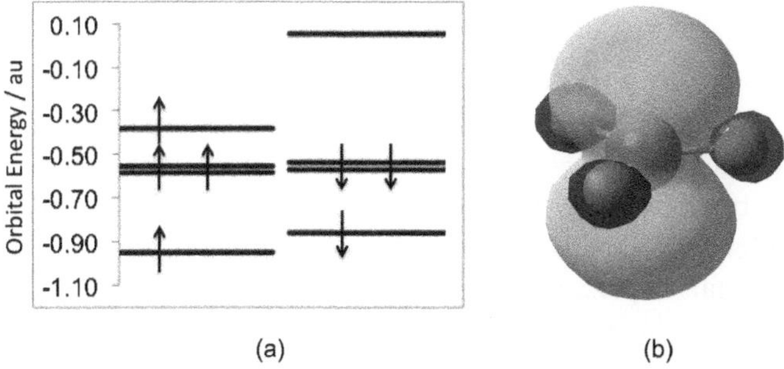

Figure 2.11   (a) UHF orbital energies, α and β sets, of the occupied valence molecular orbitals of the methyl radical. (b) Surface plot of the UHF spin density of methyl radical.

The c_{vi} and the ε_i are simply numbers and can be moved outside the integrals as

$$\sum_v^m c_{vi} \int \chi_\mu(\mathbf{r})\hat{F}\chi_v(\mathbf{r})d\mathbf{r} = \varepsilon_i \sum_v^m c_{vi} \int \chi_\mu(\mathbf{r})\chi_v(\mathbf{r})d\mathbf{r} \qquad (2.117)$$

We now have two types of integral over the basis functions $\{\chi\}$. The first of these are integrals involving the Fock operator, which we can denote as

$$F_{\mu v} = \int \chi_\mu(\mathbf{r})\hat{F}\chi_v(\mathbf{r})d\mathbf{r}$$
$$= \langle \chi_\mu|\hat{F}|\chi_v\rangle \qquad (2.118)$$

The second type of integral is simply the overlap between basis functions which we write as

$$S_{\mu v} = \int \chi_\mu(\mathbf{r})\chi_v(\mathbf{r})d\mathbf{r}$$
$$= \langle \chi_\mu|\chi_v\rangle \qquad (2.119)$$

Given m basis functions, the elements $F_{\mu v}$ and $S_{\mu v}$ will form $m \times m$ matrices. These matrices will be symmetric: $F_{v\mu} = F_{\mu v}$ and $S_{v\mu} = S_{\mu v}$. $F_{\mu v}$ is the matrix representation of the Fock operator in the basis $\{\chi\}$. Using the notation we have introduced, eqn (2.117) becomes

$$\sum_v^m F_{\mu v}c_{vi} = \varepsilon_i \sum_v^m S_{\mu v}c_{vi} \qquad (2.120)$$

There will be m such equations, $i = 1,2\cdots m$, that is the number of molecular orbitals will be equal to the number of basis functions. We can combine all m such equations together in the matrix form of the Roothaan–Hall equations

$$\mathbf{FC} = \mathbf{SCE} \qquad (2.121)$$

The form of eqn (2.121) is that of a generalised eigenvalue problem. The complication we must address is the overlap matrix $\mathbf{S}$. The molecular orbitals are orthonormal, $\langle \phi_i|\phi_j\rangle = \delta_{ij}$, but the basis functions used to expand the molecular orbital are *not* orthogonal, $\langle \chi_\mu|\chi_v\rangle = S_{\mu v}$. Collecting the molecular orbital coefficients in the columns of a matrix, $\mathbf{C}$, the orthonormality of the molecular orbitals expressed over basis functions becomes

$$\mathbf{C}^\dagger\mathbf{SC} = \mathbf{I} \qquad (2.122)$$

To solve eqn (2.121) we are free to transform the equation to any basis we wish, provided that we transform the solutions back to the original basis.

Accordingly an obvious transformation to use is one that eliminates the matrix $\mathbf{S}$. For example, if we transform $\mathbf{S}$ by $\mathbf{S}^{-\frac{1}{2}}$,

$$\mathbf{S}^{-\frac{1}{2}}\mathbf{S}\mathbf{S}^{-\frac{1}{2}} = \mathbf{I} \tag{2.123}$$

noting that both $\mathbf{S}$ and $\mathbf{S}^{-\frac{1}{2}}$ are symmetric. We can write the molecular orbital expansion coefficients as

$$\mathbf{C} = \mathbf{S}^{-\frac{1}{2}}\tilde{\mathbf{C}} \tag{2.124}$$

where $\tilde{\mathbf{C}}$ is another set of coefficients, which is related to the original set by the transformation $\mathbf{S}^{-\frac{1}{2}}$. Now we substitute this into the matrix form of the Roothaan–Hall equations

$$\mathbf{F}\mathbf{S}^{-\frac{1}{2}}\tilde{\mathbf{C}} = \mathbf{S}\mathbf{S}^{-\frac{1}{2}}\tilde{\mathbf{C}}\mathbf{E} \tag{2.125}$$

and pre-multiply by $\mathbf{S}^{-\frac{1}{2}}$ to obtain

$$\mathbf{S}^{-\frac{1}{2}}\mathbf{F}\mathbf{S}^{-\frac{1}{2}}\tilde{\mathbf{C}} = \tilde{\mathbf{C}}\mathbf{E} \tag{2.126}$$

If we write $\tilde{\mathbf{F}} = \mathbf{S}^{-\frac{1}{2}}\mathbf{F}\mathbf{S}^{-\frac{1}{2}}$ then eqn (2.126) becomes

$$\tilde{\mathbf{F}}\tilde{\mathbf{C}} = \tilde{\mathbf{C}}\mathbf{E} \tag{2.127}$$

which can be solved as a standard symmetric eigenvalue problem. The molecular orbital expansion coefficients can be obtained in the original basis using eqn (2.124). The use of the matrix $\mathbf{S}^{-\frac{1}{2}}$ corresponds to an orthogonalisation of the basis and is known as symmetric or Löwdin orthogonalisation. Some properties of symmetric orthogonalisation, and other orthogonalisation schemes, are discussed in Appendix 2B.

The overlap matrix, S, is a relatively simple matrix to form since its elements are the one-electron overlap integrals. The formation of the Fock matrix is much more complicated because of the presence of two-electron integrals. The closed-shell Fock operator $\hat{F} = \hat{h} + \sum\limits_{j}^{N/2} (2\hat{J}_j - \hat{K}_j)$ depends on the molecular orbital coefficients through the $\hat{J}$ and $\hat{K}$ operators, for example

$$\hat{J}_j(\mathbf{r}_2)\chi_v(\mathbf{r}_1) = \left(\int \phi_j(\mathbf{r}_2)\frac{1}{r_{12}}\phi_j(\mathbf{r}_2)d\mathbf{r}_2\right)\chi_v(\mathbf{r}_1)$$

$$= \left(\sum_\lambda^m c_{\lambda j}\sum_\sigma^m c_{\sigma j}\int \chi_\lambda(\mathbf{r}_2)\frac{1}{r_{12}}\chi_\sigma(\mathbf{r}_2)d\mathbf{r}_2\right)\chi_v(\mathbf{r}_1) \tag{2.128}$$

The product of the c coefficients arises from $|\phi_j|^2$ which is the probability distribution function for electron 2. Including the summation over the doubly occupied orbitals gives

$$\left(\sum_j^{N/2} 2\hat{J}_j(\mathbf{r}_2)\right)\chi_v(\mathbf{r}_1) = \left(2\sum_j^{N/2}\sum_{\lambda\sigma}^m c_{\lambda j}c_{\sigma j}\int \chi_\lambda(\mathbf{r}_2)\frac{1}{r_{12}}\chi_\sigma(\mathbf{r}_2)d\mathbf{r}_2\right)\chi_v(\mathbf{r}_1)$$

$$= \left(\sum_{\lambda\sigma}^m P_{\lambda\sigma}\int \chi_\lambda(\mathbf{r}_2)\frac{1}{r_{12}}\chi_\sigma(\mathbf{r}_2)d\mathbf{r}_2\right)\chi_v(\mathbf{r}_1) \tag{2.129}$$

where

$$P_{\lambda\sigma} = 2\sum_j^{N/2} c_{\lambda j}c_{\sigma j} \tag{2.130}$$

The elements $P_{\lambda\sigma}$ form the $m \times m$ symmetric density matrix. In terms of this density matrix the closed-shell Fock matrix becomes

$$F_{\mu v} = h_{\mu v} + \sum_{\lambda\sigma}^m P_{\lambda\sigma}\left[(\mu v|\lambda\sigma) - \frac{1}{2}(\mu\sigma|\lambda v)\right] \tag{2.131}$$

where the previously introduced notation has been adopted for the basis function integrals

$$(\mu v|\lambda\sigma) = \iint \chi_\mu(\mathbf{r}_1)\chi_v(\mathbf{r}_1)\frac{1}{r_{12}}\chi_\lambda(\mathbf{r}_2)\chi_\sigma(\mathbf{r}_2)d\mathbf{r}_1 d\mathbf{r}_2 \tag{2.132}$$

Knowing the Fock matrix, the electronic energy can be evaluated using eqn (2.90) written in terms of basis functions as

$$E_0 = \frac{1}{2}\sum_{\mu v}^m P_{\mu v}\left(h_{\mu v} + F_{\mu v}\right) \tag{2.133}$$

Using the expression for the Fock matrix, in terms of integrals over basis functions, it would seem that we can solve the Roothaan–Hall equations and obtain the optimal molecular orbital coefficients. Not quite! The density matrix, $\mathbf{P}$, is required to build the Fock matrix. Yet $\mathbf{P}$ depends on $\mathbf{C}$, which is obtained by diagonalising $\mathbf{F}$. So we need to know $\mathbf{C}$ to form $\mathbf{F}$, but we need to know $\mathbf{F}$ to obtain $\mathbf{C}$! This tells us that the Roothaan–Hall equations are non-linear and must be solved in an iterative manner.

2.4.2 Orbital Optimisation: The Self-Consistent Field (SCF) Procedure for Closed Shells

To carry out a calculation using the Roothaan–Hall formulation of the Hartree–Fock theory requires a number of steps, as shown in Figure 2.12. The initial 'guessed' set of molecular orbital coefficients can be obtained in many ways. A particularly simple method, but which often produces rather poor molecular orbital coefficients, is to diagonalise the matrix of one-electron integrals, $\mathbf{h}$. A much more successful scheme for building a molecular orbital guess involves the superposition of the atomic densities of the constituent atoms.

When the matrix of molecular orbital coefficients used to construct $\mathbf{F}$ is the same as that obtained from diagonalising $\mathbf{F}$, the procedure terminates and the orbitals are said to be self-consistent, hence the method is called the self-consistent field (SCF) procedure.

2.4.3 Basis Set Expansions in the UHF Method

The equivalent of the Roothaan–Hall equations for the case of spin-unrestricted Hartree–Fock wavefunctions are known as the "Pople–Nesbet equations". The process for obtaining the working SCF equations, starting from the definitions of the UHF Fock operators, follows closely that of the closed-shell case. The computational effort involved is almost twice that of a closed-shell calculation. In the UHF theory the number of spin–orbitals is doubled, since α and β electrons are allowed to occupy spatially distinct sets. Introducing the basis set expansion for the sets $\left\{\phi^{\alpha}\right\}$ and $\left\{\phi^{\beta}\right\}$, we write (ignoring the $\mathbf{r}$-dependence for convenience)

$$\phi_i^{\alpha} = \sum_{\mu}^{m} c_{\mu i}^{\alpha} \chi_{\mu}$$

$$\phi_i^{\beta} = \sum_{\mu}^{m} c_{\mu i}^{\beta} \chi_{\mu}$$

$$(2.134)$$

Note that the basis set, $\{\chi\}$, is identical for both α and β sets. The difference arises from two sets of expansion coefficients, $\{c^{\alpha}\}$ and $\{c^{\beta}\}$. In order to obtain $\{c^{\alpha}\}$ and $\{c^{\beta}\}$ we are now required to solve two sets of matrix equations

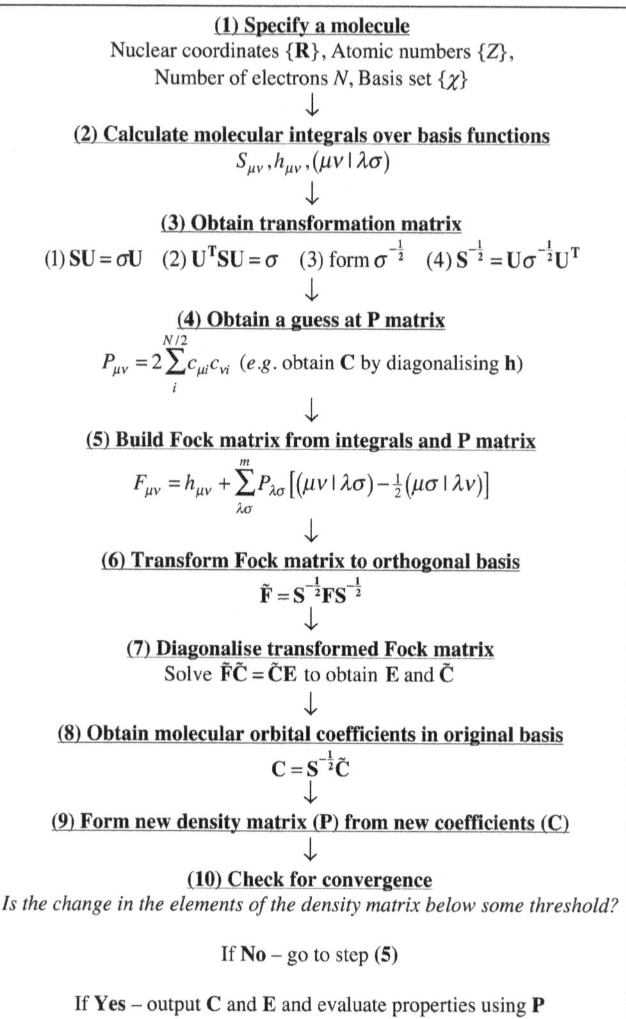

Figure 2.12 Flow chart for the SCF procedure.

$$\mathbf{F}^{\alpha}\mathbf{C}^{\alpha} = \mathbf{S}\mathbf{C}^{\alpha}\mathbf{E}^{\alpha}$$
$$\mathbf{F}^{\beta}\mathbf{C}^{\beta} = \mathbf{S}\mathbf{C}^{\beta}\mathbf{E}^{\beta}$$

$$(2.135)$$

To proceed we define density matrices, in analogy with eqn (2.130), for each spin type

$$P_{\lambda\sigma}^{\alpha} = \sum_{i}^{N_\alpha} c_{\lambda i}^{\alpha} c_{\sigma i}^{\alpha}$$

$$P_{\lambda\sigma}^{\beta} = \sum_{i}^{N_\beta} c_{\lambda i}^{\beta} c_{\sigma i}^{\beta} \tag{2.136}$$

The total density matrix, $\mathbf{P}^{\mathrm{T}}$ is the sum of these

$$\mathbf{P}^{\mathrm{T}} = \mathbf{P}^{\alpha} + \mathbf{P}^{\beta} \tag{2.137}$$

The difference, $\mathbf{P}^{\mathrm{S}}$,

$$\mathbf{P}^{\mathrm{S}} = \mathbf{P}^{\alpha} - \mathbf{P}^{\beta} \tag{2.138}$$

is the spin-density matrix. As we have discussed in Section 2.3.4, this quantity is key to calculations of spin-dependent properties. In terms of the density matrices we have introduced, the required Fock matrices can be evaluated as

$$F_{\mu v}^{\alpha} = h_{\mu v} + \sum_{\lambda\sigma}^{m} P_{\lambda\sigma}^{\alpha}[(\mu v|\lambda\sigma) - (\mu\sigma|\lambda v)] + \sum_{\lambda\sigma}^{m} P_{\lambda\sigma}^{\beta}(\mu v|\lambda\sigma)$$

$$F_{\mu v}^{\beta} = h_{\mu v} + \sum_{\lambda\sigma}^{m} P_{\lambda\sigma}^{\beta}[(\mu v|\lambda\sigma) - (\mu\sigma|\lambda v)] + \sum_{\lambda\sigma}^{m} P_{\lambda\sigma}^{\alpha}(\mu v|\lambda\sigma) \tag{2.139}$$

The α-spin Fock operator is coupled to the β-spin Fock operator through the β-spin density matrix, and *vice versa*.

A computationally more expedient form of eqn (2.139) can be obtained in terms of the total density matrix, $\mathbf{P}^{\mathrm{T}}$, as

$$F_{\mu v}^{\alpha} = h_{\mu v} + \sum_{\lambda\sigma}^{m} P_{\lambda\sigma}^{T}(\mu v|\lambda\sigma) - \sum_{\lambda\sigma}^{m} P_{\lambda\sigma}^{\alpha}(\mu\sigma|\lambda v)$$

$$F_{\mu v}^{\beta} = h_{\mu v} + \sum_{\lambda\sigma}^{m} P_{\lambda\sigma}^{T}(\mu v|\lambda\sigma) - \sum_{\lambda\sigma}^{m} P_{\lambda\sigma}^{\beta}(\mu\sigma|\lambda v) \tag{2.140}$$

The $\mathbf{C}^{\alpha}$ and $\mathbf{C}^{\beta}$ coefficient matrices are optimised in a self-consistent manner as described for the closed-shell case in Section 2.4.2. The electronic energy is evaluated as the trace of products of integrals and density matrices as

$$E_0 = \frac{1}{2} \sum_{\mu\nu}^{m} \left[P_{\mu\nu}^{T} h_{\nu\mu} + P_{\mu\nu}^{\alpha} F_{\nu\mu}^{\alpha} + P_{\mu\nu}^{\beta} F_{\nu\mu}^{\beta} \right] \tag{2.141}$$

2.4.4 Convergence of the SCF Process

The scheme of orbital optimisation we described in Section 2.4.2 is sometimes slow to converge, or even divergent. An iteration of the basic SCF process, say at cycle K, consists of constructing a new Fock matrix, $\mathbf{F}^{(K+1)}$, using the current molecular orbital coefficients, $\mathbf{C}^{(K)}$, or the density matrix constructed from them, $\mathbf{P}^{(K)}$. The matrix $\mathbf{F}^{(K+1)}$ is then diagonalised to produce a new coefficient matrix, $\mathbf{C}^{(K+1)}$,

$$\sum_{\nu}^{m} F_{\mu\nu}^{(K+1)} c_{\nu i}^{(K+1)} = \sum_{\nu}^{m} S_{\mu\nu} c_{\nu i}^{(K+1)} \varepsilon_i^{(K+1)} \tag{2.142}$$

This process is repeated until the change in the elements of the $\mathbf{C}$ or $\mathbf{P}$ matrices falls below a chosen threshold. Various schemes for accelerating convergence have been devised and here we briefly consider the three most widely used.

The first scheme is particularly simple but is often able to stabilise divergent iterations of the SCF procedure. It is based on damping of the density matrix that is used to build the Fock matrix. Suppose we have stored the density matrix of the previous iteration, $\mathbf{P}^{(K-1)}$, and have formed the density matrix for the current iteration, $\mathbf{P}^{(K)}$. A damped density matrix is defined as

$$\mathbf{P}^{(K-\text{Damped})} = \frac{\mathbf{P}^{(K)} + \kappa \mathbf{P}^{(K-1)}}{1 + \kappa} \tag{2.143}$$

κ is a positive number that determines the extent of damping. If $\kappa = 0$, the density is the unmodified density of the current iteration. As the value of κ increases the component of the density from the previous iteration, $\mathbf{P}^{(K-1)}$, is increased meaning that the change in the density between the two iterations is reduced. If the iterative process is exhibiting oscillatory behaviour, this type of damping can be very effective in stabilising the oscillations. It is easily implemented and most effective in the early stages of the SCF process. Once stability has been achieved, the damping should be switched off since it will slow the convergence of the underlying process. In sophisticated implementations, the value of κ is dynamically updated in each iteration.[3]

The second method we shall describe is that of level shifting. The details for a full implementation may be found in ref. 4. Level shifting is another way in which the size of the step taken between two iterations can be reduced. This is achieved by increasing the energies of the unoccupied orbitals. The further the unoccupied orbitals are shifted, the less they will mix with the occupied orbitals. We can write the change in the occupied orbital, ϕ_i, as

$$c_{vi}^{(K+1)} = c_{vi}^{(K)} + \sum_{a}^{unoccupied} \Delta_{ia}^{(K+1)} c_{va}^{(K)} \tag{2.144}$$

where Δ_{ia} is the amount of ϕ_a mixed into ϕ_i. Considering only Δ_{ia} and substituting into eqn (2.142) gives

$$\sum_{v}^{m} F_{\mu v}^{(K+1)} \left[c_{vi}^{(K)} + \Delta_{ia}^{(K+1)} c_{va}^{(K)} \right] = \sum_{v}^{m} S_{\mu v} \left[c_{vi}^{(K)} + \Delta_{ia}^{(K+1)} c_{va}^{(K)} \right] \varepsilon_{i}^{(K+1)} \tag{2.145}$$

Pre-multiplying by $c_{\mu a}^{(K)}$

$$\begin{aligned}
&\sum_{v}^{m} c_{\mu a}^{(K)} F_{\mu v}^{(K+1)} c_{vi}^{(K)} + c_{\mu a}^{(K)} F_{\mu v}^{(K+1)} c_{va}^{(K)} \Delta_{ia}^{(K+1)} \\
&= \left[\sum_{v}^{m} c_{\mu a}^{(K)} S_{\mu v} c_{vi}^{(K)} + \sum_{v}^{m} c_{\mu a}^{(K)} S_{\mu v} c_{va}^{(K)} \Delta_{ia}^{(K+1)} \right] \varepsilon_{i}^{(K+1)}
\end{aligned} \tag{2.146}$$

Carrying out the summations over basis functions gives the following expression in the molecular orbital basis

$$F_{ai}^{(K+1)} + \varepsilon_{a}^{(K+1)} \Delta_{ia}^{(K+1)} = \varepsilon_{i}^{(K+1)} \Delta_{ia}^{(K+1)} \tag{2.147}$$

rearranging for $\Delta_{ia}^{(K+1)}$ we obtain

$$\Delta_{ia}^{(K+1)} = - \frac{F_{ai}^{(K+1)}}{\left(\varepsilon_{a}^{(K+1)} - \varepsilon_{i}^{(K+1)} \right)} \tag{2.148}$$

The amount of mixing between ϕ_i and ϕ_a is governed by $(\varepsilon_a - \varepsilon_i)$. In the level-shifting scheme ε_a is shifted by a positive constant, b, which increases the denominator $(\varepsilon_a + b - \varepsilon_i)$ and so reduces the mixing, $\Delta_{ia}^{(K+1)}$.

The last method we shall discuss is the direct inversion in the iterative subspace (DIIS) technique devised by Pulay.[5] DIIS can be applied to any set of equations requiring an iterative solution. The cycles of any iterative process produce a series of vectors that approach the converged result. In the context of the SCF procedure, the vectors could be molecular orbital coefficients or the density matrix or the Fock matrix. In practice it has been observed that the error vector defined, at iteration K, as

$$\mathbf{e}^{(K)} = \left(\mathbf{S}^{-\frac{1}{2}} \right)^{\dagger} \left(\mathbf{F}^{(K)} \mathbf{P}^{(K)} - \mathbf{P}^{(K)} \mathbf{F}^{(K)} \right) \mathbf{S}^{-\frac{1}{2}} \tag{2.149}$$

is particularly effective. At convergence the Fock matrix and density matrix commute ensuring the vector $\mathbf{e}^{(K)}$ vanishes. The purpose of the DIIS process is to find a linear combination of error vectors that minimise the error

$$\sum_K C_K \mathbf{e}^{(K)} = \min \tag{2.150}$$

under the constraint that $\sum_K C_K = 1$. The coefficients, C_K, are determined by solution of the system of linear equations

$$\begin{bmatrix} 0 & -1 & -1 & \cdots & -1 \\ -1 & B_{11} & B_{12} & \cdots & B_{1K} \\ -1 & B_{21} & B_{22} & \cdots & B_{2K} \\ \vdots & \vdots & \vdots & \vdots & \vdots \\ -1 & B_{K1} & B_{K2} & \cdots & B_{KK} \end{bmatrix} \begin{bmatrix} -\lambda \\ C_1 \\ C_2 \\ \vdots \\ C_K \end{bmatrix} = \begin{bmatrix} -1 \\ 0 \\ 0 \\ \vdots \\ 0 \end{bmatrix} \tag{2.151}$$

The elements B_{KL} are the trace of the product of $\mathbf{e}^{(K)}$ and $\mathbf{e}^{(L)}$

$$B_{KL} = trace\left(\mathbf{e}^{(K)}\mathbf{e}^{(L)}\right) \tag{2.152}$$

Using these optimal coefficients an improved Fock matrix is constructed as

$$\mathbf{F}^{\mathrm{DIIS}} = \sum_K C_K \mathbf{F}^{(K)} \tag{2.153}$$

The DIIS method has become widely adopted as a standard technique for accelerating convergence.

2.4.5   The Direct SCF Method

To implement the SCF procedure requires access to the necessary one- and two-electron integrals. For a basis set consisting of m functions, the number of one-electron integrals, $h_{\mu\nu}$, scale as m^2. The matrix $\mathbf{h}$ is symmetric and so only the unique $m(m+1)/2$ elements are required. The evaluation and storage of $\mathbf{h}$ is simple and straightforward for even very large numbers of basis functions. The two-electron integrals, $(\mu\nu|\lambda\sigma)$, scale as m^4. We noted in Section 2.1.2 that the two-electron integrals, when evaluated over real basis functions, have an eight-fold symmetry. So the number of two-electron integrals is approximately $m^4/8$. Even for a modest number of basis functions this quantity of integrals becomes difficult to handle. The difficulty is not only the amount of disk storage needed, but also the need to move this large amount of data from disk to the computer's fast memory (RAM) in each cycle of the SCF process. To circumvent this problem, Jan Almlöf introduced the direct SCF method.[6] The speed of CPUs has increased much more quickly than that of disk access. Almlöf's idea was to evaluate the two-electron integrals without storing them. As each integral, or batch of integrals, is evaluated, their contribution to the Fock matrix is formed and the integral discarded. This means that the integrals must be re-evaluated in each SCF cycle. The cost in terms of CPU time is significant but, provided sufficient CPU time is available, the storage problem that could prevent calculations on large molecules is eliminated.

Two further ideas are used to reduce the formal m^4 scaling of the integral evaluation step. To form the Fock matrix, the density matrix is formed and the contribution, for example $P_{\lambda\sigma}(\mu v|\lambda\sigma)$, can be eliminated if the value of the integral is less than a certain threshold. To implement such a screening we must have a way of estimating the value of $(\mu v|\lambda\sigma)$ without evaluating it! Such an estimate is provided by the Schwarz inequality

$$|(\mu v|\lambda\sigma)| \leq \sqrt{(\mu\mu|vv)} \cdot \sqrt{(\lambda\lambda|\sigma\sigma)} \qquad (2.154)$$

Integrals of the form $(\mu\mu|vv)$ can be evaluated relatively easily and require only $m(m+1)/2$ storage. Hence it is possible to screen the quantity $P_{\lambda\sigma}(\mu v|\lambda\sigma)$ using $\left|P_{\lambda\sigma} \cdot \sqrt{(\mu\mu|vv)} \cdot \sqrt{(\lambda\lambda|\sigma\sigma)}\right|$. If this is below a chosen threshold, for example 10^{-10}, the integral $(\mu v|\lambda\sigma)$ need not be evaluated. In practice the test is a little more complex, since the integral $(\mu v|\lambda\sigma)$ occurs in more than one Fock matrix component.

The second idea exploits the fact that the only part of the Fock matrix that changes from iteration to iteration is the density matrix. For example, at iteration K,

$$\mathbf{F}^{(K)} = \mathbf{h} + \mathbf{GP}^{(K)}$$
$$\mathbf{F}^{(K+1)} = \mathbf{h} + \mathbf{GP}^{(K+1)} \qquad (2.155)$$

where the matrix $\mathbf{G}$ represents the two-electron integrals. Combining these two equations we can write

$$\mathbf{F}^{(K+1)} = \mathbf{F}^{(K)} + \mathbf{G} \cdot \left[\mathbf{P}^{(K+1)} - \mathbf{P}^{(K)}\right]$$
$$= \mathbf{F}^{(K)} + \mathbf{G} \cdot \Delta\mathbf{P}^{(K+1)} \qquad (2.156)$$

The change in the density matrix between two iterations determines the change in the Fock matrix. If we form the change in the density matrices, $\Delta\mathbf{P}^{(K+1)}$, and use it to screen the integral evaluation using the Schwarz inequality we will make large savings in computational effort, since as the process converges $\Delta\mathbf{P}^{(K+1)}$ tends to zero!

These two ideas, combined, serve to reduce the steep m^4 scaling of the Hartree–Fock SCF process. The extent of the reduction in effort depends on the type of molecule being studied. Long, thin molecules scale better than short, broad molecules. In general the scaling is reduced from m^4 to between m^3 and m^2. More precise integral estimates than the Schwarz inequality have been devised and, along with further subtleties of implementation, it is possible for large molecules to obtain linear scaling (order m) of the SCF procedure.

2.5 Molecular Basis Sets

By now it should be apparent that the basis set, $\{\chi\}$, to which we have repeatedly referred, plays a central role in determining the quality of quantum chemical calculations, see Figure 1.15. The majority of modern calculations use gaussian type functions that have a radial dependence of $e^{-\alpha r^2}$, where α is an orbital exponent. We are free to choose the form of the basis functions we use. The adoption of gaussian type functions rests on a balance of relatively easy integral evaluation and good representation of the electronic distribution in molecules. More appropriate basis functions are available in the form of Slater type functions, which are better suited to the description of the electron distribution in the nuclear region. The problem with Slater type functions is that the evaluation of integrals is sufficiently more demanding, computationally, that it becomes preferable to use a larger number of gaussian type functions. In this section our discussion will cover gaussian type functions only. We shall not deal with the evaluation of molecular integrals here, since that is an advanced topic beyond the scope of this text. The interested reader may consult refs. 7 and 8. Ref. 7 provides a very fine modern tutorial on integral evaluation over gaussian type functions, and ref. 8 extends the discussion to the evaluation of property integrals.

2.5.1 Gaussian Type Functions

Un-normalised gaussian type basis functions have the general form

$$g(\alpha,i,j,k,\mathbf{r}-\mathbf{r}_C) = (x-x_C)^i (y-y_C)^j (z-z_C)^k e^{-\alpha|\mathbf{r}-\mathbf{r}_C|^2} \quad (2.157)$$

The function is centred at a point $\mathbf{r}_C$ with coordinates (x_C,y_C,z_C). Typically $\mathbf{r}_C$ is taken as the position of the nucleus to which the basis function is attached. The radial dependence is modulated by the exponent, α, which is always a positive number. Figure 2.13 shows the radial factor for $\alpha = 1$, 2 and 0.5. A larger value of α contracts the function, concentrating it towards the origin, $\mathbf{r}_C$, while smaller values extend the function making it more diffuse in space. Molecular integrals involve integration of this basis function over the electronic position vector, $\mathbf{r}$, with coordinates (x,y,z). The numbers i, j and k are non-negative integers that determine the angular form of the basis function. The sum $i+j+k$ is related to the angular momentum quantum number, l. For example when $i+j+k=0$, eqn (2.157) represents an s-type basis function, or when $i+j+k=2$ a d-type basis function. The members of the angular momentum shell are specified by i, j and k individually. For example $i+j+k=1$ is a p-type shell, and $i=1$, $j=0$, $k=0$ represents a p_x function, or $i=0$, $j=0$, $k=1$ a p_z function.

The property of gaussian type functions that makes them attractive for use in molecular calculations is that a product of two gaussian functions centred at different locations is itself a gaussian function located between the two functions. An example of this gaussian product property is shown in Figure 2.14 for two s-type gaussian functions. Consequently, molecular integrals involving products of two gaussian functions can be reduced to an integral over a single centre. Two normalised s-type gaussian functions, centred at $\mathbf{R}_A$ and $\mathbf{R}_B$ with exponents α_A and α_B, have the form

$$g_A(\mathbf{r}) = \left(\frac{\pi}{2\alpha_A}\right)^{-\frac{3}{4}} e^{-\alpha_A|\mathbf{r}-\mathbf{R}_A|^2}$$

$$g_B(\mathbf{r}) = \left(\frac{\pi}{2\alpha_B}\right)^{-\frac{3}{4}} e^{-\alpha_B|\mathbf{r}-\mathbf{R}_B|^2}$$

$$(2.158)$$

Their product and its properties are specified by

$$g_A(\mathbf{r})g_B(\mathbf{r}) = Ke^{-\alpha_P|r-R_P|^2}$$

$$\alpha_P = \alpha_A + \alpha_B$$

$$\mathbf{R}_P = \frac{\alpha_A\mathbf{R}_A + \alpha_B\mathbf{R}_B}{\alpha_A + \alpha_B}$$

$$K = \left(\frac{\pi}{2}\right)^{-\frac{3}{2}} (\alpha_A\alpha_B)^{\frac{3}{4}} e^{-\frac{\alpha_A\alpha_B}{\alpha_P}|\mathbf{R}_A-\mathbf{R}_B|^2}$$

$$(2.159)$$

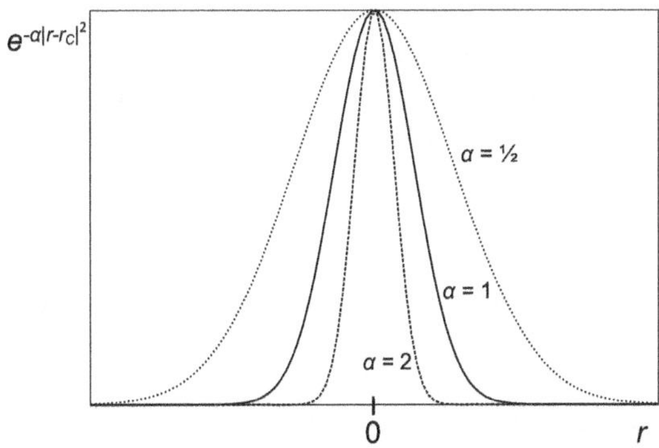

Figure 2.13 The behaviour of gaussian type functions as the orbital exponent is varied.

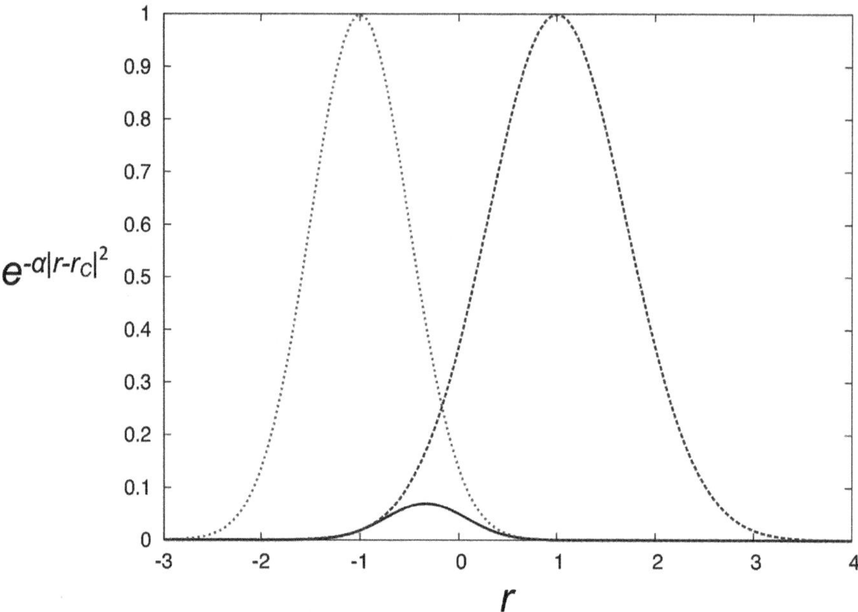

Figure 2.14 Two gaussian type functions, centred at $r = -1$ and $r = +1$, and their product which is centred in between.

This is known as the "gaussian product theorem" and becomes very significant when we must deal with two-electron integrals involving three or four centres. The gaussian product theorem allows these integrals to be simplified and easily handled.

We know from elementary quantum mechanics that atomic shells of angular momentum, l, have $2l+1$ sub-shells. For example, in a p shell, $l = 1$, and we have $2l+1=3$, corresponding to the three sub-shells. Now consider the case, $l = 2$, where we have $2l+1=5$ sub-shells, which we are familiar with as the five d orbitals. If we look at the sum of the individual components in eqn (2.157), we can form $i+j+k=2$ in *six* ways, rather than five, see Table 2.2. We can continue by looking at the case of $l = 3$, where we find that rather than the familiar seven f orbitals, there are 10 gaussian type functions of the form given in eqn (2.157). It can be quickly deduced that the number of sub-shells that can be formed from the angular factor $(x-x_C)^i(y-y_C)^j(z-z_C)^k$ for a total angular momentum, l, is $(l+1)(l+2)/2$. This form of the angular factor defines the function in eqn (2.157) as a cartesian gaussian type function. To produce the familiar five d type or seven f orbitals, requires that we use spherical harmonic angular factors. In most modern implementations, the molecular integrals are evaluated over the $(l+1)(l+2)/2$ cartesian functions and then transformed to the 'pure' or spherical harmonic representation. This is easily done, for example the six cartesian d-type functions can be transformed into the real, orthogonal, spherical harmonic representation using the non-square transformation

$$
\begin{bmatrix}
-\frac{1}{2} & -\frac{1}{2} & 1 & 0 & 0 & 0 \\
\frac{\sqrt{3}}{2} & -\frac{\sqrt{3}}{2} & 0 & 0 & 0 & 0 \\
0 & 0 & 0 & 1 & 0 & 0 \\
0 & 0 & 0 & 0 & 1 & 0 \\
0 & 0 & 0 & 0 & 0 & 1
\end{bmatrix}
\begin{bmatrix}
d_{x^2} \\
d_{y^2} \\
d_{z^2} \\
d_{xy} \\
d_{xz} \\
d_{yz}
\end{bmatrix}
=
\begin{bmatrix}
d_{3z^2 - r^2} \\
d_{x^2 - y^2} \\
d_{xy} \\
d_{xz} \\
d_{yz}
\end{bmatrix}
\qquad (2.160)
$$

A discussion and general formulae for arbitrary l can be found in ref. 9.

The single gaussian function in eqn (2.157) is referred to as a "primitive cartesian gaussian function". We have already seen that the physically more correct Slater type functions, Figure 1.3, are very different to the gaussian functions. The difference is particularly acute near the centre of the function and also at large distances from the centre. A way to improve the description in these areas is to use more than one primitive function. In fact a large set of primitive functions can be used, including large and small exponents. The variational optimisation of the orbital, as in the SCF method, will result in an optimum combination of the primitive gaussian functions. While the primitives individually give a poor representation of the molecular orbitals, a linear combination of many primitives can give almost arbitrary flexibility to their form. Of course the cost of using such a large basis set will result in a substantial increase in the computational cost of any calculations.

A way to improve the representation of orbitals without using very large sets of primitive functions is to take a linear combination of the primitive functions to obtain contracted gaussian functions. For example, consider the closed-shell neon atom. Huzinaga[10] devised a large set of primitive functions, consisting of nine s functions and five p functions. The exponents of this $9s5p$ set are shown in Table 2.3 (upper table), where each type of function is numbered sequentially in order of decreasing exponent. A Hartree–Fock SCF calculation in this basis yields five doubly occupied orbitals with the orbital coefficients shown in Table 2.3 (lower table). The electronic energy obtained is -128.52674 au. The lowest orbital must correspond to the $1s$ atomic orbital and the second to the atomic $2s$ orbital. Inspection of the orbital coefficients shows that the

Table 2.2 The six cartesian gaussian type functions that can be formed for $i + j + k = 2$.

$l = 2$	i	j	k	Sub-shell
	1	1	0	d_{xy}
	1	0	1	d_{xz}
	0	1	1	d_{yz}
	2	0	0	d_{x^2}
	0	2	0	d_{y^2}
	0	0	2	d_{z^2}

seven functions with the largest exponents contribute more to orbital 1 than to orbital 2. The $1s$ and $2s$ atomic orbitals of neon are approximately represented in this basis as

$$\phi_{1s}(\mathbf{r}) = 0.00120\chi_{1s}(\mathbf{r}) + 0.00909\chi_{2s}(\mathbf{r}) + 0.04131\chi_{3s}(\mathbf{r})$$
$$+ 0.13790\chi_{4s}(\mathbf{r}) + 0.36251\chi_{5s}(\mathbf{r}) + 0.13006\chi_{6s}(\mathbf{r})$$
$$+ 0.47267\chi_{7s}(\mathbf{r})$$
$$\phi_{2s}(\mathbf{r}) = 0.61430\chi_{8s}(\mathbf{r}) + 0.50208\chi_{9s}(\mathbf{r})$$

$$(2.161)$$

Note that these expressions are not properly normalised since the coefficients refer to the use of all nine s primitive functions. Rather than dealing with nine primitive s functions we can choose to reduce the set into a smaller number of contracted functions. For example we can split each of the combinations in eqn (2.161) for the $1s$ and $2s$ orbitals into two contracted s functions. Similarly the p shells can be contracted into two sets. The form of these contracted functions, with properly normalised contraction coefficients, is shown in Table 2.4 (upper table) and the optimised SCF orbitals obtained in this contracted basis are given in Table 2.4 (lower table). The energy obtained with this contracted basis set is -128.52235 au, which is a little higher than that obtained with the full uncontracted set of primitive functions. However the number of basis functions has been reduced from 24 to 10. This is a significant saving, if we recall the scaling properties of the Hartree–Fock SCF method (see Table 1.2). The atomic $1s$ orbital is now composed of the two contracted functions but with much more equal weight. Similarly the $2s$ atomic orbital is essentially just a combination of the two more diffuse contracted s functions. We could choose several different contraction schemes, for example we could contract the s-type primitive functions into only two contracted functions, one to represent each of the atomic $1s$ and $2s$ orbitals. Similarly, the p-type primitive functions could be contracted down to a single function. This would reduce the number of basis function further to only five, with a corresponding reduction in the computational effort. However, the energy would again increase, as we will have restricted the variational space yet further. Typically, such an extreme level of contraction will not produce a basis set useful for anything other than the most qualitative calculations.

The contraction scheme we have described is usually referred to as "segmented contraction". This alludes to the fact that each contracted gaussian function includes a set of primitives unique to that function. An alternative contraction scheme would be to include the same set of primitive functions in each contracted shell of a given angular momentum. In the example in Table 2.3, we could use the contractions

Table 2.3 The $9s5p$ basis set; the orbital exponents (upper table) and the occupied molecular orbital coefficients (lower table) of neon obtained from an SCF calculation using this basis set.

No.	Shell	Exponent
1	*s*	12100
2	*s*	1821
3	*s*	432.8
4	*s*	132.5
5	*s*	43.77
6	*s*	5.127
7	*s*	14.91
8	*s*	1.491
9	*s*	0.4468
1	*p*	56.45
2	*p*	12.92
3	*p*	3.865
4	*p*	1.203
5	*p*	0.3444

No.	Function	Molecular orbital 1	2	3	4	5
1	$1s$	0.00120	−0.00028	0	0	0
2	$2s$	0.00909	−0.00217	0	0	0
3	$3s$	0.04131	−0.00977	0	0	0
4	$4s$	0.13790	−0.03537	0	0	0
5	$5s$	0.36251	−0.10136	0	0	0
6	$6s$	0.13006	−0.01925	0	0	0
7	$7s$	0.47267	−0.20769	0	0	0
8	$8s$	−0.00217	0.61430	0	0	0
9	$9s$	0.00185	0.50208	0	0	0
10	$1p_x$	0	0	0	0.01632	0
11	$1p_y$	0	0	0	0	0.01632
12	$1p_z$	0	0	0.01632	0	0
13	$2p_x$	0	0	0	0.10164	0
14	$2p_y$	0	0	0	0	0.10164
15	$2p_z$	0	0	0.10164	0	0
16	$3p_x$	0	0	0	0.30929	0
17	$3p_y$	0	0	0	0	0.30929
18	$3p_z$	0	0	0.30929	0	0
19	$4p_x$	0	0	0	0.48577	0
20	$4p_y$	0	0	0	0	0.48577
21	$4p_z$	0	0	0.48577	0	0
22	$5p_x$	0	0	0	0.34966	0
23	$5p_y$	0	0	0	0	0.34966
24	$5p_z$	0	0	0.34966	0	0

Table 2.4 The $9s5p$ basis set contracted down to a $4s2p$ set; the orbital exponents and contraction coefficients (upper table) and the occupied molecular orbital coefficients (lower table) of neon obtained from an SCF calculation using this basis set.

No.	Contracted shell	No.	Primitive shell	Exponent	Contraction coefficient
1	s	1	s	12100	0.00209307
	s	2	s	1821	0.0158585
	s	3	s	432.8	0.0720450
	s	4	s	132.5	0.240471
	s	5	s	43.77	0.632163
	s	6	s	5.127	0.226810
2	s	7	s	14.91	-1.0
3	s	8	s	1.491	1.0
4	s	9	s	0.4468	1.0
1	p	1	p	56.45	0.0208750
	p	2	p	12.92	0.130032
	p	3	p	3.865	0.395679
	p	4	p	1.203	0.621450
2	p	5	p	0.3444	1.0

No.	Function	Molecular orbital 1	2	3	4	5
1	$1s$	0.57347	-0.14954	0	0	0
2	$2s$	-0.47266	0.20737	0	0	0
3	$3s$	-0.00221	0.62981	0	0	0
4	$4s$	0.00186	0.49488	0	0	0
5	$1p_x$	0	0	0	0	0.78202
6	$1p_y$	0	0	0	0.78202	0
7	$1p_z$	0	0	0.78202	0	0
8	$2p_x$	0	0	0	0	0.3492
9	$2p_y$	0	0	0	0.3492	0
10	$2p_z$	0	0	0.3492	0	0

$$\phi_{1s}(\mathbf{r}) = 0.00120\chi_{1s}(\mathbf{r}) + 0.00909\chi_{2s}(\mathbf{r}) + 0.04131\chi_{3s}(\mathbf{r})$$

$$+ 0.13790\chi_{4s}(\mathbf{r}) + 0.36251\chi_{5s}(\mathbf{r}) + 0.13006\chi_{6s}(\mathbf{r})$$

$$+ 0.47267\chi_{7s}(\mathbf{r}) - 0.00217\chi_{8s}(\mathbf{r}) + 0.00185\chi_{9s}(\mathbf{r})$$

$$\phi_{2s}(\mathbf{r}) = -0.00028\chi_{1s}(\mathbf{r}) - 0.00217\chi_{2s}(\mathbf{r}) - 0.00977\chi_{3s}(\mathbf{r})$$

$$- 0.03537\chi_{4s}(\mathbf{r}) - 0.10136\chi_{5s}(\mathbf{r}) - 0.01925\chi_{6s}(\mathbf{r})$$

$$- 0.20769\chi_{7s}(\mathbf{r}) + 0.61430\chi_{8s}(\mathbf{r}) + 0.50208\chi_{9s}(\mathbf{r})$$

$$(2.162)$$

Here the primitives used to represent the contracted function are identical, but have distinct contraction coefficients. This is called a "general contraction". The basis sets we shall discuss are of the segmented type.

2.5.2 Types of Contracted Basis Sets

A minimal basis set is the simplest type of basis set. A single contracted basis function is used to represent each of the atomic basis functions for all shells up to and including the valence shell of an atom. A minimal basis set would include:

1 basis function for H, He ($1s$)
5 basis functions for Li–Ne ($1s$, $2s$, $2p_x$, $2p_y$, $2p_z$)
9 basis functions for Na–Ar ($1s$, $2s$, $2p_x$, $2p_y$, $2p_z$, $3p_x$, $3p_y$, $3p_z$)
and so on.

For example, a minimal basis set for water consists of seven basis functions: $1s$ on the two hydrogen atoms and $1s$, $2s$, $2p_x$, $2p_y$, $2p_z$, on the oxygen atom. Table 2.5 shows the five doubly occupied molecular orbitals obtained from a Hartree–Fock SCF calculations using the STO-3G minimal basis. The STO-3G basis is obtained by fitting an optimised Slater type orbital with a linear combination of three gaussian type functions. As we have suggested above, minimal basis sets are generally inadequate for anything beyond qualitative work.

An improvement on the minimal basis set can be made by adopting a double-zeta (DZ) basis set. Zeta, ζ, in this name refers to the exponents of the functions. In the earlier quantum chemical literature, orbital exponents were usually referred to as ζ. Hence DZ implies two exponents. Each basis function in the minimal basis set is replaced by two contracted basis functions. Compared to a minimal basis set, the number of basis functions is doubled. The occupied molecular orbitals obtained from a Hartree–Fock SCF calculation on water using a DZ basis are given in Table 2.6. A number of variations of this type of basis set are used. In a triple-zeta basis set (TZ), three basis functions are used to represent each of the orbitals in all shells up to and including the valence shell of an atom. In a split-valence basis set (SV), the splitting of the basis functions is only applied to the valence shell basis orbitals. Each valence atomic orbital is represented by two basis functions while each inner-shell atomic orbital is represented by a single contracted basis function. For example, in water the oxygen atom will have the $1s$ atomic orbital represented by a single contracted function, but the valence shell $2s$ atomic orbital will be represented by two contracted functions. The $2p$ shell will be represented by two contracted basis functions.

So far our emphasis has been on accurate representations of atomic basis functions, yet our main concern in this book is with chemistry and molecules. When molecules form, the atomic orbitals must be allowed to distort, or polarise, to provide optimal overlap between atomic constituents. (A weak

Table 2.5 The occupied molecular orbital coefficients of water obtained from an SCF calculation using the STO-3G basis set.

No.	Atom	Function	1	2	3	4	5
1	O	$1s$	0.99418	−0.23370	0	−0.10194	0
2		$2s$	0.02609	0.84192	0	0.52920	0
3		$2p_x$	0	0	0	0	1.0
4		$2p_y$	0	0	0.60574	0	0
5		$2p_z$	−0.00413	−0.12180	0	0.77174	0
6	H	$1s$	−0.00572	0.15702	0.44649	−0.28684	0
7	H	$2s$	−0.00572	0.15702	−0.44649	−0.28684	0

analogy with qualitative chemical ideas is that of mixing an s and three p orbitals to produce four orbitals directed to the corners of a tetrahedron. This analogy should not be taken too literally since the molecular orbital model is quite distinct from the localised hybrid description of bonding, but this analogy does illustrate the idea that mixing orbitals allows them to distort in essentially any direction.) The basis sets we have described have included basis functions of angular momentum up to and including the valence shell of each atom. Adding further such basis functions will allow greater flexibility in the radial extension of the orbitals, but to describe angular distortions we must include basis functions that represent orbitals for which the value of the angular momentum quantum number, l, is larger than the maximum value encountered in the valence shell. For example, the inclusion of p-type basis functions will allow an s-type function to distort its spherical distribution along any of the axes of the p basis functions. Similarly, d-type functions enable the distortion of p functions away from the axes they are aligned along. The addition of polarisation functions to a DZ basis set results in a double-zeta

Table 2.6 The occupied molecular orbital coefficients of water obtained from an SCF calculation using a DZ basis set.

No.	Atom	Function	1	2	3	4	5
1	O	$1s$	0.58102	−0.13062	0	−0.04681	0
2		$2s$	0.46124	−0.18108	0	−0.06531	0
3		$3s$	−0.00021	0.51021	0	0.19817	0
4		$4s$	0.00188	0.45939	0	0.26405	0
5		$5p_x$	0	0	0	0	0.72908
6		$5p_y$	0	0	0.57136	0	0
7		$5p_z$	−0.00164	−0.12309	0	0.62345	0
8		$6p_x$	0	0	0	0	0.40860
9		$6p_y$	0	0	0.18410	0	0
10		$6p_z$	0.00046	−0.03435	0	0.32715	0
11	H	$1s$	0.00002	0.13257	0.25171	−0.14096	0
12		$2s$	−0.00017	0.00772	0.13460	−0.06579	0
13	H	$1s$	0.00002	0.13257	−0.25171	−0.14096	0
14		$2s$	−0.00017	0.00772	−0.13460	−0.06579	0

plus polarisation basis set (DZP). For water, a DZP includes a set of three *p*-type functions on each hydrogen atom and a set of five *d*-type functions on the oxygen atom. Table 2.7 gives the molecular orbitals obtained for water using a DZP basis set. Higher polarisation functions are often needed for accurate work. A widely used family of basis sets, developed by Dunning,[11] are the correlation consistent basis sets, denoted cc-pV*X*Z. These are split-valence basis sets, with $X = 2, 3, 4, 5, 6 \dots$ denoting valence (V) *double, triple, quadruple … zeta.*, with increasing shells of polarisation functions. These large basis sets were designed to allow systematic extrapolation to the complete basis set limit. For the oxygen atom, the sets of polarisation functions included as a function of X are: $1d$ (cc-pVDZ); $2d\,1f$ (cc-pVTZ); $2d\,2f\,1g$ (cc-pVQZ); $4d\,3f\,2g\,1h$ (cc-pV5Z); $5d\,4f\,3g\,2h\,1i$ (cc-pV6Z).

In addition to polarisation functions, when dealing with systems that are negative ions or contain a large number of lone pairs, it is necessary to include diffuse functions in the basis set. These functions are often of the same angular momentum as the valence shell orbitals, but have small exponents, which give the orbitals a large radial extension. This provides a mechanism whereby a molecular orbital containing a negative ion or a lone pair can expand its spatial extent to minimise electron–electron repulsions. In principle, for such systems, diffuse functions should be included for all types of basis function in the set. This is done in the augmented, aug-cc-pV*X*Z, basis sets.[12]

The range of basis sets currently in use is vast. There are many 'families' of basis sets. Some of the best known being the older Pople-style split-valence basis sets. These basis sets have the constraint that the exponents of *s* and *p* functions are the same, but they have different contraction coefficients. This provides savings in the computational effort associated with integral evaluation, as the *s* and *p* shells share much common data. Examples of Pople-style basis sets are 3-21G, 6-31G, 6-311G. In these designations the first number indicates the number of gaussian primitives that are contracted to represent the core shell basis functions. For example, 6-31G implies that for an oxygen atom six primitive gaussian functions are contracted to represent the 1*s* atomic basis function. For the sulphur atom, the 1*s*, 2*s*, and 2*p* basis functions would be treated with a contraction of six primitive gaussian functions each. The valence shell basis functions, 2*s* 2*p*, are split into two functions with the tighter function being a contraction of three primitive gaussian functions and the outer function being uncontracted. Similarly, 6-311G describes a triple split of the valence shell basis functions. General polarisation functions are given in parentheses with the first entry referring to non-hydrogen atoms and the second to hydrogen. For example, 6-311G(2df,2pd) implies that two sets of *d* and one set of *f* polarisation functions are to be added to all atoms other than hydrogen and 2 sets of *p* and one set of *d* polarisation functions are to be added to hydrogen atoms. In the older literature (d, p) was denoted as **. Diffuse sets can also be included using the "+" nomenclature, for example 6-311++G(2df,2pd). Again, the first "+" refers to non-hydrogen atoms and the second to hydrogen.

Table 2.7 The occupied molecular orbital coefficients of water obtained from an SCF calculation using a DZP basis set, the five d functions are labelled with their m_l value.

No.	Atom	Function	1	2	3	4	5
1	O	$1s$	0.58106	−0.13102	0	−0.04337	0
2		$2s$	0.46123	−0.18106	0	−0.06066	0
3		$3s$	0.00035	0.51453	0	0.18341	0
4		$4s$	0.00227	0.41828	0	0.26491	0
5		$5p_x$	0	0	0	0	0.72419
6		$5p_y$	0	0	0.56789	0	0
7		$5p_z$	−0.00130	−0.08970	0	0.62587	0
8		$6p_x$	0	0	0	0	0.39972
9		$6p_y$	0	0	0.19638	0	0
10		$6p_z$	0.00013	−0.01390	0	0.31575	0
11		$7d_0$	0.00005	0.00285	0	−0.02617	0
12		$7d_{+1}$	0	0	0	0	−0.02519
13		$7d_{-1}$	0	0	−0.03665	0	0
14		$7d_{+2}$	0.00001	−0.00430	0	0.00653	0
15		$7d_{-2}$	0	0	0	0	0
16	H	$1s$	0	0.13500	0.22982	−0.14170	0
17		$2s$	−0.00034	0.02784	0.13718	−0.07287	0
18		$3p_x$	0	0	0	0	0.02177
19		$3p_y$	0.00015	−0.02100	−0.00888	0.01578	0
20		$3p_z$	−0.00003	0.01299	0.01674	0.00629	0
21	H	$1s$	0	0.13500	−0.22982	−0.14170	0
22		$2s$	−0.00034	0.02784	−0.13718	−0.07287	0
23		$3p_x$	0	0	0	0	0.02177
24		$3p_y$	−0.00015	0.02100	−0.00888	−0.01578	0
25		$3p_z$	−0.00003	0.01299	−0.01674	0.00629	0

Another very useful, systematic, family of basis sets is that of Ahlrichs and coworkers.[13] These can be accessed *via* the basis set library of the Turbomole program at http://bases.turbo-forum.com/TBL/tbl.html. Another site that collates a wide range of basis sets for general access is available at the Basis Set Exchange, https://bse.pnl.gov/bse/portal.

2.5.3 Basis Set Superposition Error: Non-Covalent Interactions

The treatment of non-covalent interactions poses a challenge for computational quantum chemistry. Its conventional, *ab initio*, treatment requires large basis sets and correlated methods. Non-covalent interactions are present in a large range of systems. The gas-phase dimers of small molecules, for example water or hydrogen fluoride, are held together by non-covalent forces such as hydrogen bonding, dispersion and dipole–dipole interactions. These are also the key interactions in determining the structures of many biomolecules. The basis set superposition error arises in the calculation of weakly bound systems.

Its origin is simply the use of finite basis sets. In the limit of a complete basis set, the basis set superposition error would vanish.

For example, suppose we were interested in calculating the binding energy of two water molecules combining to form a dimer, Figure 2.15. Let us denote the first water molecule as A, with basis set $\{\chi_A\}$ and the second as B, with basis set $\{\chi_B\}$. The structure of the dimer corresponds to a minimum on the potential energy surface and there will be more than one such structure. Its binding energy could be calculated as the energy of the dimer minus the energies of the two infinitely separated monomers.

$$E_{\text{Binding}} = E_{AB}^{\chi_A + \chi_B} - E_A^{\chi_A} - E_B^{\chi_B} \tag{2.163}$$

However, the energy of an individual water molecule is computed using the basis functions on that monomer alone. In the dimer, the basis set is the combination of the basis sets of the two monomers. The effect of this is that each monomer is effectively described by a larger basis in the dimer than in the monomer and if we apply eqn (2.163) to calculate the binding energy we can expect that the binding energy will be too large.

A simple method used to reduce the basis set superposition error is known as the "counterpoise correction". In this scheme the monomer energies are calculated in the dimer basis, that is

$$E_{\text{Binding}} = E_{AB}^{\chi_A + \chi_B} - E_A^{\chi_A + \chi_B} - E_B^{\chi_A + \chi_B} \tag{2.164}$$

This has proved to be an effective correction, but it must be emphasised that it is not a complete cure for the basis set superposition error. It should be borne in mind that calculating the energy of the monomer in the basis of the dimer allows the monomer to use the whole dimer basis set. In the dimer, each monomer can effectively only access the unoccupied orbitals of the other monomer, hence in the counterpoise correction additional flexibility is being given to each monomer.

Table 2.8 shows the counterpoise correction for a water dimer configuration using the family of correlation consistent basis sets and the MP2 method, see Section 2.8.1. To place these numbers in perspective we note that experimentally[14] the binding energy of the water dimer is 13–15 kJ mol^{-1} and *ab initio* computed energies are typically 21–25 kJ mol^{-1}. The magnitude of the basis set superposition error is very significant and must be addressed in the study of non-covalent interactions. Table 2.8 also informs us that the choice of basis function is very important, simply enlarging the basis set in the cc-pVXZ family produces a decrease in the basis set superposition error but the decrease is quite slow. Adding diffuse functions to the basis set, with a large spatial extent, is much more effective.

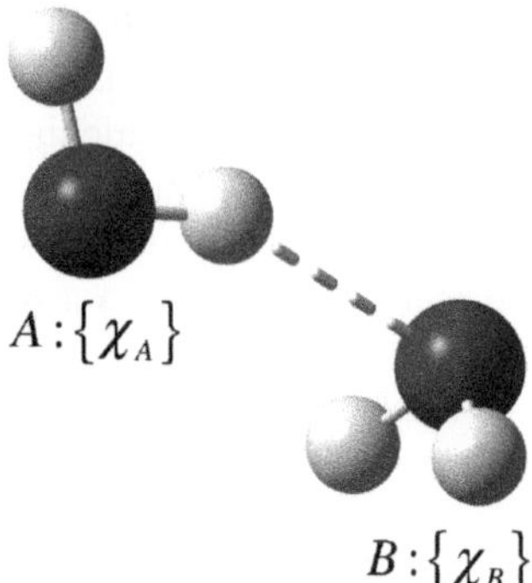

Figure 2.15   Non-covalently bound water dimer consisting of monomers *A* and *B* with their associated basis sets.

2.6   Electron Correlation: Background to Post Hartree–Fock Methods

In the following sections we shall discuss some of the common methods used to treat the electron correlation problem. We introduced some aspects of the electron correlation problem in Section 1.4 and saw that we can choose to tackle it using wavefunction-based methods or density functional theory. We shall address density functional theory at the end of this chapter. The remaining sections of this chapter will be concerned with wavefunction-based techniques and this section introduces some key ideas that we shall need.

Recall that the exact wavefunction, within a given basis set, can be written as a linear combination of Slater determinants. The leading term is taken to be the Hartree–Fock reference determinant, $|\Psi_0\rangle$, and all the substituted determinants that can be formed from it

$$|\Psi\rangle = C_0|\Psi_0\rangle + \sum_{i}^{\text{occupied}} \sum_{a}^{\text{unoccupied}} C_i^a |\Psi_i^a\rangle + \sum_{i<j}^{\text{occupied}} \sum_{a<b}^{\text{unoccupied}} C_{ij}^{ab} |\Psi_{ij}^{ab}\rangle$$

$$+ \sum_{i<j<k}^{\text{occupied}} \sum_{a<b<c}^{\text{unoccupied}} C_{ijk}^{abc} |\Psi_{ijk}^{abc}\rangle + \sum_{i<j<k<l}^{\text{occupied}} \sum_{a<b<c<d}^{\text{unoccupied}} C_{ijkl}^{abcd} |\Psi_{ijkl}^{abcd}\rangle + \cdots \tag{2.165}$$

This expansion is written in terms of spin–orbitals, with i, j, k, l labelling orbitals that are occupied in $|\Psi_0\rangle$ and a, b, c, d those that are unoccupied. For an *N*-electron system, up to *N*-fold excitations must be included. To obtain the energy corresponding to this wavefunction requires that we form the quantity

$$E = \frac{\langle\Psi|H|\Psi\rangle}{\langle\Psi|\Psi\rangle} = \langle\Psi|H|\Psi\rangle \quad [\langle\Psi|\Psi\rangle = 1] \tag{2.166}$$

It can be seen that this will require the evaluation of matrix elements between

Table 2.8 The basis set superposition error for the water dimer using the cc-pVXZ basis sets. As the basis set becomes more complete, the basis set superposition error is reduced.

Basis set	Size of dimer basis set	Basis set superposition error / kJ mol^{-1}
cc-pVDZ	48	18.6
cc-pVTZ	116	8.7
cc-pVQZ	230	4.1
aug-cc-pVDZ	82	3.9
aug-cc-pVTZ	184	1.9

different Slater determinants. If the wavefunction in eqn (2.165) consists of N_{Det} Slater determinants, it will be necessary to evaluate $(N_{\text{Det}})^2$ matrix elements to obtain the energy. The matrix element rules given in Section 2.1.1 all rely on the availability of the molecular one- and two-electron integrals over the set of molecular orbitals, $\{\phi\}$. In the previous section we saw that the one- and two-electron integrals are evaluated over the atomic basis functions $\{\chi\}$. The corresponding integrals over molecular orbitals are obtained by transformation. For the one-electron integrals the transformation maps the atomic basis function integrals, $h_{\mu\nu}$, to the molecular orbital integrals, h_{pq}. This transformation involves only two indices and may be written

$$h_{pq} = \sum_{\mu\nu}^{m} c_{\mu p} h_{\mu\nu} c_{\nu q} \tag{2.167}$$

Here p and q denote arbitrary molecular orbitals and the labels μ, ν range over the full set of basis functions. There are m^2 elements, h_{pq}, and each of these requires m^2 operations to form. So, as written, the transformation of the one-electron integrals to the molecular orbital basis requires m^4 operations. More efficiently, we could form the partially transformed integrals, $h_{p\nu}$,

$$h_{p\nu} = \sum_{\mu}^{m} c_{\mu p} h_{\mu\nu} \tag{2.168}$$

which requires m^3 effort, and then complete the transformation to h_{pq} as

$$h_{pq} = \sum_{\nu}^{m} c_{\nu q} h_{p\nu} \tag{2.169}$$

This last transformation also requires m^3 effort. Hence, the m^4 demand of eqn (2.167) has been ameliorated by using two partial transformations, each of m^3 effort. The only disadvantage of the two-step transformation is the need to store the intermediate integrals, h_{pv}. This disadvantage is outweighed by the benefit of a significantly reduced number of operations. The storage of m^2 elements is not too demanding, even for quite large values of m.

The two-electron integrals range over four indices and their direct transformation scales very steeply as m^8

$$(pq|rs) = \sum_{\mu\nu\lambda\sigma}^{m} c_{\mu p} c_{\nu q} (\mu\nu|\lambda\sigma) c_{\lambda r} c_{\sigma s} \tag{2.170}$$

Again, we can reduce the effort required by using four partial transformations, each requiring m^5 computational effort,

$$(pv|\lambda\sigma) = \sum_{\mu}^{m} c_{\mu p} (\mu\nu|\lambda\sigma) \quad (i)$$

$$(pq|\lambda\sigma) = \sum_{\nu}^{m} c_{\nu q} (pv|\lambda\sigma) \quad (ii)$$

$$(pq|r\sigma) = \sum_{\lambda}^{m} c_{\lambda r} (pq|\lambda\sigma) \quad (iii) \tag{2.171}$$

$$(pq|rs) = \sum_{\sigma}^{m} c_{\sigma s} (pq|r\sigma) \quad (iv)$$

We can reduce the effort even further by noting the permutational equivalences between integrals, as noted in Section 2.1.2. Rather than forming all m^4 integrals we need only form integrals, $(pq|rs)$, satisfying the restrictions

$$p \geq q$$

$$r \geq s \tag{2.172}$$

$$pq \geq rs$$

This reduces the number of integrals to approximately $m^4/8$. In principle, if there is sufficient fast memory available on our computer, we can store all the molecular orbital integrals one after the other and access them using the addressing

$$pq = p(p-1)/2 + q$$

$$rs = r(r-1)/2 + s \tag{2.173}$$

$$pqrs = pq(pq-1)/2 + rs$$

For example, if $m = 2$, the restrictions in eqn (2.172) and the canonical order specified by eqn (2.173) implies the storage of the six permutationally unique integrals $(11|11); (21|11); (21|21); (22|11); (22|21); (22|22)$, rather than the full, $2^4 = 16$, set of integrals.

In practice, the size of the basis set, m, for which it is possible to hold the canonical list of integrals in memory is quite limited. Typically each integral requires eight bytes to store, and there are exactly $\frac{1}{2}[m(m+1)/2]$ $[m(m+1)/2+1]$ integrals in the canonical list. For example, 8 GB (8589934592 bytes) of fast memory can accommodate the canonical list for up to $m = 181$ only. Doubling the memory to 16 GB will only increase this up to $m = 215$. Yet, we often wish to perform calculations with several times this number of basis functions. The handling of such large amounts of data provides one of the many significant quantum chemical challenges to the computer programmer's ingenuity. It is well to remember that the extraordinary developments in quantum chemistry, which have made it possible to perform calculations with thousands of basis functions, have come about as much through the development of algorithms and new methods as they have through the improvements in computational hardware, Figure 1.1.

2.6.1 Brillouin's Theorem

Eqn (2.165) expresses the exact wavefunction, $|\Psi\rangle$, as a linear combination of classes of substituted determinants generated from the Hartree–Fock reference determinant. These classes refer to single, double, triple, ... substitutions of occupied orbitals with unoccupied orbitals. We know from the rules for matrix elements between determinants that the Hartree–Fock reference determinant will not interact with another determinant that differs from $|\Psi_0\rangle$ by more than two substitutions. A further property of $|\Psi_0\rangle$ is that, provided the orbitals from which it is formed have been optimised, it will not interact with any determinants that differ from it by a single substitution. To see that this is so, consider the expression for the closed-shell Fock matrix

$$F_{\mu v} = h_{\mu v} + \sum_{j}^{N/2} \sum_{\lambda\sigma}^{m} c_{\lambda j} c_{\sigma j} [2(\mu v|\lambda\sigma) - (\mu\sigma|\lambda v)] \tag{2.174}$$

Now let us transform this matrix into the molecular orbital basis

$$\mathbf{C}^{\dagger} \mathbf{F}^{AO} \mathbf{C} = \mathbf{F}^{MO} \tag{2.175}$$

When the orbitals are optimised, $\mathbf{F}^{MO}$ will be block diagonal. In the molecular orbital basis the off-diagonal blocks of $\mathbf{F}^{MO}$, corresponding to occupied–unoccupied orbital pairs, are related to the gradient of the Hartree–Fock energy with respect to the molecular orbital coefficients. Hence, at convergence, these blocks must be zero. Now consider the specific case of the occupied orbital, ϕ_i, and the unoccupied orbital, ϕ_a,

$$F_{ia} = \sum_{\mu\nu}^{m} c_{\mu i} F_{\mu\nu} c_{\nu a}$$

$$= \sum_{\mu\nu}^{m} c_{\mu i} c_{\nu a} \left[h_{\mu\nu} + \sum_{j}^{N/2} \sum_{\lambda\sigma}^{m} c_{\lambda j} c_{\sigma j} [2(\mu\nu|\lambda\sigma) - (\mu\sigma|\lambda\nu)] \right] \quad (2.176)$$

$$= h_{ia} + \sum_{j}^{N/2} [2(ia|jj) - (jj|ja)]$$

Now consider a four-electron system as depicted in Figure 2.16. The Fock matrix element, F_{ia}, will consist of the integrals

$$F_{ia} = h_{ia} + 2(ia|ii) - (ii|ia) + 2(ia|jj) - (ij|ja)$$
$$= h_{ia} + (ia|ii) + 2(ia|jj) - (ij|ja) \quad (2.177)$$

The second line is obtained by noting that the permutational symmetry of the two-electron integrals implies that $(ia|ii) = (ii|ia)$. Now let us use eqn (2.20) to evaluate the matrix element between the configurations shown in Figure 2.16(a) and (b)

$$\left\langle \begin{matrix} b & - \\ a & - \\ j & -\uparrow\downarrow- \\ i & -\uparrow\downarrow- \end{matrix} \right| H \left| \begin{matrix} b & - \\ a & -\uparrow- \\ j & -\uparrow\downarrow- \\ i & -\downarrow- \end{matrix} \right\rangle = h_{ia} + (ia|\bar{i}\bar{i}) + (ia|jj) - (ij|ja) + (ia|\bar{j}\bar{j}) \quad (2.178)$$

Note that eqn (2.20) applies to spin–orbitals, while the closed-shell Fock matrix has spin eliminated from it by integration. In terms of spatial integrals it is clear that eqns (2.177) and (2.178) are equivalent. We know that for optimal orbitals eqn (2.177) yields zero. This implies that the matrix element between $|\Psi_0\rangle$ and a singly substituted determinant, $|\Psi_i^a\rangle$, is also zero. This is Brillouin's theorem. We have illustrated it for the closed-shell case only, but the theorem is quite general and applies to the UHF and ROHF cases as well as to multi-configurational reference wavefunctions.

2.6.2 Spin Eigenfunctions/Configuration State Functions

In Section 2.3 we introduced the idea of spin symmetry in the context of the ROHF and UHF wavefunctions. In general Slater determinants are not eigenfunctions of $\hat{S}^2$, although they are eigenfunctions of $\hat{S}_z$. In special cases, for example all doubly occupied or all singly occupied orbitals, Slater determinants can be spin eigenfunctions. The difficulty arises in situations involving singly occupied orbitals that are spin-paired. For example, consider the determinant $|\phi_1\bar{\phi}_2\rangle$ for which

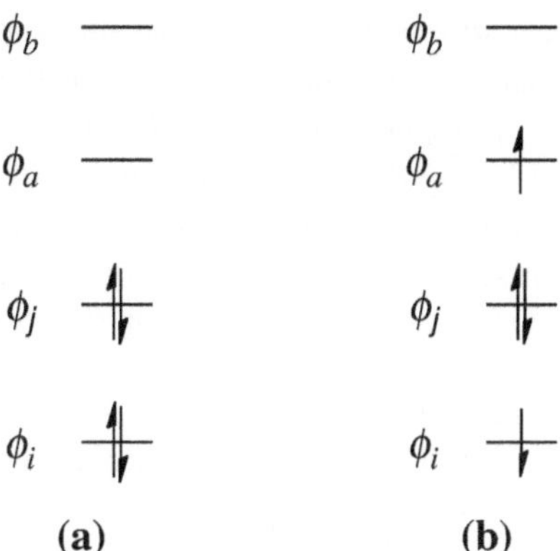

Figure 2.16 A single substitution in which an orbital, ϕ_i, occupied in the Hartree–Fock determinant (a) is substituted by an unoccupied orbital ϕ_a (b).

$\langle \hat{S}_z \rangle = 0$. Now consider the action of $\hat{S}^2$ on this determinant, using eqn (2.37),

$$\hat{S}^2 |\phi_1 \bar{\phi}_2\rangle = \left\{ \sum_{\substack{\alpha\beta \\ \text{permuations}}} \hat{P}_{\alpha\beta} + \frac{1}{4}\left[(N_\alpha - N_\beta)^2 2N_\alpha + 2N_\beta \right] \right\} |\phi_1 \bar{\phi}_2\rangle$$

$$= \left\{ \sum_{\substack{\alpha\beta \\ \text{permuations}}} \hat{P}_{\alpha\beta} + 1 \right\} |\phi_1 \bar{\phi}_2\rangle \qquad (2.179)$$

$$= |\bar{\phi}_1 \phi_2\rangle + |\phi_1 \bar{\phi}_2\rangle$$

A state with $S = 0$ should return an eigenvalue of zero, $S(S+1) = 0$. Clearly $|\phi_1 \bar{\phi}_2\rangle$ does not satisfy this condition. However if we take a linear combination, $|\phi_1 \bar{\phi}_2\rangle \pm |\bar{\phi}_1 \phi_2\rangle$, we obtain

$$\hat{S}^2 \left\{ |\phi_1 \bar{\phi}_2\rangle \pm |\bar{\phi}_1 \phi_2\rangle \right\} = \left\{ \sum_{\substack{\alpha\beta \\ \text{permuations}}} \hat{P}_{\alpha\beta} + 1 \right\} \left\{ |\phi_1 \bar{\phi}_2\rangle \pm |\bar{\phi}_1 \phi_2\rangle \right\}$$

$$= \left\{ |\bar{\phi}_1 \phi_2\rangle \pm |\phi_1 \bar{\phi}_2\rangle \right\} + \left\{ |\phi_1 \bar{\phi}_2\rangle \pm |\bar{\phi}_1 \phi_2\rangle \right\} \qquad (2.180)$$

$$= 2(|\phi_1 \bar{\phi}_2\rangle + |\bar{\phi}_1 \phi_2\rangle), \text{ or } 0$$

If $S(S+1) = 2$, then $S = 1$, while if $S(S+1) = 0$ then $S = 0$. So a linear combination of Slater determinants does produce spin eigenfunctions. In fact it produces spin eigenfunctions corresponding to $S = 0, 1$. The $S = 1$ case corresponds to a triplet spin multiplicity, $2S+1 = 3$. In terms of spin angular momentum vector diagrams, the triplet state has three components, which in the absence of a magnetic field are degenerate. The components are usually represented as their projection on the z-axis (taken as the axis of spin quantisation), as shown in Figure 2.17. The $S = 0$ state is obtained by the $|\phi_1\bar{\phi}_2\rangle - |\bar{\phi}_1\phi_2\rangle$ combination. It is always possible to combine Slater determinants to produce spin eigenfunctions. Such combinations are termed "configuration state functions". Typically, the number of configuration state functions, for a given value of S, is smaller than the corresponding number of Slater determinants that can be formed. For example, for two orbitals and two electrons, with $S = 0$, there are three configuration state functions but four Slater determinants, see Figure 2.18. The number of Slater determinants is given by

$$\binom{m}{N_\alpha}\binom{m}{N_\beta} \tag{2.181}$$

and the number of configuration state functions is given by Weyl's formula

$$\frac{2S+1}{m+1}\binom{m+1}{N/2-S}\binom{m+1}{N/2+S+1} \tag{2.182}$$

As we have seen, the evaluation of matrix elements between Slater determinants is simply done. We could form configuration state functions as linear combinations of Slater determinants and then combine the matrix element contributions accordingly. However this is rather inefficient, and many schemes have been developed for the direct calculation of matrix elements between configuration state functions. In this text, we shall only deal with Slater determinants.

2.6.3 Integral Approximations: Density Fitting

Two-electron integrals over the molecular orbitals are very numerous. In order to use many of the techniques we have discussed it is essential to have the molecular orbital integrals available in the computer's fast memory. To avoid having to store all the two-electron integrals we must look for suitable

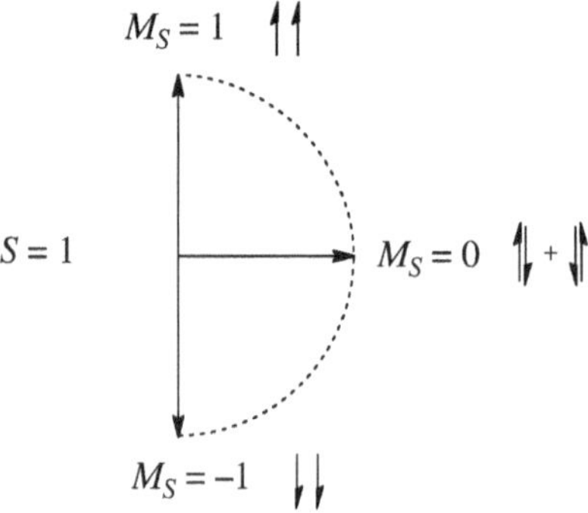

Figure 2.17   The triplet M_S spin components corresponding to $S = 1$.

approximations. Returning to the form of the two-electron integral over basis functions

$$(\mu v | \lambda \sigma) = \iint \chi_\mu(\mathbf{r}_1)\chi_v(\mathbf{r}_1)\frac{1}{r_{12}}\chi_\lambda(\mathbf{r}_2)\chi_\sigma(\mathbf{r}_2)d\mathbf{r}_1 d\mathbf{r}_2 \qquad (2.183)$$

we note that the integral involves products of the basis functions, such as $\chi_\mu(\mathbf{r}_1)\chi_v(\mathbf{r}_1)$. As the size of the basis set increases, the products of basis functions will become more and more numerically linearly dependent. This means that a given product of basis functions is more and more likely to be expressible in terms of a linear combination of other such products of basis functions. For a basis set of dimension m, there are $m(m+1)/2$ unique pairs. Suppose that we form a fit, in a least-squares sense, to the product $\chi_\mu(\mathbf{r}_1)\chi_v(\mathbf{r}_1)$ as

$$\chi_\mu(\mathbf{r}_1)\chi_v(\mathbf{r}_1) = \sum_Q^{m_{\text{fit}}} c_{\mu v}^Q \chi_Q(\mathbf{r}_1) \qquad (2.184)$$

$\chi_Q(\mathbf{r}_1)$ is a member of an auxilliary set of basis functions, $\{\chi_Q\}$, which is used to fit the products $\chi_\mu(\mathbf{r}_1)\chi_v(\mathbf{r}_1)$. The number of auxillary functions is m_{fit}, and provided that it is smaller than $m(m+1)/2$ some efficiency may be gained in the integral evaluation step. We can obtain a measure of the error in the fitting by defining a residuum as

$$R_{\mu v}(\mathbf{r}_1) = \chi_\mu(\mathbf{r}_1)\chi_v(\mathbf{r}_1) - \sum_Q^{m_{\text{fit}}} c_{\mu v}^Q \chi_Q(\mathbf{r}_1) \qquad (2.185)$$

This in turn provides an estimate of the error in the two-electron integral

$$(R_{\mu v} | R_{\lambda \sigma}) = \iint R_{\mu v}(\mathbf{r}_1)\frac{1}{r_{12}}R_{\lambda\sigma}(\mathbf{r}_2)d\mathbf{r}_1 d\mathbf{r}_2 \qquad (2.186)$$

Minimisation of the error in the two-electron integral leads to a set of linear equations from which the fitting coefficients, $c_{\mu v}^Q$, can be obtained. Defining the integrals

$$(\mu v | Q) = \iint \chi_\mu(\mathbf{r}_1)\chi_v(\mathbf{r}_1)\frac{1}{r_{12}}\chi_Q(\mathbf{r}_2)d\mathbf{r}_1 d\mathbf{r}_2$$

$$(P|Q) = \iint \chi_P(\mathbf{r}_1)\frac{1}{r_{12}}\chi_Q(\mathbf{r}_2)d\mathbf{r}_1 d\mathbf{r}_2 \qquad (2.187)$$

and the matrix, $\mathbf{V}$, with elements

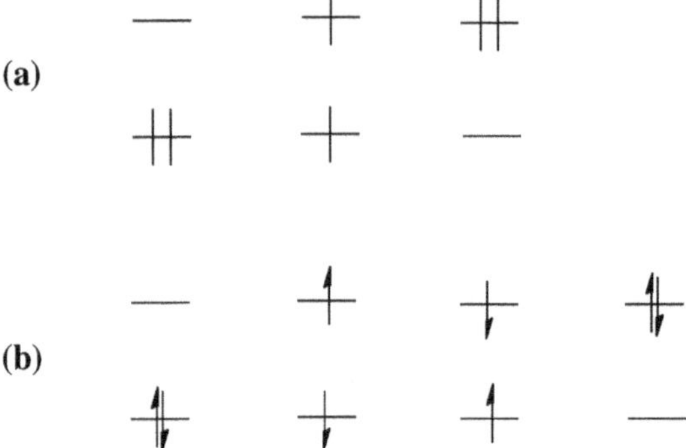

Figure 2.18   (a) Configuration states function with overall spin $S = 0$ that can be formed from two electrons distributed in two orbitals. (b) The Slater determinants that can be formed from two electrons distributed in two orbitals with $S = 0$.

$$V_{PQ} = (P|Q) \tag{2.188}$$

The optimal fitting coefficients can be obtained as

$$c_{\mu\nu}^{Q} = \sum_{P}^{m_{\text{fit}}} (\mu\nu|P) V_{PQ}^{-1} \tag{2.189}$$

accordingly, the two-electron integral may be obtained as

$$(\mu\nu|\lambda\sigma)^{\text{DF}} = \sum_{PQ}^{m_{\text{fit}}} (\mu\nu|P) V_{PQ}^{-1} (Q|\lambda\sigma) \tag{2.190}$$

The superscript "DF" is to indicate that this has been obtained in the density fitting approximation, also referred to as the "resolution of the identity approximation". In the limit that the set $\{\chi_Q\}$ becomes complete, the expression in eqn (2.190) becomes exact.

So far no reduction in computational cost has been achieved! If we were to proceed in the usual manner and now transform $(\mu\nu|\lambda\sigma)^{\text{DF}}$ into the molecular orbital basis we would still have the regular computational effort of performing the steps in eqn (2.171). The advantage of the method is to use the three- and two-index quantities directly rather than the four-index electron repulsion integral. By forming the array

$$B_{pq}^{P} = \sum_{Q}^{m_{\text{fit}}} V_{PQ}^{-\frac{1}{2}} \sum_{\mu}^{m} c_{\mu p} \sum_{\nu}^{m} c_{\nu q} (Q|\mu\nu) \tag{2.191}$$

the two-electron integral in the molecular orbital basis, $(pq|rs)$, can be obtained as

$$(pq|rs)^{\text{DF}} = \sum_P^{m_{\text{fit}}} B_{pq}^P B_{rs}^P \qquad (2.192)$$

The storage of the **B** array requires $m^2 m_{\text{fit}}$ space, which is much more manageable than $\approx m^4/8$. The formation of the integral requires summation over m_{fit} auxillary functions and this incurs a cost. However, the two-electron integrals can be formed in the computer's fast memory and make it feasible to perform calculations on much larger systems than would be possible using the conventional canonical list of integrals.

Finally we note that very efficient standard fitting basis sets of auxillary functions are available at http://bases.turbo-forum.com/TBL/tbl.html.

2.7 Configuration Interaction Methods

The task of the configuration interaction (CI) method is to find the minimum in the energy

$$E = \frac{\langle \Psi_{\text{CI}}|H|\Psi_{\text{CI}}\rangle}{\langle \Psi_{\text{CI}}|\Psi_{\text{CI}}\rangle} \qquad (2.193)$$

by varying the coefficients that occur in the determinantal expansion of the wavefunction. The CI wavefunction is

$$|\Psi_{\text{CI}}\rangle = \sum_M C_M |\Psi_M\rangle \qquad (2.194)$$

where the index M ranges over all the various types of substitution given in eqn (2.165). Additionally we must satisfy the wavefunction normalisation condition. Hence the variation of the energy is not a free variation but is constrained by the requirement $\langle \Psi_{\text{CI}}|\Psi_{\text{CI}}\rangle = 1$. We have met the problem of optimising energy functionals in the presence of constraints when we dealt with the Hartree–Fock method. Here the situation is a little simpler since the variation is linear, unlike the variation of the Hartree–Fock energy with respect to the molecular orbital coefficients.

With proper normalisation, the CI energy is given by $E_{\text{CI}} = \langle \Psi_{\text{CI}}|H|\Psi_{\text{CI}}\rangle$ and the Lagrange function is

$$L = E_{\text{CI}} - \lambda[\langle \Psi_{\text{CI}}|\Psi_{\text{CI}}\rangle - 1] \qquad (2.195)$$

Substituting the CI expansion yields

$$L = \sum_{MN} C_M C_N \langle \Psi_M | H | \Psi_N \rangle - \lambda \left[\sum_{MN} C_M C_N \langle \Psi_M | \Psi_N \rangle - 1 \right] \qquad (2.196)$$

The condition to be satisfied is that $\delta L = 0$ with respect to variation of the coefficients

$$\delta L = \sum_{MN} \delta C_M C_N \langle \Psi_M | H | \Psi_N \rangle + \sum_{MN} C_M \delta C_N \langle \Psi_M | H | \Psi_N \rangle$$

$$- \lambda \sum_{MN} \delta C_M C_N \langle \Psi_M | \Psi_N \rangle - \lambda \sum_{MN} C_M \delta C_N \langle \Psi_M | \Psi_N \rangle \qquad (2.197)$$

$$= 0$$

As previously we deal with real functions only, which implies that $\langle \Psi_{\text{CI}} |$ is equivalent to $| \Psi_{\text{CI}} \rangle$, but in the general case the variations of $| \Psi_{\text{CI}} \rangle$ and $\langle \Psi_{\text{CI}} |$ provide two conditions

$$\delta L' = \sum_{M} \delta C_M \left\{ \sum_{N} C_N \langle \Psi_M | H | \Psi_N \rangle - \lambda \sum_{N} C_N \langle \Psi_M | \Psi_N \rangle \right\} = 0$$

$$\delta L'' = \sum_{N} \delta C_N \left\{ \sum_{M} C_M \langle \Psi_M | H | \Psi_N \rangle - \lambda \sum_{M} C_M \langle \Psi_M | \Psi_N \rangle \right\} = 0$$

$$(2.198)$$

and $\delta L = \delta L' + \delta L''$. Since the variations δC_M or δC_N are arbitrary, the quantities in curly brackets above must vanish. This gives

$$\sum_{N} C_N \langle \Psi_M | H | \Psi_N \rangle = \lambda \sum_{N} C_N \langle \Psi_M | \Psi_N \rangle \qquad (2.199)$$

for δC_M. Including the variation for all M, we can see that the Lagrange multiplier, λ, is equal to the energy, E_{CI},

$$\lambda = \frac{\displaystyle\sum_{MN} C_M C_N \langle \Psi_M | H | \Psi_N \rangle}{\displaystyle\sum_{MN} C_M C_N \langle \Psi_M | \Psi_N \rangle} = E_{\text{CI}} \qquad (2.200)$$

or in matrix form

$$\mathbf{Hc} = E_{\text{CI}} \mathbf{Sc} \qquad (2.201)$$

where $\mathbf{c}$ is the column vector containing the expansion coefficients corresponding to the energy, E_{CI}. We know from our previous discussion of Slater determinants that they are orthonormal, $\langle \Psi_M | \Psi_N \rangle = \delta_{MN}$, which allows us to simplify eqn (2.201) to

$$\mathbf{H}\mathbf{c} = E_{CI}\mathbf{c} \tag{2.202}$$

The lowest energy, E_{CI}, and its corresponding eigenvector, $\mathbf{c}$, represent the ground-state energy and wavefunction, respectively. The variational principle tells us that $E_{CI} \geq E_{Exact}$. Since the matrix $\mathbf{H}$ is of dimension $N_{Det} \times N_{Det}$, with N_{Det} given by eqn (2.181), we can obtain N_{Det} energies and corresponding wavefunctions. There exists a theorem that extends the variational principle to states above the ground state. The MacDonald–Hylleraas–Undheim theorem establishes that any eigenvalue of $\mathbf{H}$ is an upper bound to the corresponding exact excited state energy. Accordingly, it is useful to write eqn (2.202) as a matrix eigenvalue equation that can furnish all N_{Det} energies and wavefunctions

$$\mathbf{H}\mathbf{C} = \mathbf{C}\mathbf{E} \tag{2.203}$$

The $\mathbf{H}$ matrix we have considered corresponds to eqn (2.165) and within a given basis set is exact. The form of $\mathbf{H}$ is governed by Slater's rules for matrix elements, which tell us that determinants differing in more than two spin–orbitals will have a matrix element of zero. Additionally, if the molecular orbitals from which the determinants are built have been optimised according to the Hartree–Fock principle, then Brillouin's theorem holds. These factors impose a structure on the form of $\mathbf{H}$. If we denote the complete class of single, double, triple, ... substituted determinants by the abbreviations $|\mathbf{S}\rangle, |\mathbf{D}\rangle, |\mathbf{T}\rangle, |\mathbf{Q}\rangle, \cdots$ where

$$|\mathbf{S}\rangle = \sum_{i}^{\text{occupied}} \sum_{a}^{\text{unoccupied}} \hat{T}_i^a |\Psi_0\rangle$$

$$|\mathbf{D}\rangle = \sum_{i<j}^{\text{occupied}} \sum_{a<b}^{\text{unoccupied}} \hat{T}_{ij}^{ab} |\Psi_0\rangle \tag{2.204}$$

$$\vdots$$

and $\hat{T}$ represents a general substitution operator, then $\mathbf{H}$ has the form

$$\mathbf{H} = \begin{bmatrix}
\langle\Psi_0|H|\Psi_0\rangle & 0 & \langle\Psi_0|H|\mathbf{D}\rangle & 0 & 0 & \cdots \\
0 & \langle\mathbf{S}|H|\mathbf{S}\rangle & \langle\mathbf{S}|H|\mathbf{D}\rangle & \langle\mathbf{S}|H|\mathbf{T}\rangle & 0 & \cdots \\
\langle\mathbf{D}|H|\Psi_0\rangle & \langle\mathbf{D}|H|\mathbf{S}\rangle & \langle\mathbf{D}|H|\mathbf{D}\rangle & \langle\mathbf{D}|H|\mathbf{T}\rangle & \langle\mathbf{D}|H|\mathbf{Q}\rangle & \cdots \\
0 & \langle\mathbf{T}|H|\mathbf{S}\rangle & \langle\mathbf{T}|H|\mathbf{D}\rangle & \langle\mathbf{T}|H|\mathbf{T}\rangle & \langle\mathbf{T}|H|\mathbf{Q}\rangle & \cdots \\
0 & 0 & \langle\mathbf{Q}|H|\mathbf{D}\rangle & \langle\mathbf{Q}|H|\mathbf{T}\rangle & \langle\mathbf{Q}|H|\mathbf{Q}\rangle & \cdots \\
\vdots & \vdots & \vdots & \vdots & \vdots & \cdots
\end{bmatrix} \tag{2.205}$$

The matrix elements of $\mathbf{H}$ are obtained using the one- and two-electron integrals transformed into the molecular orbital basis. The simple form of Slater's rules implies that the integrals need only be multiplied by ± 1 before being summed into the total matrix element. Whether the value is $+1$ or -1 depends on the form of the two determinants, $|\Psi_M\rangle$ and $|\Psi_N\rangle$. Before applying eqns (2.18), (2.20) or (2.22), the two determinants must be permuted into maximum coincidence and the change in phase following each permutation accumulated. The fixed factors of ± 1 are called "structure constants" or "coupling coefficients". It is possible to write a general form for all cases covered in eqns (2.18), (2.20) and (2.22),

$$\langle \Psi_M | H | \Psi_N \rangle = \sum_{ij}^{\text{occupied}} h_{ij}\gamma_{ij}^{MN} + \sum_{ijkl}^{\text{occupied}} (ij|kl)\Gamma_{ijkl}^{MN} \tag{2.206}$$

For example, consider the two determinants

$$|\Psi_M\rangle = |\phi_1\bar{\phi}_1\phi_2\bar{\phi}_2\phi_4\rangle$$
$$|\Psi_N\rangle = |\phi_1\phi_2\bar{\phi}_2\phi_3\bar{\phi}_3\rangle$$
$$|\phi_1\phi_2\bar{\phi}_2\phi_3\bar{\phi}_3\rangle = -|\phi_1\phi_2\bar{\phi}_2\bar{\phi}_3\phi_3\rangle \tag{2.207}$$
$$= +|\phi_1\phi_2\bar{\phi}_3\bar{\phi}_2\phi_3\rangle$$
$$= -|\phi_1\bar{\phi}_3\phi_2\bar{\phi}_2\phi_3\rangle$$

There are two spin–orbital differences and eqn (2.22) tells us that

$$\langle \Psi_M | H | \Psi_N \rangle = -\langle \phi_1\bar{\phi}_1\phi_2\bar{\phi}_2\phi_4 | H | \phi_1\bar{\phi}_3\phi_2\bar{\phi}_2\phi_3 \rangle$$
$$= -(\bar{1}\bar{3}|43) \tag{2.208}$$

or using the canonical list of integrals $(43|31)$, which implies that in eqn (2.206) $\Gamma_{4331}^{MN} = -1$, with all other terms being zero. We could proceed to evaluate all matrix elements by having the canonical list of integrals available and working out the structure constants to form the matrix elements, determinant by determinant. This is often referred to as a "configuration-driven scheme". Alternatively we could evaluate all the γ_{ij}^{MN} and Γ_{ijkl}^{MN} values and order them by the integral label $(ij|kl)$. Then we could obtain the contribution of $(ij|kl)$ to all matrix elements at once. This is termed an "integral-driven scheme" and is particularly useful when the canonical list of integrals cannot be held in the computer's fast memory. The manipulation of determinants is greatly simplified using the idea of α and β strings introduced by Handy.[15]

The dimension of $\mathbf{H}$ being $N_{\text{Det}} \times N_{\text{Det}}$ means that it is only possible to store the matrix for cases up to $N_{\text{Det}} \approx 10^3$. In Appendix 2C we discuss how to calculate a few eigenvalues and eigenvectors of very large matrices. Here we briefly describe the principles that allow the eigenvalues and eigenvectors to be found without the need to store the matrix $\mathbf{H}$. Suppose we have an estimate of the eigenvalue, $E^{(0)}$. We can obtain this by choosing a normalised starting vector of $c^{(0)} = (1,0,0,\cdots)$ in which case

$$E^{(0)} = \mathbf{c}^{(0)\dagger}\mathbf{H}\mathbf{c}^{(0)} \qquad (2.209)$$

which is simply the Hartree–Fock energy. We now need to improve the starting vector as

$$\mathbf{c}^{(1)} = \mathbf{c}^{(0)} + \delta\mathbf{c}^{(0)} \qquad (2.210)$$

the elements of $\delta\mathbf{c}$ are given by

$$\delta C_M^{(0)} = \frac{r_M^{(0)}}{E^{(0)} - \langle \Psi_M | H | \Psi_M \rangle}$$
$$r_M^{(0)} = \sum_N \left(\langle \Psi_M | H | \Psi_N \rangle - E^{(0)}\delta_{MN} \right) C_N^{(0)} \qquad (2.211)$$

The vector $\mathbf{r}$, with elements r_M, is related to the first derivative of the energy with respect to the CI coefficients, see eqn (2.198), and will vanish when the optimal coefficients are obtained. Note that to form r_M only requires one row of the matrix $\mathbf{H}$. The elements $\langle \Psi_M | H | \Psi_N \rangle$ can be formed and summed to produce r_M without the need to access all of $\mathbf{H}$. Having obtained the coefficient vector $\mathbf{c}^{(1)}$, we must normalise it such that $\mathbf{c}^{(1)\dagger}\mathbf{c}^{(1)} = 1$, and then an update is obtained as

$$E^{(1)} = \mathbf{c}^{(1)\dagger}\mathbf{H}\mathbf{c}^{(1)}$$
$$r_M^{(1)} = \sum_N \left(\langle \Psi_M | H | \Psi_N \rangle - E^{(1)}\delta_{MN} \right) C_N^{(1)}$$
$$\delta c_M^{(1)} = \frac{r_M^{(1)}}{E^{(1)} - \langle \Psi_M | H | \Psi_M \rangle} \qquad (2.212)$$
$$\mathbf{c}^{(2)} = \mathbf{c}^{(1)} + \delta\mathbf{c}^{(1)}$$

The process is iterated until the elements of the displacement vector fall below a chosen threshold, such as $\delta C_M < 10^{-6}$ for all M. The procedure we have just described is known as the "Cooper–Nesbet method" and has been superseded by more recent techniques. However, all methods for large matrices use essentially the same quantities as above, and so are able to avoid the need to store or access all the elements of $\mathbf{H}$ at once.

2.7.1 Density Matrices and Natural Orbitals

CI expansions based on Hartree–Fock molecular orbitals obey Brillouin's theorem and provide a well-defined starting point. Convergence of the CI expansion is generally very slow when based on Hartree–Fock orbitals. That is, many highly substituted determinants are required. The CI method can be used with any set of molecular orbitals. In the limit of full CI, the energy is independent of the orbital set used to calculate it. A well-chosen set of orbitals will lead to a compact CI expansion in which a minimal number of orbitals will have significant populations, while a poorly chosen set will produce a large CI expansion in which many orbitals will have significant populations.

  Löwdin showed that the most rapidly convergent CI expansions are obtained when the orbitals are eigenfunctions of the one-electron density matrix, and he called these "natural orbitals". We have met the one-electron density for the case of the closed-shell Hartree–Fock wavefunction in eqn (2.130). For CI wavefunctions the one-electron density in the molecular orbital basis is given by

$$\rho(\mathbf{r}) = \sum_{ij}^{\substack{\text{molecular}\\\text{orbitals}}} \gamma_{ij}\phi_i(\mathbf{r})\phi_j(\mathbf{r}) \tag{2.213}$$

The one-electron density matrix, γ_{ij}, is related to the structure constants of eqn (2.206) and the CI expansion coefficients by

$$\gamma_{ij} = \sum_{MN}^{N_{\text{Det}}} C_M \gamma_{ij}^{MN} C_N \tag{2.214}$$

γ can be diagonalised using a transformation, $\mathbf{U}$,

$$\mathbf{U}^{\dagger}\gamma\mathbf{U} = \mathbf{n} \tag{2.215}$$

where $\mathbf{n}$ is a diagonal matrix. The elements of $\mathbf{n}$ are occupation numbers or orbital populations. If Φ is the matrix of molecular orbitals from which the CI expansion was built, then the natural orbitals are defined by

$$\mathbf{\Phi}^{\text{NO}} = \mathbf{\Phi}\mathbf{U} \tag{2.216}$$

In terms of the natural orbitals γ is diagonal and the one-electron density can be written as

$$\rho(\mathbf{r}) = \sum_{i}^{\substack{\text{molecular}\\\text{orbitals}}} n_i \phi_i^{\text{NO}}(\mathbf{r})\phi_i^{\text{NO}}(\mathbf{r}) \tag{2.217}$$

To connect the one-electron density matrix in the molecular orbital basis, γ, with that in the atomic orbital basis, $\mathbf{P}$, we can substitute the molecular orbital expansion into eqn (2.217)

$$\rho(\mathbf{r}) = \sum_{i}^{\substack{\text{molecular}\\\text{orbitals}}} \sum_{\mu\nu}^{m} c_{\mu i} c_{\nu j} \gamma_{ij} \chi_\mu(\mathbf{r}) \chi_\nu(\mathbf{r}) \tag{2.218}$$

and form

$$P_{\mu\nu} = \sum_{ij}^{\substack{\text{molecular}\\\text{orbitals}}} c_{\mu i} \gamma_{ij} c_{\nu j} \tag{2.219}$$

Note that for a closed-shell Hartree–Fock wavefunction $\gamma_{ij} = 2\delta_{ij}$ $(i, j$ are occupied) and eqn (2.219) reduces to the form we met in eqn (2.130). For completeness we note that the two-electron density matrix in the molecular orbital basis can be obtained from the structure constants in a similar fashion as

$$\Gamma_{ijkl} = \sum_{MN}^{N_{\text{Det}}} C_M C_N \Gamma_{ijkl}^{MN} \tag{2.220}$$

To obtain the natural orbitals from $\mathbf{P}$ we must solve the eigenvalue problem

$$\left(\mathbf{S}^{\frac{1}{2}}\mathbf{P}\mathbf{S}^{\frac{1}{2}}\right)\left(\mathbf{S}^{\frac{1}{2}}\mathbf{C}^{\text{NO}}\right) = \left(\mathbf{S}^{\frac{1}{2}}\mathbf{C}^{\text{NO}}\right)\mathbf{n} \tag{2.221}$$

or equivalently

$$(\mathbf{SPS})\mathbf{C}^{\text{NO}} = \mathbf{SC}^{\text{NO}}\mathbf{n} \tag{2.222}$$

The factors of $\mathbf{S}^{\frac{1}{2}}$ (or $\mathbf{S}$) enter to ensure that all orbitals obey the correct normalisation, $\mathbf{C}^{\dagger}\mathbf{SC} = \mathbf{I}$.

2.7.2 Truncated Configuration Interaction Methods

Brillouin's theorem states that there will be no direct interaction between the Hartree–Fock determinant and any singly substituted determinant derived from it. So a natural starting point for truncated CI methods is to limit the expansion to double substitutions (CID),

$$|\Psi_{\text{CID}}\rangle = C_0|\Psi_0\rangle + \sum_{i<j}^{\text{occupied}} \sum_{a<b}^{\text{unoccupied}} C_{ij}^{ab}|\Psi_{ij}^{ab}\rangle \tag{2.223}$$

We know from the discussion in Section 1.4.2 that the CID wavefunction will be variational but will not satisfy the size-consistency requirement. In the case of two helium atoms at large separation, the size-consistency error was traced to the omission from the wavefunction of a quadruple substitution, in fact a simultaneous double excitation on each atom. A simple scheme to correct for the missing quadruple substitutions was proposed by Davidson, it requires knowledge of the ground-state energy at the CID and Hartree–Fock levels and also the coefficient of the Hartree–Fock determinant, C_0, in the normalised CID wavefunction

$$\Delta E_Q^{\text{Davidson}} = \left(1 - C_0^2\right)\left(E_{\text{CID}} - E_0\right) \tag{2.224}$$

Many other forms of size-consistency correction have been investigated[16] and have proven useful. Care should be taken when employing these corrections to ensure that C_0 remains large, meaning that the correction remains small. The inclusion of the Davidson correction removes the variational bound on the energy. If employed in situations where C_0 is small, which implies that the Hartree–Fock determinant is not a reliable reference wavefunction, a large $\Delta E_Q^{\text{Davidson}}$ will be obtained which can easily fall below the exact energy.

Often the same size-consistency corrections are employed when singly substituted determinants are included in the wavefunction to produce a CISD calculation

$$|\Psi_{\text{CISD}}\rangle = C_0|\Psi_0\rangle + \sum_i^{\text{occupied}} \sum_a^{\text{unoccupied}} C_i^a|\Psi_i^a\rangle + \sum_{i<j}^{\text{occupied}} \sum_{a<b}^{\text{unoccupied}} C_{ij}^{ab}|\Psi_{ij}^{ab}\rangle \tag{2.225}$$

Table 2.9 gives energies obtained for water, using two small basis sets. Note that while the single substitutions do not interact directly with the Hartree–Fock determinant, their inclusion in the wavefunctions does provide a small improvement in the energy over the CID calculation. The singly substituted determinants mix with the doubly substituted determinants, which in turn interact with the Hartree–Fock determinant, see eqn (2.205). A much more significant improvement in the energy is given by Davidson's correction to the CID energy amounting to $\approx 4.8 \times 10^{-3}$ au. The addition of $\Delta E_Q^{\text{Davidson}}$ to the CISD calculation improves the energy without falling below the exact energy. Of course, the water molecule only contains 10 electrons and the magnitude of the error in the CID/CISD calculations will increase with larger numbers of electrons.

Because of the size-consistency error and the steep computational scaling, truncated CI methods such as CID/CISD are not so widely used for treating the correlation problem. Extension beyond the level of double substitutions leads to very large numbers of determinants making the computations prohibitively expensive, except for very small molecules. That said, truncated CI methods that include single substitutions are useful for the

Table 2.9 Comparison of energies obtained with approximate CI calculations with full CI for water, in 6-31G and 6-311G basis sets. C_o is the coefficient of the Hartree–Fock determinant in the CID and CISD wavefunction.

Method	6-31G	C_o	6-311G	C_o
HF	−75.985359	1.0	−76.010918	1.0
CID	−76.113788	0.981027	−76.164563	0.980713
CISD	−76.114401	0.980545	−76.166104	0.979966
CID + $\Delta E_Q^{\text{Davidson}}$	−76.118615	—	−76.170337	—
CISD + $\Delta E_Q^{\text{Davidson}}$	−76.119373	—	−76.172259	—
Full CI	−76.121023	—	−76.175179	—

study of excited states. The MacDonald–Hylleraas–Undheim theorem establishes the eigenvalues of the electronic hamiltonian as upper bounds to exact excited state energies. At the simplest level this involves a CI expansion including only singly substituted determinants and is termed "CIS",

$$|\Psi_{\text{CIS}}\rangle = C_0|\Psi_0\rangle + \sum_i^{\text{occupied}} \sum_a^{\text{unoccupied}} C_i^a|\Psi_i^a\rangle \qquad (2.226)$$

The CIS wavefunction is both variational and size-consistent. It is also relatively compact, and so can be applied to the study of the excited states of quite large molecular systems. We shall formulate the CIS method briefly here, although it should be noted that since it provides no information on the ground state beyond that contained in the Hartree–Fock determinant it is usual to exclude $|\Psi_0\rangle$ from the formulation. Because of the relationship between elements of the Fock matrix, in the molecular orbital basis, and the matrix elements between determinants that differ by a single spin–orbital it is possible to obtain very efficient formulations of the necessary equations. The matrix diagonalisation step must now be able to obtain many eigenvalues, typically a few tens, of the hamiltonian matrix. A commonly used strategy is described in Appendix 2C.

Starting from the Hartree–Fock calculation on the ground state, we need only consider two types of matrix element, namely $\langle\Psi_0|H|\Psi_i^a\rangle$ and $\langle\Psi_i^a|H|\Psi_j^b\rangle$. The first of these is particularly simple, see Section 2.6.1,

$$\langle\Psi_0|H|\Psi_i^a\rangle = F_{ia} \qquad (2.227)$$

where F_{ia} is an element of the occupied (i)–unoccupied (a) block of the Fock matrix transformed to the molecular orbital basis. The other type of matrix element, $\langle\Psi_i^a|H|\Psi_j^b\rangle$, is a little more involved since we must consider all possible cases: $i = j$ and $a = b$; $i \neq j$ and $a = b$; $i = j$ and $a \neq b$; $i \neq j$ and $a \neq b$. The final form is[17]

$$\langle \Psi_i^a | H | \Psi_j^b \rangle = E_0 \delta_{ij} \delta_{ab} + F_{ab} \delta_{ij} - F_{ij} \delta_{ab} + (ai|jb) - (ab|ji) \qquad (2.228)$$

The structure of the hamiltonian matrix is shown below

$$\mathbf{H}^{\mathrm{CIS}} = \begin{bmatrix} E_0 & F_{ia} & F_{jb} & \cdots \\ F_{ia} & \langle \Psi_i^a | H | \Psi_j^b \rangle & \langle \Psi_i^a | H | \Psi_j^b \rangle & \cdots \\ F_{jb} & \langle \Psi_i^a | H | \Psi_j^b \rangle & \langle \Psi_i^a | H | \Psi_j^b \rangle & \cdots \\ \vdots & \vdots & \vdots & \vdots \end{bmatrix} \qquad (2.229)$$

If all the F_{ia} values are zero, as will be the case for Hartree–Fock optimised orbitals, the reference determinant will not mix with any of the singly substituted determinants. The eigenvalues of this matrix represent the energies of the electronic states of the system. Eigenvalue I represents the energy of the $(I - 1)$ excited state and has eigenvector elements $C_i^{a(I)}$. It is also possible to solve this eigenvalue problem for the transition energies directly by subtracting E_0 from all diagonal elements to form

$$\bar{\mathbf{H}}^{\mathrm{CIS}} = \mathbf{H}^{\mathrm{CIS}} - E_0 \mathbf{I} \qquad (2.230)$$

The corresponding energy can be written in terms of the expansion coefficients as

$$E_{\mathrm{CIS}} = E_0 + 2 \sum_i^{\mathrm{occupied}} \sum_a^{\mathrm{unoccupied}} C_0 C_i^a F_{ia} + \sum_i^{\mathrm{occupied}} \sum_{ab}^{\mathrm{unoccupied}} C_i^a C_i^b F_{ab}$$

$$- \sum_{ij}^{\mathrm{occupied}} \sum_a^{\mathrm{unoccupied}} C_i^a C_j^a F_{ij} + \sum_{ij}^{\mathrm{occupied}} \sum_{ab}^{\mathrm{unoccupied}} C_i^a C_j^b [(ai|jb) - (ab|ji)] \qquad (2.231)$$

The closed-shell Fock matrix in the molecular orbital basis is block-diagonal for optimised orbitals, and the energy expression is simplified to

$$E_{\mathrm{CIS}} = E_0 + \sum_i^{\mathrm{occupied}} \sum_a^{\mathrm{unoccupied}} \left(C_i^a \right)^2 (\varepsilon_a - \varepsilon_i)$$

$$+ \sum_{ij}^{\mathrm{occupied}} \sum_{ab}^{\mathrm{unoccupied}} C_i^a C_j^b [(ai|jb) - (ab|ji)] \qquad (2.232)$$

In obtaining these equations, the permutational symmetry of the two-electron integrals has been used. Slightly different, but equivalent, forms can be derived. The case of UHF-type reference determinants can be handled in a similar fashion. The α- and β-spin excitations must be treated separately since the spatial orbitals of the two sets will not be equivalent. We shall not dwell on the details here, they can be found in ref. 17. We shall return to a more practical formulation in our discussion of excited-state methods in Chapter 3.

2.7.3 The Frozen Core Approximation

In the CI methods we have discussed we have assumed that substituted determinants are formed from the full molecular orbital set. Yet many chemical questions concern the electronic behaviour in the valence region. Generating determinants in which very low energy molecular orbitals are substituted is not always a constructive thing to do. Deep lying orbitals may not influence the valence molecular properties significantly and so their inclusion may offer no improvement in calculated properties. Additionally, low-lying orbitals require special basis functions for the accurate description of the electron correlation associated with them. Unless such functions are included in the basis set, then a balanced description of correlation effects will not be obtained. If these two conditions hold, it may be preferable to freeze very low lying orbitals and not allow any determinants to be generated by substitutions from them. If there are N_C electrons in the frozen core there will be $N_C/2$ orbitals that remain doubly occupied in all determinants of the CI expansion. A slightly modified "frozen core" hamiltonian must be used

$$\hat{H}_{\text{FCore}} = E_{\text{FCore}} + \sum_{i=N_C+1}^{N} \hat{h}_{\text{FCore}}(i) + \sum_{i<j=N_C+1}^{N} \frac{1}{r_{ij}} \qquad (2.233)$$

The E_{FCore} term is simply the energy of the $N_C/2$ doubly occupied orbitals

$$E_{\text{FCore}} = 2 \sum_{i}^{N_C/2} h_{ii} + \sum_{ij}^{N_C/2} [2(ii|jj) - (ij|ji)] \qquad (2.234)$$

The one-electron operator, $\hat{h}_{\text{FCore}}(i)$, includes the interaction of electron i with the N_C electrons of the frozen core

$$\hat{h}_{\text{FCore}}(i) = \hat{h}(i) + \sum_{j}^{N_C/2} [2\hat{J}_j(i) - \hat{K}_j(i)] \qquad (2.235)$$

where $\hat{J}$ and $\hat{K}$ are the familiar Coulomb and exchange operators.

To appreciate the effect of the frozen core approximation on the computational effort required, consider the full CI calculation on water reported in Table 2.9. Ignoring spatial symmetry, the full CI expansion in the 6-31G basis set consists of 1,656,369 Slater determinants. If the lowest molecular orbital, corresponding to the $1s$ orbital on oxygen is frozen, then the full CI expansion is reduced to 245,025 Slater determinants. A reduction of more than one million determinants! The effect is even more marked in the slightly larger 6-311G basis set, where the full CI expansion contains 135,210,384 determinants and freezing the core $1s$ orbital on oxygen reduces this to 9,363,600 determinants. The frozen core approximation can be profitably applied to any correlation technique, since it not only reduces the number of determinants that must be calculated but also reduces the size of the two-electron integral transformation. For the example of water with the 6-31G basis set, the single frozen core orbital reduces the length of the canonical list of molecular orbital integrals from 4186 to 3086, a saving of some 26%. For the 6-311G basis set, the saving is about 21%. The saving in the number of floating point operations is even more dramatic, since the transformation scales as m^5, providing speedups of about 3–4.

2.8   Perturbation Methods

We have stressed that size-consistency is an important requirement in an electronic structure method. Truncated CI methods are not size-consistent and have a substantial computational cost that scales as an iterative m^6. Ideally we want methods that have a lower scaling of cost with molecular size, and are size-consistent from the outset. Perturbation theory can provide a route to such methods. In this chapter we shall specifically consider the Rayleigh–Schrödinger perturbation theory, which is size-consistent in all orders.

The starting point is the recognition that the full electronic problem is difficult to solve, so we look for a related model problem, which is easily solved, and then describe the real problem as a relatively small 'perturbation' of the model problem. In mathematical terms we split the real hamiltonian, $\hat{H}$, into a model hamiltonian, $\hat{H}_0$, and a perturbation, $\hat{V}$,

$$\hat{H} = \hat{H}_0 + \hat{V} \tag{2.236}$$

We also introduce an order parameter, λ, which can take any value between 0 and 1. λ allows us to move systematically from the model problem to the real problem

$$\hat{H} = \hat{H}_0 + \lambda \hat{V} \tag{2.237}$$

The exact problem has the usual electronic Schrödinger equation, $\hat{H}|\Psi_A\rangle = E_A|\Psi_A\rangle$ ($A = 1,2,\cdots$), for all electronic states A. The model problem satisfies

$$\hat{H}_0 \left| \Psi_A^{(0)} \right\rangle = E_A^{(0)} \left| \Psi_A^{(0)} \right\rangle \quad (A = 1, 2, \cdots) \tag{2.238}$$

We assume that a complete set of solutions, $\left\{ \Psi^{(0)} \right\}$, is known and that this set is orthonormal,

$$\left\langle \Psi_A^{(0)} \middle| \Psi_B^{(0)} \right\rangle = \delta_{AB} \tag{2.239}$$

Additionally, we assume that the states in eqn (2.238) are non-degenerate. Substituting the λ-dependent hamiltonian into the Schrödinger equation gives

$$\left(\hat{H}_0 + \lambda \hat{V} \right) |\Psi_A(\lambda)\rangle = E_A(\lambda) |\Psi_A(\lambda)\rangle \tag{2.240}$$

This form of the hamiltonian makes the energy and the corresponding wavefunction dependent on λ. We can expand $E_A(\lambda)$ and $|\Psi_A(\lambda)\rangle$ in a Taylor series around the point $\lambda = 0$,

$$E_A(\lambda) = E_A(0) + \frac{dE_A(0)}{d\lambda}\lambda + \frac{1}{2!}\frac{d^2 E_A(0)}{d\lambda^2}\lambda^2 + \cdots$$

$$|\Psi_A(\lambda)\rangle = |\Psi_A(0)\rangle + \frac{d|\Psi_A(0)\rangle}{d\lambda}\lambda + \frac{1}{2!}\frac{d^2|\Psi_A(0)\rangle}{d\lambda^2}\lambda^2 + \cdots \tag{2.241}$$

When $\lambda = 0$, $E_A(\lambda) = E_A^{(0)}$ and $|\Psi_A(\lambda)\rangle = \left| \Psi_A^{(0)} \right\rangle$. We can adopt a simpler notation for the derivatives and numerical factors in eqn (2.241) by writing

$$X^{(n)} = \frac{1}{n!}\frac{d^n X(0)}{d\lambda^n} \tag{2.242}$$

Using this notation, the Taylor expansions of $E_A(\lambda)$, and $|\Psi_A(\lambda)\rangle$ can be written as

$$E_A(\lambda) = E_A^{(0)} + \lambda E_A^{(1)} + \lambda^2 E_A^{(2)} + \cdots$$

$$|\Psi_A(\lambda)\rangle = \left| \Psi_A^{(0)} \right\rangle + \lambda \left| \Psi_A^{(1)} \right\rangle + \lambda^2 \left| \Psi_A^{(2)} \right\rangle + \cdots \tag{2.243}$$

Before proceeding further we must impose normalisation on $\left| \Psi_A^{(0)} \right\rangle$ and $|\Psi_A(\lambda)\rangle$. The solutions of the model problem are orthonormal, eqn (2.239). $|\Psi_A(\lambda)\rangle$ is required to satisfy

$$\left\langle \Psi_A(\lambda) \middle| \Psi_A^{(0)} \right\rangle = 1 \tag{2.244}$$

This is termed "intermediate normalisation" and amounts to imposing orthogonality between $|\Psi_A^{(0)}\rangle$ and all $|\Psi_A^{(n)}\rangle$. To see this, we substitute $|\Psi_A(\lambda)\rangle$ from eqn (2.243) into eqn (2.244)

$$\langle\Psi_A(\lambda)|\Psi_A^{(0)}\rangle=1=\langle\Psi_A^{(0)}|\Psi_A^{(0)}\rangle+\lambda\langle\Psi_A^{(1)}|\Psi_A^{(0)}\rangle+\lambda^2\langle\Psi_A^{(2)}|\Psi_A^{(0)}\rangle+\cdots \qquad (2.245)$$

Given eqn (2.239) and the fact that λ can take any value between 0 and 1 implies that

$$\langle\Psi_A^{(n)}|\Psi_A^{(0)}\rangle=0 \qquad (2.246)$$

for all orders n. Substituting the expansions of $E_A(\lambda)$ and $|\Psi_A(\lambda)\rangle$ into the Schrödinger equation yields

$$\begin{aligned}
(\hat{H}_0+\lambda\hat{V})&\left(|\Psi_A^{(0)}\rangle+\lambda|\Psi_A^{(1)}\rangle+\lambda^2|\Psi_A^{(2)}\rangle+\cdots\right)\\
&=\left(E_A^{(0)}+\lambda E_A^{(1)}+\lambda^2 E_A^{(2)}+\right)\left(|\Psi_A^{(0)}\rangle+\lambda|\Psi_A^{(1)}\rangle+\lambda^2|\Psi_A^{(1)}\rangle+\cdots\right)
\end{aligned} \qquad (2.247)$$

We multiply out the terms and collect together those of a given order in λ. For example,

$$\hat{H}_0|\Psi_A^{(0)}\rangle=E_A^{(0)}|\Psi_A^{(0)}\rangle \quad (\lambda^0)$$

$$\hat{H}_0|\Psi_A^{(1)}\rangle+\hat{V}|\Psi_A^{(0)}\rangle=E_A^{(0)}|\Psi_A^{(1)}\rangle+E_A^{(1)}|\Psi_A^{(0)}\rangle \quad (\lambda^1) \qquad (2.248)$$

$$\hat{H}_0|\Psi_A^{(2)}\rangle+\hat{V}|\Psi_A^{(1)}\rangle=E_A^{(0)}|\Psi_A^{(2)}\rangle+E_A^{(1)}|\Psi_A^{(1)}\rangle+E_A^{(2)}|\Psi_A^{(0)}\rangle \quad (\lambda^2)$$

These equations are obtained by setting $\lambda=1$, corresponding to the exact hamiltonian. Now each equation is pre-multiplied by $\langle\Psi_A^{(0)}|$ and the orthogonality between the model wavefunctions and the exact wavefunction, eqn (2.246), yields

$$E_A^{(0)}=\langle\Psi_A^{(0)}|\hat{H}_0|\Psi_A^{(0)}\rangle$$

$$E_A^{(1)}=\langle\Psi_A^{(0)}|\hat{V}|\Psi_A^{(0)}\rangle \qquad (2.249)$$

$$E_A^{(2)}=\langle\Psi_A^{(0)}|\hat{V}|\Psi_A^{(1)}\rangle$$

Note that to obtain the second-order energy, we need only know the first-order wavefunction. In fact there is a rule, established by Eugene Wigner, which states that knowledge of the wavefunction to order n, $|\Psi_A^{(n)}\rangle$, is sufficient to obtain the energies to order $2n+1$. We shall not pursue the formal subtleties of perturbation theory beyond second order. The interested reader should consult ref. 18. The remaining task for us is to turn the equations in eqn (2.248) into workable forms, from which we can evaluate the perturbed energies and wavefunctions.

Rearranging the second line of eqn (2.248) and using the second line of eqn (2.249),

$$\left(E_A^{(0)} - \hat{H}_0\right)|\Psi_A^{(1)}\rangle = \left(\hat{V} - E_A^{(1)}\right)|\Psi_A^{(0)}\rangle$$
$$= \left(\hat{V} - \langle\Psi_A^{(0)}|\hat{V}|\Psi_A^{(0)}\rangle\right)|\Psi_A^{(0)}\rangle \qquad (2.250)$$

This is clearly not an eigenvalue equation. From the definition of $|\Psi_A^{(1)}\rangle$ in eqn (2.242) we can see that this is a complicated inhomogeneous differential equation, and one way of finding a solution is to expand $|\Psi_A^{(1)}\rangle$ in terms of the complete set of solutions of $\hat{H}_0$, that is

$$|\Psi_A^{(1)}\rangle = \sum_B C_B^{(1)}|\Psi_B^{(0)}\rangle \qquad (2.251)$$

Using eqn (2.239), pre-multiplying by $\langle\Psi_D^{(0)}|$ and integrating yields

$$\langle\Psi_D^{(0)}|\Psi_A^{(1)}\rangle = \sum_B C_B^{(1)}\langle\Psi_D^{(0)}|\Psi_B^{(0)}\rangle$$
$$= C_D^{(1)} \qquad (2.252)$$

The orthogonality of the model state, $|\Psi_A^{(0)}\rangle$, to all orders of corrections, $|\Psi_A^{(n)}\rangle$, provides the condition $\langle\Psi_A^{(0)}|\Psi_A^{(1)}\rangle = 0$, which implies that $C_A^{(1)} = 0$. The first-order wavefunction is thus

$$|\Psi_A^{(1)}\rangle = \sum_{B \neq A} |\Psi_B^{(0)}\rangle\langle\Psi_B^{(0)}|\Psi_A^{(1)}\rangle \qquad (2.253)$$

Returning to eqn (2.250), pre-multiplying by $\langle\Psi_B^{(0)}|$ and integrating gives

$$E_A^{(0)}\langle\Psi_B^{(0)}|\Psi_A^{(1)}\rangle - \langle\Psi_B^{(0)}|\hat{H}_0|\Psi_A^{(1)}\rangle = \langle\Psi_B^{(0)}|\hat{V}|\Psi_A^{(0)}\rangle \qquad (2.254)$$

The second term on the left-hand side can be reworked by noting that

$$\langle \Psi_B^{(0)} | \hat{H}_0 | \Psi_A^{(1)} \rangle = \langle \Psi_A^{(1)} | \hat{H}_0 | \Psi_B^{(0)} \rangle^*$$

$$\hat{H}_0 | \Psi_B^{(0)} \rangle = E_B^{(0)} | \Psi_B^{(0)} \rangle$$

$$\therefore \tag{2.255}$$

$$\langle \Psi_B^{(0)} | \hat{H}_0 | \Psi_A^{(1)} \rangle = \langle \Psi_A^{(1)} | \Psi_B^{(0)} \rangle^* E_B^{(0)}$$

$$= \langle \Psi_B^{(0)} | \Psi_A^{(1)} \rangle E_B^{(0)}$$

Eqn (2.254) now becomes

$$\left(E_A^{(0)} - E_B^{(0)} \right) \langle \Psi_B^{(0)} | \Psi_A^{(1)} \rangle = \langle \Psi_B^{(0)} | \hat{V} | \Psi_A^{(0)} \rangle \tag{2.256}$$

which gives

$$\langle \Psi_B^{(0)} | \Psi_A^{(1)} \rangle = C_B^{(1)} = \frac{\langle \Psi_B^{(0)} | \hat{V} | \Psi_A^{(0)} \rangle}{\left(E_A^{(0)} - E_B^{(0)} \right)} \tag{2.257}$$

and $| \Psi_A^{(1)} \rangle$ is

$$| \Psi_A^{(1)} \rangle = \sum_{B \neq A} | \Psi_B^{(0)} \rangle \frac{\langle \Psi_B^{(0)} | \hat{V} | \Psi_A^{(0)} \rangle}{\left(E_A^{(0)} - E_B^{(0)} \right)} \tag{2.258}$$

The second-order energy can now be obtained as

$$E_A^{(2)} = \langle \Psi_A^{(0)} | \hat{V} | \Psi_A^{(1)} \rangle = \sum_{B \neq A} \frac{\langle \Psi_A^{(0)} | \hat{V} | \Psi_B^{(0)} \rangle \langle \Psi_B^{(0)} | \hat{V} | \Psi_A^{(0)} \rangle}{\left(E_A^{(0)} - E_B^{(0)} \right)} \tag{2.259}$$

This concludes our consideration of the Rayleigh–Schrödinger perturbation theory. This particular variety of perturbation theory is just one of several. An equivalent variational formulation exists, due to Hylleraas. It provides a variational form that leads to the same first-order wavefunction and second-order energy as we have obtained above. Assuming an arbitrary trial wavefunction $| \bar{\Psi} \rangle$, containing variable parameters, the second-order energy can be obtained from the Hylleraas functional

$$J_2[\bar{\Psi}] = 2\langle\bar{\Psi}|\hat{V}|\Psi_A^{(0)}\rangle + \langle\bar{\Psi}|\hat{H}_0 - E_A^{(0)}|\Psi_A^{(0)}\rangle \qquad (2.260)$$

The variational condition that provides the optimal $|\bar{\Psi}\rangle$ is

$$\hat{V}|\Psi_A^{(0)}\rangle + \left(\hat{H}_0 - E_A^{(0)}\right)|\bar{\Psi}\rangle = 0 \qquad (2.261)$$

with the assumption that $\langle\bar{\Psi}\mid\Psi_A^{(0)}\rangle = 0$. This is equivalent to the first-order equation in eqn (2.248). If we make the substitution $|\bar{\Psi}\rangle = \left|\Psi_A^{(1)}\right\rangle$, then it can be established that $J_2\left[\Psi_A^{(1)}\right] = E_A^{(2)}$. It can further be shown that $J_2[\bar{\Psi}] \geq E_A^{(2)}$, which provides us with a variational condition for obtaining $\Psi_A^{(1)}$. The Hylleraas functional can be extended to higher even-orders of perturbation theory. Its advantages will become clear in subsequent sections.

2.8.1 Møller–Plesset Perturbation Theory

In the preceding section we have obtained some of the formal results of Rayleigh–Schrödinger perturbation theory. To obtain a usable electronic structure method we must specify the form of the model hamiltonian, $\hat{H}_0$, and its eigenfunctions, $\left\{\Psi^{(0)}\right\}$. The most widely used definition of $\hat{H}_0$ is that which was first introduced by Møller and Plesset in 1934, and developed in the 1970s by Pople and coworkers[19] and also Bartlett and coworkers.[20]

The Møller–Plesset approach takes $\hat{H}_0$ to be the Hartree–Fock hamiltonian

$$\hat{H}_0 = \sum_i^N \hat{F}(i) \qquad (2.262)$$

which is the sum of Fock operators. Since the exact hamiltonian is written as $\hat{H} = \hat{H}_0 + \lambda\hat{V}$, the perturbation can be written as

$$\hat{V} = \hat{H} - \hat{H}_0 = \sum_i^N \hat{h}(i) + \sum_{i<j}^N \frac{1}{r_{ij}} - \sum_i^N \hat{h}(i) - \sum_{ij}^N \left(\hat{J}_j(i) - \hat{K}_j(i)\right)$$

$$= \sum_{i<j}^N \frac{1}{r_{ij}} - \sum_{ij}^N \left(\hat{J}_j(i) - \hat{K}_j(i)\right) \qquad (2.263)$$

The Hartree–Fock wavefunction $|\Psi_0\rangle$ is an eigenfunction of $\hat{H}_0$ and satisfies the equation

$$\hat{H}_0|\Psi_0\rangle = E^{(0)}|\Psi_0\rangle \tag{2.264}$$

$E^{(0)}$ is not the Hartree–Fock energy but rather the sum of occupied spin–orbital energies

$$E^{(0)} = \sum_i^N \varepsilon_i \tag{2.265}$$

Using the results of Rayleigh–Schrödinger perturbation theory, the first-order energy, eqn (2.249), is

$$E^{(1)} = \langle \Psi_0 | V | \Psi_0 \rangle$$
$$= \langle \Psi_0 | \sum_{i<j}^N \frac{1}{r_{ij}} | \Psi_0 \rangle - \langle \Psi_0 | \sum_{ij}^N \left(\hat{J}_j(i) - \hat{K}_j(i) \right) | \Psi_0 \rangle \tag{2.266}$$

Applying Slater's rules, eqn (2.18), the first term is

$$\langle \Psi_0 | \sum_{i<j}^N \frac{1}{r_{ij}} | \Psi_0 \rangle = \frac{1}{2} \sum_{ij}^N \left[(ii|jj) - (ij|ji) \right] \tag{2.267}$$

the summation here is over spin–orbitals. The second term in eqn (2.266) can be evaluated using the definitions of the $\hat{J}$ and $\hat{K}$ operators as

$$\langle \Psi_0 | \sum_{ij}^N \hat{J}_j(i) - \hat{K}_j(i) | \Psi_0 \rangle = \sum_{ij}^N \left[(ii|jj) - (ij|ji) \right] \tag{2.268}$$

Consequently

$$E^{(1)} = -\frac{1}{2} \sum_{ij}^N \left[(ii|jj) - (ij|ji) \right] \tag{2.269}$$

Hence the total energy to first order is simply the Hartree–Fock energy, see Section 2.2.5,

$$E^{(0)} + E^{(1)} = \sum_i^N \varepsilon_i - \frac{1}{2} \sum_{ij}^N \left[(ii|jj) - (ij|ji) \right] \tag{2.270}$$
$$= E_0$$

At this point it is useful to introduce a notation for the integrals $(ii|jj) - (ij|ji)$ as they occur frequently in perturbation theory. We shall denote this antisymmetrised pair of integrals using a double vertical bar as

$$(ii\|jj) = (ii|jj) - (ij|ji) \tag{2.271}$$

Eqn (2.270) can be written in terms of the double-bar integrals as

$$E^{(0)} + E^{(1)} = \sum_i^N \varepsilon_i - \frac{1}{2} \sum_{ij}^N (ii\|jj) \tag{2.272}$$

Proceeding to the second-order energy, we use eqn (2.259) to write

$$E^{(2)} = \sum_{B \neq 0} \frac{\left| \langle \Psi_0 | \hat{V} | \Psi_B^{(0)} \rangle \right|^2}{E^{(0)} - E_B^{(0)}} \tag{2.273}$$

Recalling the definition of the perturbation, $\hat{V}$, the numerator can be written as

$$\langle \Psi_0 | \hat{V} | \Psi_B^{(0)} \rangle = \langle \Psi_0 | \hat{H} | \Psi_B^{(0)} \rangle - \langle \Psi_0 | \hat{H}_0 | \Psi_B^{(0)} \rangle \tag{2.274}$$

We know that if $\left| \Psi_B^{(0)} \right\rangle$ is a single substitution relative to $|\Psi_0\rangle$, then the first matrix element is zero by Brillouin's theorem. The second term is also zero since $\hat{H}_0 | \Psi_B^{(0)} \rangle = E_B^{(0)} | \Psi_B^{(0)} \rangle$ and so $\langle \Psi_0 | \hat{H}_0 | \Psi_B^{(0)} \rangle = \langle \Psi_0 | \Psi_B^{(0)} \rangle E_B^{(0)} = 0$. We can also exclude all $\left| \Psi_B^{(0)} \right\rangle$ that differ from $|\Psi_0\rangle$ by three or more substitutions because of Slater's rules. This leaves only doubly substituted determinants to consider, which using the familiar notation

$$\sum_{B \neq 0} \left| \Psi_B^{(0)} \right\rangle = \overset{\text{occupied}}{\underset{i<j}{\sum}} \; \overset{\text{unoccupied}}{\underset{a<b}{\sum}} \left| \Psi_{ij}^{ab} \right\rangle \tag{2.275}$$

allows all unique doubly substituted determinants to be written as a restricted sum over the spin–orbitals. The term $\langle \Psi_0 | \hat{H}_0 | \Psi_{ij}^{ab} \rangle$ vanishes since

$$\langle \Psi_0 | \hat{H}_0 | \Psi_{ij}^{ab} \rangle = \langle \Psi_0 | E^{(0)} - \varepsilon_i - \varepsilon_j + \varepsilon_a + \varepsilon_b | \Psi_{ij}^{ab} \rangle$$

$$= \langle \Psi_0 | \Psi_{ij}^{ab} \rangle \left(E^{(0)} - \varepsilon_i - \varepsilon_j + \varepsilon_a + \varepsilon_b \right) \tag{2.276}$$

$$= 0$$

The final form for $E^{(2)}$ becomes

$$
\begin{aligned}
E^{(2)} &= \overset{\text{occupied}}{\underset{i<j}{\sum}}\ \overset{\text{unoccupied}}{\underset{a<b}{\sum}}\ \frac{\left|\langle\Psi_0|\hat{H}|\Psi_{ij}^{ab}\rangle\right|^2}{\varepsilon_i+\varepsilon_j-\varepsilon_a-\varepsilon_b} \\[2mm]
&= \overset{\text{occupied}}{\underset{i<j}{\sum}}\ \overset{\text{unoccupied}}{\underset{a<b}{\sum}}\ \frac{|(ia\|jb)|^2}{\varepsilon_i+\varepsilon_j-\varepsilon_a-\varepsilon_b} \\[2mm]
&= \overset{\text{occupied}}{\underset{i<j}{\sum}}\ \overset{\text{unoccupied}}{\underset{a<b}{\sum}}\ (ia\|jb)t_{ij}^{ab}
\end{aligned}
\tag{2.277}
$$

where $t_{ij}^{ab}=\dfrac{(ia\|jb)}{\varepsilon_i+\varepsilon_j-\varepsilon_a-\varepsilon_b}$. The final line of eqn (2.277) is often written with the restrictions on the summations removed as

$$
E^{(2)} = \frac{1}{4}\overset{\text{occupied}}{\underset{ij}{\sum}}\ \overset{\text{unoccupied}}{\underset{ab}{\sum}}\ (ia\|jb)t_{ij}^{ab}
\tag{2.278}
$$

This expression for the correlation energy at the second order of the Møller–Plesset theory is referred to as "MP2". It is the most computationally economical *ab initio* correlation method since it only requires a sub-set of the transformed two-electron integrals. These can be obtained using the partial transformations described in eqn (2.171) as

$$
\begin{aligned}
(iv|\lambda\sigma) &= \sum_{\mu}^{m} c_{\mu i}(\mu v|\lambda\sigma) \quad \left[m_{\text{occ}}m^4\right] \\[2mm]
(ia|\lambda\sigma) &= \sum_{v}^{m} c_{va}(iv|\lambda\sigma) \quad \left[m_{\text{occ}}m_{\text{unocc}}m^3\right] \\[2mm]
(ia|j\sigma) &= \sum_{\lambda}^{m} c_{\lambda j}(ia|\lambda\sigma) \quad \left[m_{\text{occ}}^2 m_{\text{unocc}}m^2\right] \\[2mm]
(ia|jb) &= \sum_{\sigma}^{m} c_{\sigma b}(ia|j\sigma) \quad \left[m_{\text{occ}}^2 m_{\text{unocc}}^2 m\right]
\end{aligned}
\tag{2.279}
$$

The scaling of each partial transformation is given in square brackets in terms of the basis set dimension, m, the number of occupied orbitals, m_{occ}, and the number of unoccupied orbitals, m_{unocc}. m is the largest of these numbers and so the first partial transformation is the most demanding step.

The computational requirements can be significantly reduced for large molecules using the density-fitted approximate integrals of Section 2.6.3. Using the B_{pq}^P arrays of eqn (2.191), the transformed integrals can be obtained as

$$(ia|jb)^{\mathrm{DF}} = \sum_P^{m_{\mathrm{fit}}} B_{ia}^P B_{jb}^P \quad [m_{\mathrm{occ}}^2 m_{\mathrm{unocc}}^2 m_{\mathrm{fit}}] \tag{2.280}$$

Much development of this technique has been carried out by Manby and coworkers[21] that has enabled density-fitted MP2 calculations to be performed on large molecular systems, achieving a linear scaling with system size.

2.8.2 Improvements in Low-Order Perturbation Theory: Spin Component Scaling and Orbital Optimisation

The double-substituted determinants that enter the MP2 wavefunction are formed from three components. The first corresponds to a double substitution where both electrons are of α-spin, $(\alpha\alpha)$. The second set is the β-spin counterpart, $(\beta\beta)$. The third set involves a simultaneous single substitution of each type, $(\alpha\beta)$. From our discussion of the Fermi hole in Section 1.4.1 we know that the Hartree–Fock wavefunction contains correlation for electrons of the same spin, but lacks correlation for pairs of opposite spin. MP2, being based on a Hartree–Fock reference state, is prone to overestimate the same spin correlation while yielding too little of the opposite spin correlation. By comparing MP2 energies with higher level *ab initio* calculations, Grimme[22] developed the spin-component-scaled MP2 method, SCS-MP2. No additional computational cost is entailed since the contributions of the different spin components: $\alpha\alpha;\ \beta\beta;\ \alpha\beta$, are usually evaluated separately. That is, the summation in eqn (2.277) is broken up over sets of spin–orbitals. Thus the second-order energy can be written as

$$E^{(2)} = E_{\alpha\alpha}^{(2)} + E_{\beta\beta}^{(2)} + E_{\alpha\beta}^{(2)} \tag{2.281}$$

In the SCS-MP2 method

$$E_{\mathrm{SCS}}^{(2)} = f_{\mathrm{SS}}\left(E_{\alpha\alpha}^{(2)} + E_{\beta\beta}^{(2)}\right) + f_{\mathrm{OS}} E_{\alpha\beta}^{(2)} \tag{2.282}$$

and Grimme proposed the use of the values $f_{\mathrm{SS}} = \frac{1}{3}$ and $f_{\mathrm{OS}} = \frac{6}{5}$. The SCS-MP2 method provides improved molecular properties, such as bond lengths and vibrational frequencies, but fails for long range van der Waals systems.[23] The improvements provided by the SCS-MP2 method are significant for closed-shell molecules but less effective for open-shell systems. This situation can be improved by optimising the orbitals from which the MP2 wavefunction is built. The conventional MP2 method relies on the molecular orbitals being eigenfunctions of the Fock operator. The use of optimised orbitals can be addressed using the variational Hylleraas perturbation theory. The MP2 amplitudes, t_{ij}^{ab} in eqn (2.277), and the orbitals must be simultaneously optimised. This is done by adding to the Hylleraas expression for the

second-order energy the energy of the reference determinant. This is no longer
the Hartree–Fock determinant, since the orbitals will not in general satisfy the
Hartree–Fock equation. The Hartree–Fock determinant by definition corresponds to the best single determinant energy and so the energy of the reference
determinant will increase. The change in the orbitals is described by a rotation
matrix, $\mathbf{R}$. The functional to be minimised depends on $\mathbf{R}$ and the amplitudes, $\mathbf{t}$,

$$L(\mathbf{t},\mathbf{R}) = \langle \Psi_0(\mathbf{t},\mathbf{R})|\hat{H}|\Psi_0(\mathbf{t},\mathbf{R})\rangle + 2\langle \Psi^{(1)}(\mathbf{t},\mathbf{R})|\hat{V}|\Psi_0(\mathbf{t},\mathbf{R})\rangle$$
$$+ \langle \Psi^{(1)}(\mathbf{t},\mathbf{R})|\hat{H}_0 - \hat{V}|\Psi^{(1)}(\mathbf{t},\mathbf{R})\rangle \tag{2.283}$$

The optimisation of $L(\mathbf{t},\mathbf{R})$ is complicated[23] and we shall not consider the
details here. Clearly the simple non-iterative MP2 scheme is lost and a
substantial computational cost is incurred. However the description of
molecular properties for open-shell systems is improved.

Table 2.10 compares the MP2 and SCS-MP2 energies for water with the
results in Table 2.9. Larger basis sets including polarisation functions are
included. Note that the variational Hartree–Fock energies converge with the
size of the basis set more rapidly than the correlated energies. For the MP2
method, the spin component energies show a different rate of convergence with
basis set size. The same spin correlation energies are smaller in magnitude and
converge more quickly than the opposite spin component.

Perturbation theory methods overcome the problem of size-consistency
errors associated with truncated CI methods. However, the MP2 method is of
limited accuracy. It is always possible to go to higher orders of perturbation
theory, MP3, MP4 ..., but these techniques are more demanding and scale as:
MP3 (m^6); MP4 (m^7). The MP3 method does not provide significant
improvements over MP2. In fact MP3 results are often poorer than MP2.
MP4 does generally provide a significant improvement, but at a large
computational cost. Additionally the formalism of the MPn series, when $n \geq 4$,
is quite complicated and best handled using diagrammatic techniques.[18] More
reliable techniques are needed. Experience, obtained since the 1980s, has
suggested that higher order MP methods are less useful than coupled-cluster
methods of similar computational cost.

2.9 Coupled-Cluster Methods

The coupled-cluster method currently provides some of the most accurate
wavefunction-based calculations of energies and molecular properties. The full
coupled-cluster method is equivalent to full CI. The excitation operators
introduced in eqn (1.68) are applied to a reference wavefunction, usually the
Hartree–Fock determinant, in the form of an exponential operator

$$|\Psi_{\text{CC}}\rangle = e^{\hat{T}}|\Psi_0\rangle \tag{2.284}$$

Table 2.10 Comparison of energies obtained with MP2 and SCS-MP2 methods for water. Numbers from variational CI methods are included for comparison.

Method	*6-31G*	*6-311G*	*6-311G(d)*	*6-311G(d,p)*
HF	-75.985359	-76.010918	-76.031634	-76.045811
MP2	-76.113092	-76.166185	-76.253018	-76.280992
$E^{(2)}(\alpha\alpha)$	-0.015023	-0.017145	-0.026478	-0.028394
$E^{(2)}(\alpha\beta)$	-0.097685	-0.120975	-0.168426	-0.178392
$E^{(2)}(\beta\beta)$	-0.015023	-0.017145	-0.026478	-0.028394
SCS$-$MP2	-76.112596	-76.167518	-76.251397	-76.278810
CID	-76.113788	-76.164563	-76.251540	-76.278328
CISD	-76.114401	-76.166104	-76.252538	-76.279348
Full CI	-76.121023	-76.175179	—	—

We saw in Section 1.4.4 that the expansion of the exponential generates an infinite series in which the $\hat{T}$ operator occurs in linear form (connected clusters) and also in products (disconnected clusters). The product terms suffice to introduce size-consistency in the resultant energies, regardless of the level of truncation of $\hat{T}$. For example, restricting to double substitutions, $\hat{T} \approx \hat{T}_2$,

$$e^{\hat{T}_2}|\Psi_0\rangle = |\Psi_0\rangle + \overbrace{\sum_{i<j}}^{\text{occupied}}\overbrace{\sum_{a<b}}^{\text{unoccupied}} t_{ij}^{ab}|\Psi_{ij}^{ab}\rangle + \frac{1}{2}\overbrace{\sum_{i<j<k<l}}^{\text{occupied}}\overbrace{\sum_{a<b<c<d}}^{\text{unoccupied}} t_{ij}^{ab} t_{kl}^{cd}|\Psi_{ijkl}^{abcd}\rangle + \cdots \tag{2.285}$$

generates the quadruple-substituted determinant $\left|\Psi_{ijkl}^{abcd}\right\rangle$, but rather than being allowed its own variable coefficient, its contribution is determined by the product of double substitutions, see Figure 2.19. If $\hat{T}_4$ was included then the quadruple-substituted determinant could take on its own, independent, amplitude. Using $\hat{T} \approx \hat{T}_2$, the coefficient of the quadruple-substituted determinant is given as the product of double excitations, see Figure 2.19(b) and (c).

2.9.1 The Coupled-Cluster Doubles Equations

The coupled-cluster method is not variational, except in the limit that the complete $\hat{T}$ operator is used. The energy and coupled-cluster amplitudes are obtained by projection rather than variational minimisation. Let us illustrate this for the coupled-cluster doubles (CCD) model in which $\hat{T} \approx \hat{T}_2$. The coupled-cluster wavefunction, eqn (2.284), is substituted into the Schrödinger equation

$$\hat{H}e^{\hat{T}_2}|\Psi_0\rangle = E_{\text{CCD}}e^{\hat{T}_2}|\Psi_0\rangle \tag{2.286}$$

and the exponential is expanded using eqn (1.81)

$$\hat{H}\left(1+\hat{T}_2+\frac{1}{2}\hat{T}_2^2+\cdots\right)|\Psi_0\rangle = E_{\text{CCD}}\left(1+\hat{T}_2+\frac{1}{2}\hat{T}_2^2+\cdots\right)|\Psi_0\rangle \qquad (2.287)$$

The effect of the $\hat{T}_2$ operators is to generate the double- and quadruple-substituted determinants in eqn (2.285). We need not consider higher terms since they cannot interact with the doubly substituted determinants or with $|\Psi_0\rangle$. Including the doubles cluster amplitudes, t_{ij}^{ab}, gives

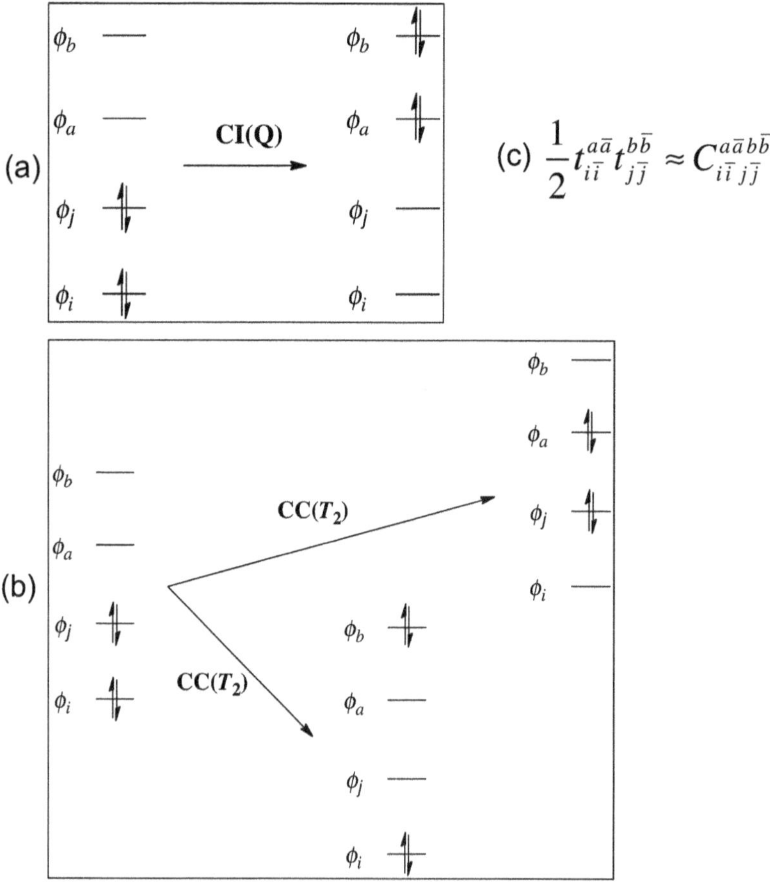

Figure 2.19 (a) A quadruple substitution with its own independent coefficient, as in a CI calculation. (b) Double substitutions in the coupled-cluster method that have independent coefficients. (c) In the CCD method, the coefficient given to the excitation in (a) is a product of the coefficients corresponding to the substitutions in (b).

$$\hat{H}|\Psi_0\rangle + \sum_{i<j}^{\text{occupied}} \sum_{a<b}^{\text{unoccupied}} t_{ij}^{ab}\hat{H}|\Psi_{ij}^{ab}\rangle + \frac{1}{2}\sum_{i<j<k<l}^{\text{occupied}} \sum_{a<b<c<d}^{\text{unoccupied}} t_{ij}^{ab} t_{kl}^{cd}\hat{H}|\Psi_{ijkl}^{abcd}\rangle =$$

$$E_{\text{CCD}}|\Psi_0\rangle + E_{\text{CCD}}\sum_{i<j}^{\text{occupied}} \sum_{a<b}^{\text{unoccupied}} t_{ij}^{ab}|\Psi_{ij}^{ab}\rangle + \frac{1}{2}E_{\text{CCD}}\sum_{i<j<k<l}^{\text{occupied}} \sum_{a<b<c<d}^{\text{unoccupied}} t_{ij}^{ab} t_{kl}^{cd}|\Psi_{ijkl}^{abcd}\rangle \tag{2.288}$$

To obtain the energy we pre-multiply this equation by $\langle\Psi_0|$ and integrate

$$\langle\Psi_0|\hat{H}|\Psi_0\rangle + \sum_{i<j}^{\text{occupied}} \sum_{a<b}^{\text{unoccupied}} t_{ij}^{ab}\langle\Psi_0|\hat{H}|\Psi_{ij}^{ab}\rangle$$

$$+ \frac{1}{2}\sum_{i<j<k<l}^{\text{occupied}} \sum_{a<b<c<d}^{\text{unoccupied}} t_{ij}^{ab} t_{kl}^{cd}\langle\Psi_0|\hat{H}|\Psi_{ijkl}^{abcd}\rangle$$

$$= E_{\text{CCD}}\langle\Psi_0|\Psi_0\rangle + E_{\text{CCD}}\sum_{i<j}^{\text{occupied}} \sum_{a<b}^{\text{unoccupied}} t_{ij}^{ab}\langle\Psi_0|\Psi_{ij}^{ab}\rangle \tag{2.289}$$

$$+ \frac{1}{2}E_{\text{CCD}}\sum_{i<j<k<l}^{\text{occupied}} \sum_{a<b<c<d}^{\text{unoccupied}} t_{ij}^{ab} t_{kl}^{cd}\langle\Psi_0|\Psi_{ijkl}^{abcd}\rangle$$

Using the orthonormality of the determinants and Slater's rules, this reduces to

$$E_0 + \sum_{i<j}^{\text{occupied}} \sum_{a<b}^{\text{unoccupied}} t_{ij}^{ab}\langle\Psi_0|\hat{H}|\Psi_{ij}^{ab}\rangle = E_{\text{CCD}} \tag{2.290}$$

To obtain the amplitudes we follow the same procedure, but now we pre-multiply eqn (2.288) by $\langle\Psi_{mn}^{ef}|$ to obtain

$$\langle\Psi_{mn}^{ef}|\hat{H}|\Psi_0\rangle + \sum_{i<j}^{\text{occupied}} \sum_{a<b}^{\text{unoccupied}} t_{ij}^{ab}\langle\Psi_{mn}^{ef}|\hat{H}|\Psi_{ij}^{ab}\rangle$$

$$+ \frac{1}{2}\sum_{i<j<k<l}^{\text{occupied}} \sum_{a<b<c<d}^{\text{unoccupied}} t_{ij}^{ab} t_{kl}^{cd}\langle\Psi_{mn}^{ef}|\hat{H}|\Psi_{ijkl}^{abcd}\rangle$$

$$= E_{\text{CCD}}\langle\Psi_{mn}^{ef}|\Psi_0\rangle + E_{\text{CCD}}\sum_{i<j}^{\text{occupied}} \sum_{a<b}^{\text{unoccupied}} t_{ij}^{ab}\langle\Psi_{mn}^{ef}|\Psi_{ij}^{ab}\rangle \tag{2.291}$$

$$+ \frac{1}{2}E_{\text{CCD}}\sum_{i<j<k<l}^{\text{occupied}} \sum_{a<b<c<d}^{\text{unoccupied}} t_{ij}^{ab} t_{kl}^{cd}\langle\Psi_{mn}^{ef}|\Psi_{ijkl}^{abcd}\rangle$$

Again this reduces to

$$
\langle \Psi_{mn}^{ef} | \hat{H} | \Psi_0 \rangle + \sum_{i<j}^{\text{occupied}} \sum_{a<b}^{\text{unoccupied}} t_{ij}^{ab} \langle \Psi_{mn}^{ef} | \hat{H} | \Psi_{ij}^{ab} \rangle
$$

$$
+ \frac{1}{2} \sum_{i<j<k<l}^{\text{occupied}} \sum_{a<b<c<d}^{\text{unoccupied}} t_{ij}^{ab} t_{kl}^{cd} \langle \Psi_{mn}^{ef} | \hat{H} | \Psi_{ijkl}^{abcd} \rangle \tag{2.292}
$$

$$
= E_{\text{CCD}} t_{mn}^{ef}
$$

It is customary to subtract E_0 from both sides of the equation and obtain an expression in terms of the correlation energy $\Delta E_{\text{CCD}} = E_{\text{CCD}} - E_0$. To proceed we note that there will be one such equation for each doubles amplitude, t_{mn}^{ef}. Eqn (2.292) is quadratic in the **t** coefficients and so must be solved iteratively. At the simplest level a scheme might be to take all **t** coefficients on the left-hand side of eqn (2.292) to be zero and $E^{(0)} = E_0$, after which an approximate set of **t** coefficients is easily obtained. From these coefficients an approximate energy can be obtained from eqn (2.290) and the left-hand side of eqn (2.292) formed. This gives a new set of **t** coefficients and the process is repeated until the changes in the **t** coefficients fall below some chosen threshold, signalling convergence of the process. Much more robust numerical procedures exist but we shall not go into details here.

A similar, but more involved, approach can be used to derive the coupled-cluster singles and doubles (CCSD) equations. The CCSD energy is given by

$$
E_{\text{CCSD}} = E_0 + \sum_{i<j}^{\text{occupied}} \sum_{a<b}^{\text{unoccupied}} \left(\frac{1}{2} t_i^a t_j^b + t_{ij}^{ab} \right) \langle \Psi_0 | \hat{H} | \Psi_{ij}^{ab} \rangle \tag{2.293}
$$

and now clearly depends on the single-cluster amplitudes, t_i^a. CCSD is a very widely used computational technique, it has the same formal computational demands as the CISD method and scales as an iterative m^6 method. In practice it is more expensive than CISD and much efficiency can be gained by formulating the equation directly over the atomic basis functions.[24]

2.9.2   Higher Order Methods

Proceeding beyond CCSD to CCSDT and further has been achieved[25] but the application of such methods is limited to quite small molecules. A very successful approach has been to carry out a CCSD calculation and then use the optimised amplitudes to form an estimate of the effect of triple substitutions. The contribution of triple substitutions is calculated using fourth-order Møller–Plesset perturbation theory, but using the CCSD amplitudes in place of the coefficients of the perturbed second-order wavefunction. Finally the coupling between single and triple substitutions is included from fifth-order perturbation theory. This method is denoted CCSD(T) and has been shown to provide very accurate energies and molecular properties, provided the Hartree–Fock reference wavefunction is reliable.

An alternative coupled-cluster technique is Brueckner theory. Here the orbitals are allowed to change from their Hartree–Fock form so as to eliminate all amplitudes of the $\hat{T}_1$ operator from the wavefunction. In analogy with eqn (2.292), the condition that satisfies this requirement is

$$\langle \Psi_i^a | \hat{H}(1 + \hat{T}_2) | \Psi_0 \rangle = 0 \qquad (2.294)$$

This technique is denoted "Brueckner doubles" (BD). It is also possible to add a perturbative triples correction, as done in the CCSD(T) method, to yield the BD(T) technique. It has been observed that BD and CCSD give results of very similar quality. Table 2.11 compares energies obtained by the coupled-cluster techniques.

2.10 Localised Orbital Formulations of Post Hartree–Fock Techniques

The extension of the correlation methods we have discussed so far to ever-larger molecules poses a serious challenge in terms of computational resources. It is very limiting if all we can do is to await the arrival of more powerful computers! The m^6 scaling of the CISD method is severe and in a sense also 'unphysical'. To understand this let us consider an example, hexanoic acid, depicted in Figure 2.20. The highest occupied molecular orbital, obtained from the canonical Hartree–Fock procedure, is delocalised over the whole frame of the molecule, see Figure 2.21. The phases of the orbitals match between every pair of bonded carbon atoms. There are six carbon atoms and so $6(6-1)/2 = 15$ pairs of carbon atoms. The adjacent pairs are spatially close, but some of the pairs are quite remote from each other. We can imagine that if we replaced hexanoic acid with octanoic acid, containing two more carbon atoms, the adjacent pairs would look quite similar. The distant pairs would now stretch even further away from each other. Increasing the molecular size by adding two more $-CH_2-$ units, to form octanoic acid, we would still expect the corresponding canonical molecular orbital of octanoic acid to be delocalised over the whole molecular frame. Yet we know from the transferability of many molecular properties, for example bond dissociation energies and group vibrational wavenumbers, that their features must be determined by the local electronic structure and not the structure of the whole molecule. Similarly the dynamic electron correlation associated with a pair of electrons should fall off as the electrons move apart. It appears 'unphysical' that the orbitals located at the carbonyl end of hexanoic acid should have a strong influence on the correlation energy of electrons at the methyl group end. The correlation space of such pairs should be separate. The canonical

Table 2.11   Comparison of energies obtained with coupled-cluster methods for
water. Numbers from full CI with smaller basis sets are included.

Method	6-31G	6-311G	6-311G(d)	6-311G(d,p)
CCD	−76.118912	−76.170643	−76.260226	−76.287666
CCSD	−76.119548	−76.172162	−76.261239	−76.288723
BD	−76.119436	−76.171854	−76.261031	−76.288506
CCSD(T)	−76.120512	−76.174942	−76.264908	−76.293424
BD(T)	−76.120469	−76.174781	−76.264810	−76.293321
Full CI	−76.121023	−76.175179	—	—

molecular orbitals do not allow any separation as they traverse the whole
molecular frame.

It is possible to mix the occupied molecular orbitals amongst themselves
without changing the total energy of the Hartree–Fock wavefunction from its
variational minimum. Obviously we can also mix the unoccupied orbitals
amongst themselves without changing the energy. Denoting the canonical
molecular orbitals as $\mathbf{\Phi}^{\text{CMO}}$, we can choose a rotation matrix $\mathbf{U}^{\text{Loc}}$ such that

$$\mathbf{\Phi}^{\text{LMO}} = \mathbf{\Phi}^{\text{CMO}}\mathbf{U}^{\text{Loc}} \tag{2.295}$$

where $\mathbf{\Phi}^{\text{LMO}}$ represents a localised set of molecular orbitals. We shall address
how we can obtain $\mathbf{U}^{\text{Loc}}$ in Chapter 3. Figures 2.22(a) and (b) show two
localised occupied molecular orbitals of hexanoic acid. These bonding orbitals
are essentially fully localised between two carbon atoms. If we changed from
hexanoic acid to octanoic acid these highly localised and doubly occupied
orbitals will not change qualitatively. Localising the unoccupied orbitals
produces a corresponding pair of highly localised antibonding orbitals,
Figures 2.22(c) and (d). The first antibonding orbital, Figure 2.22(c) should
provide the most important correlating orbital for the first bonding orbital
Figure 2.22(a). Similarly for the second bonding and antibonding pair,
Figure 2.22(b) and (d). The first bonding orbital, Figure 2.22(a), will have
some mixing with the second antibonding orbital, Figure 2.22(d), but its
magnitude will be much less than with the spatially closer orbital in
Figure 2.22(c). We would not expect any significant interaction between

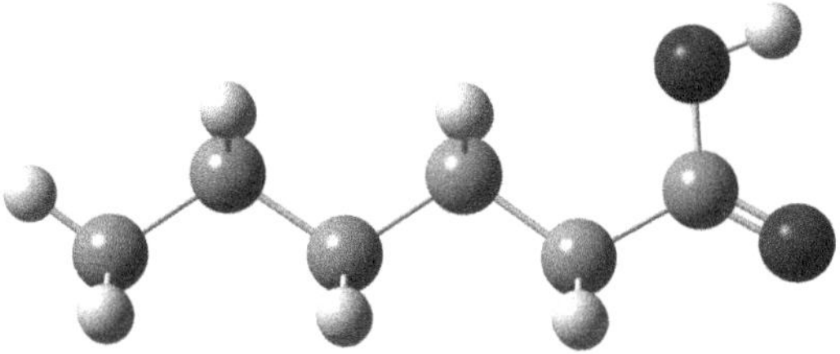

Figure 2.20   The molecular structure of hexanoic acid with its chain of $-CH_2-$ units.

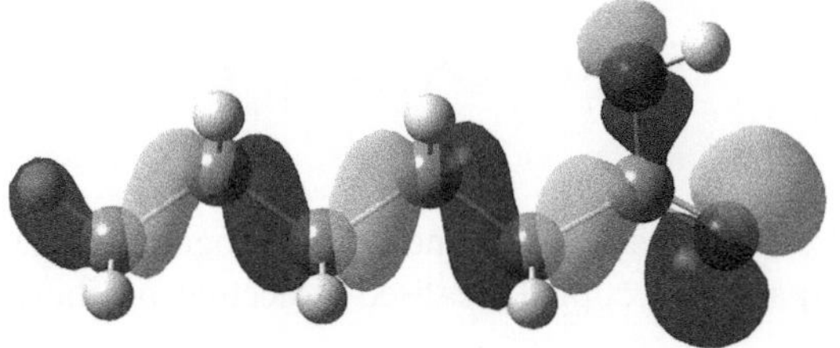

Figure 2.21 A surface plot of the highest occupied molecular orbital in hexanoic acid.

bonding and antibonding pairs further distant from each other. Hence the correlating orbitals for the bonding orbital, Figure 2.22(a), become essentially independent of the size of the molecule, unlike the situation with the canonical molecular orbitals. The use of localised orbitals allows a more 'physical' scaling of the CI procedure with molecular size.

A number of groups have developed correlation methods based on localised molecular orbitals.[26-28] Local versions of CI, MP2 and CCSD have been successfully implemented. The occupied canonical Hartree–Fock molecular orbitals, in many cases, localise easily. The unoccupied orbitals tend to be diffuse and their localisation can be problematic. To avoid the problem associated with localising unoccupied orbitals, projected atomic orbitals are used for the unoccupied orbital space. Atomic orbitals being, by definition, already localised. The localised occupied molecular orbitals are given as

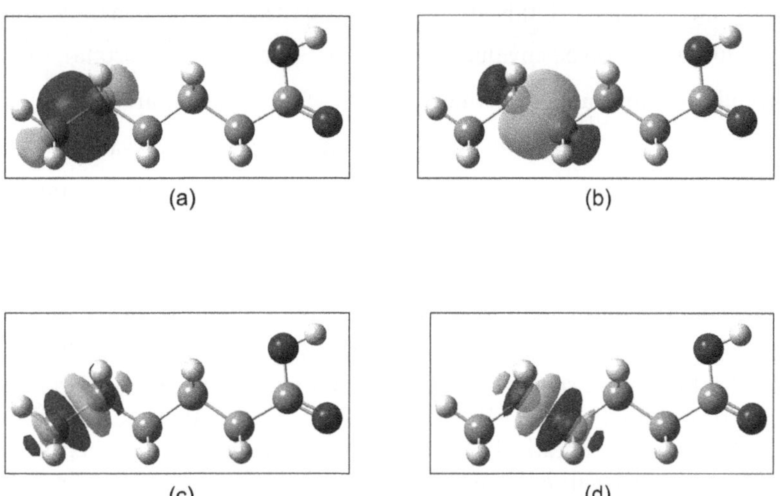

Figure 2.22 Surface plots of two localised bonding molecular orbitals, (a) and (b), in hexanoic acid and their antibonding counterparts, (c) and (d).

$$\left|\phi_i^{\mathrm{LMO}}\right\rangle = \sum_{\mu}^{m} c_{\mu i}^{\mathrm{LMO}}\left|\chi_\mu\right\rangle \tag{2.296}$$

The unoccupied orbital space is obtained by orthogonalising the atomic basis functions to the space of occupied localised molecular orbitals. These are given as

$$\left|\widetilde{\chi}_\mu\right\rangle = \left(1 - \sum_{i}^{\mathrm{occupied}} \left|\phi_i^{\mathrm{LMO}}\right\rangle\left\langle\phi_i^{\mathrm{LMO}}\right|\right)\left|\chi_\mu\right\rangle$$
$$= \sum_{v}^{m} c_{v\mu}\left|\chi_v\right\rangle \tag{2.297}$$

or in matrix form

$$\widetilde{\mathbf{C}} = \mathbf{I} - \mathbf{PS} \tag{2.298}$$

where $P_{\mu v} = \sum_{i}^{\mathrm{occupied}} c_{\mu i}^{\mathrm{LMO}} c_{v i}^{\mathrm{LMO}}$ and $\mathbf{S}$ is the basis function overlap matrix. The projected atomic orbitals are localised but are not orthonormal amongst themselves. The overlap matrix of the projected atomic orbitals is

$$\widetilde{\mathbf{S}} = \widetilde{\mathbf{C}}^{\dagger}\mathbf{S}\widetilde{\mathbf{C}} \tag{2.299}$$

The projected atomic orbitals are linearly dependent and diagonalisation of $\widetilde{\mathbf{S}}$ will produce m_{occ} zero eigenvalues. To each occupied localised orbital, $\left|\phi_i^{\mathrm{LMO}}\right\rangle$, an orbital domain is assigned which consists of all $\left|\widetilde{\chi}_\mu\right\rangle$ that are spatially close to $\left|\phi_i^{\mathrm{LMO}}\right\rangle$.[29] The details of how domains are defined varies between implementations and consequently the energies computed by any local correlation technique depend slightly on the details of the implementation. Care must be taken in the definition of orbital domains to ensure that there are no discontinuities in the potential energy surface. The technique of domain merging[30] has been shown to be effective in this respect. The use of localised molecular orbitals means that an orbital energy cannot be associated with the localised orbital in the same way as for the canonical Hartree–Fock orbitals. In the case of localised orbital MP2 methods this requires that the second-order energy be obtained iteratively using the variational Hylleraas perturbation theory, since the localised molecular orbitals are not eigenfunctions of the model hamiltonian, the Fock operator.

2.11 Non-Dynamic Electron Correlation and Multi-configurational Reference Wavefunctions

The correlation methods we have described so far work well provided the reference wavefunction, the Hartree–Fock determinant, gives a qualitatively correct description of the electronic structure. In Section 1.4.5 we discussed a common situation where this requirement is not fulfilled, namely the stretching of a bond. As the bond distance is increased it becomes necessary to introduce more than a single determinant to prevent the wavefunction from describing a non-physical, spurious, situation.

We may not be too concerned with the extreme dissociation limit of bonds. However, if we wish to study for example, transition state structures, then the same problem can easily arise. Transition state structures are typically intermediate geometries between that of the reactants and the products. Hence we can imagine that often these will involve stretched bonds, which are broken in the course of the reaction or incipient stretched bonds that are forming as the reaction proceeds. So we must take care that the techniques we use to study transition state structures are suitable for this purpose.

Many molecules require a multi-configurational description, even at their equilibrium geometries. Their electronic structure is such that there is only a small energy gap between occupied and unoccupied orbitals, which means that more than a single Slater determinant will be necessary to describe the non-dynamic electron correlation. We can sometimes deal with such situations by using high-order single reference techniques, such as the coupled-cluster methods. If the non-dynamic correlation is centred around the description of a single bond, for example the dissociation of molecular hydrogen, then the CCSD method will be adequate. For cases where more than a single bond is involved, for example the dissociation of molecular nitrogen, such an approach will fail. To correct for the structure-dependent non-dynamic electron correlation often requires the addition of only a small number of determinants to produce a qualitatively correct description of the electronic structure. For example, consider the ozone molecule in a cc-pVTZ basis. At the Hartree–Fock level, the predicted geometry gives excessively short O–O bonds, as we might expect. We can move to some of the techniques we have described to see how the geometry is improved. Table 2.12 lists some of these. The CISD calculation, based on Hartree–Fock optimised orbitals, increases the bond distance by 0.03 Å but is still considerably shorter than the experimental bond length. The unoccupied orbitals produced by the Hartree–Fock procedure tend to be too diffuse and so the weights of substituted determinants built from these orbitals tend to be underestimated. However, the CISD calculation produces a coefficient $C_0 = 0.935$ for the Hartree–Fock determinant, this implies that some 87% of the CISD wavefunction is accounted for by $|\Psi_0\rangle$. The only other significant coefficient, $C = -0.104$, corresponds to a double substitution of the highest occupied molecular orbital (HOMO) with the lowest unoccupied molecular orbital (LUMO), see Figure 2.23. This only supplies a further 1% of the CISD

Table 2.12 Geometric parameters of ozone obtained using the cc-pVTZ basis set.

Method	R / Å	θ / °	Energy / au
HF	1.194	119.26	−224.35671
CISD	1.224	118.19	−224.99927[a]
MP2	1.284	116.64	−224.11897[a]
CCSD	1.250	117.58	−225.08665[a]
CCSD(T)	1.276	116.94	−225.13278[a]
CASSCF(2,2)	1.253	115.74	−224.45302
CASSCF(FVS)	1.285	116.67	−224.57831
Experiment	1.272	116.78	—

[a]A frozen core approximation was applied to the lowest three molecular orbitals.

wavefunction and means that the remaining ∼12% of the CISD wavefunction comes from many small contributions of other substituted determinants.

Moving to the CCSD level, the geometry is slightly improved. The contribution of the Hartree–Fock determinant is reduced to ∼81% of the CCSD wavefunction and the HOMO → LUMO doubly substituted determinant now accounts for ∼3% of the wavefunction. Proceeding to the CCSD(T) level, we obtain a reasonable estimate of the geometry with errors of 0.004 Å and 0.16° in the bond length and angle, respectively.

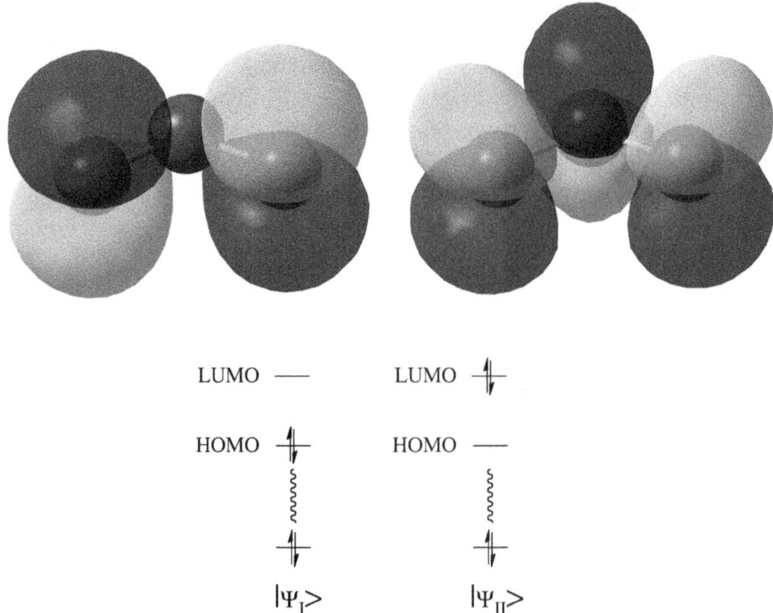

Figure 2.23 The highest occupied and lowest unoccupied molecular orbitals in ozone and the two determinants, and , in which they are substituted with each other.

All these high-order methods include many determinants. The weights of all these determinants, except for two of them, are individually extremely small but collectively add up to >10% of the wavefunction. The computational cost of these high-order methods, see Table 1.2, is very high. Since we have only found two significant determinants contributing, we might ask what the effect of including just these two determinants in the reference wavefunction would be. The outcome of variationally optimising the mixing coefficient of the determinants $|\Psi_I\rangle$ and $|\Psi_{II}\rangle$ of Figure 2.23 and also the orbitals from which they are built gives the result indicated as CASSCF(2,2) in Table 2.12. In terms of bond length, the CASSCF(2,2) goes more than half way from the Hartree–Fock value to the CCSD(T) value, although the angle is reduced too much. However, this calculation contains only two determinants! It is an example of a multi-configurational self-consistent field (MCSCF) calculation. The coefficients of the two determinants are $C_I = 0.903$ and $C_{II} = -0.430$, the doubly substituted determinant now counts for >18% of the wavefunction. Of course we have been able to choose the determinants required based on much more extensive calculations. How might we choose the important determinants in the absence of such information? It turns out that there is no single, foolproof, way to do this. Rather one must use one's chemical and physical insight, and a good deal of experience, and proceed very cautiously. Rather than choosing individual determinants to include, we can choose a small number of 'active' orbitals and electrons, as described in Section 1.4.5, and generate all possible determinants of the correct spin for that set of orbitals and electrons. This constitutes the complete active space (CASSCF) variant of the MCSCF method.

An alternative way of choosing the active orbitals and electrons for ozone, is to use the idea of the full valence shell (FVS) active space. Each oxygen atom possesses six valence electrons and four valence orbitals, $2s\,2p$. This gives a full valence shell for ozone of 18 electrons in 12 orbitals. A full CI in this small space will contain 48,400 determinants, which is a small calculation by current standards. The outcome of this calculation is denoted "CASSCF(FVS)" in Table 2.12. The angle is in good agreement with the CCSD(T) result and experiment, but the bond length is too long. Determinant $|\Psi_{II}\rangle$ accounts for ~9% of the full valence shell CASSCF wavefunction. The FVS is easily arrived at from simple physical principles but cannot be applied to systems with more than three or four atoms. For example adding another, hypothetical, oxygen atom to the active space would increase the size to 24 electrons in 16 orbitals and, in the absence of symmetry, 3,312,400 determinants!

An alternative approach would be to use a UHF calculation and look at the natural orbital (UNO) occupation numbers. This method can be very effective, but is very geometry dependent and must be used with caution. For example, a Hartree–Fock calculation using the experimental geometry of ozone converges to a UHF solution easily. The UNOs formed from the total density, see Sections 2.4.3 and 2.7.1, show occupancies of 1.822 and 0.178 for two orbitals, with all other orbitals having occupancies >1.99 or <0.01. Hence such a calculation appears to choose the two-electron two-orbital space we have discussed. However, a similar calculation at the closed-shell Hartree–Fock geometry, in which the bond length is considerably shorter, shows very little tendency to converge to a UHF solution and typically converges to the RHF solution, implying that the RHF determinant is a good reference wavefunction. The non-dynamic correlation is structure dependent and its balanced treatment requires considerable care. Multi-configurational methods cannot be viewed as black box procedures, they require a good deal of skill and experience to use correctly.

Having defined a suitable active space, and there may be more than one choice, it should be borne in mind that the CASSCF method yields a qualitatively correct reference wavefunction. To obtain quantitative accuracy, more extended treatment of the correlation problem is needed. This can, very inefficiently, be achieved by expanding the active space but is better carried out with the multi-reference analogues of the CI, MP2 and CC methods we have discussed.

2.11.1 The MCSCF Method and Associated Optimisation Problems

The MCSCF methods require a coupled optimisation of configuration mixing coefficients as well as orbital expansion coefficients. A number of methods, which are quite different from each other in their details, have been developed for dealing with these tasks. Here we shall only give a brief outline of some of the considerations involved.

First of all the orbital variation is usually performed rather differently than in the RHF or UHF approaches. As we have said, the orbital space in a MCSCF calculation is divided into inactive, active and unoccupied or virtual sets. The inactive orbitals are doubly occupied in all determinants and the virtual orbitals are empty in all determinants. The active orbitals are allowed to have all possible occupancies, for a complete active space. We need to consider the mixing of orbitals between these spaces, see Figure 2.24. Specifically we must consider: inactive–active; inactive–virtual; active–active and active–virtual mixings. If a complete active space is used, that is a full CI in the active space, the active–active mixings will not change the energy and should be omitted. The conventional SCF method involves mixing the occupied orbitals with the unoccupied and it is possible to define a Fock operator for this purpose. The difficulty in the MCSCF methods is that the orbital mixings depicted in Figure 2.24 cannot be written in terms of a single Fock operator. Hence it is preferable to consider the orbital optimisation problem from a different perspective. Starting from an orthonormal set of orbitals, $\mathbf{\Phi}$, we mix the orbitals to lower the energy under the constraint of maintaining orthonormality. A unitary transformation, by definition, preserves the orthonormality of vectors. Hence we can consider the orbital variation problem to be that of finding a unitary transformation, $\mathbf{U}$, that transforms the orbitals

$$\mathbf{\Phi}' = \mathbf{U}\mathbf{\Phi} \tag{2.300}$$

such that the energy is lowered. The transformation matrix $\mathbf{U}$ is written as the exponential of an antisymmetric matrix, $\mathbf{X}$,

$$\mathbf{U} = e^{\mathbf{X}} \quad [\mathbf{X}^+ = -\mathbf{X}] \tag{2.301}$$

The advantage of such a parameterisation of $\mathbf{U}$ is that $e^{\mathbf{X}}$ describes the simultaneous rotation of all required pairs of orbitals rather than sequential pairwise rotations. For an m orbital problem, $\mathbf{X}$ will *not* have the full dimension $m \times m$, because the elements x_{pq} will be zero when p, q refers to inactive–inactive pairs or virtual–virtual pairs, also active–active pairs for a complete active space. $\mathbf{X}$ has the form,

$$X = \begin{pmatrix} 0 & x_{pq} & x_{pr} \\ -x_{pq} & \ddots & x_{qr} \\ -x_{pr} & -x_{qr} & 0 \end{pmatrix} \tag{2.302}$$

and denoting the dimension of the orbital sub-spaces as, m_{inactive}, m_{active}, m_{virtual}, a maximum dimension of $m_{\text{inactive}} \times m_{\text{active}} + m_{\text{inactive}} \times m_{\text{virtual}} + m_{\text{active}} \times m_{\text{active}} + m_{\text{active}} \times m_{\text{virtual}}$. $e^{\mathbf{X}}$ can be evaluated in several ways. At the simplest level the expansion of the exponential gives

$$\mathbf{U} = e^{\mathbf{X}} = \mathbf{I} + \mathbf{X} + \frac{\mathbf{X}^2}{2} + \frac{\mathbf{X}^3}{6} + \cdots \tag{2.303}$$

This expansion is often truncated at the third term and can provide a stable representation of $e^{\mathbf{X}}$ for the purpose of orbital optimisation. Other forms are

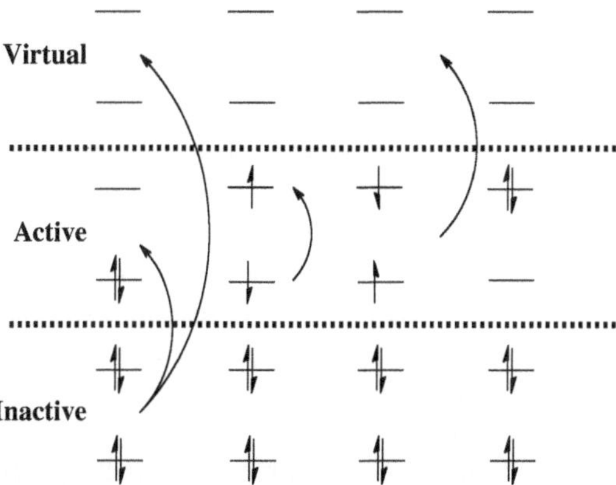

Figure 2.24 The three different orbital spaces in a MCSCF calculation. The arrows indicate the orbital mixings that must be allowed to optimise the energy.

possible, accurate to higher orders, which involve evaluating matrix functions. For example the Cayley transform

$$\mathbf{U} = e^{\mathbf{X}} = (\mathbf{I} - \mathbf{X})(\mathbf{I} + \mathbf{X})^{-1} \tag{2.304}$$

The MCSCF wavefunction is a linear combination of Slater determinants, that is a limited CI type wavefunction

$$|\Psi_{\mathrm{MCSCF}}\rangle = \sum_M C_M |\Psi_M\rangle \tag{2.305}$$

In Section 2.7 we noted that the matrix elements between determinants can be written in terms of one- and two-electron structure constants,

$$\langle \Psi_M | H | \Psi_N \rangle = \sum_{pq}^{\text{occupied}} h_{pq} \gamma_{pq}^{MN} + \sum_{pqrs}^{\text{occupied}} (pq|rs) \Gamma_{pqrs}^{MN} \tag{2.306}$$

The density matrices

$$\gamma_{pq} = \sum_{MN}^{N_{\text{Det}}} C_M \gamma_{pq}^{MN} C_N$$

$$\Gamma_{pqrs} = \sum_{MN}^{N_{\text{Det}}} C_M{}_{pqrs}^{MN} C_N \tag{2.307}$$

allow the energy to be written as

$$E = \sum_{pq}^{\text{occupied}} h_{pq} \gamma_{pq} + \sum_{pqrs}^{\text{occupied}} (pq|rs) \Gamma_{pqrs} \tag{2.308}$$

We now wish to vary the orbitals so as to minimise the energy, given the density matrices γ and Γ. These density matrices depend on the configuration mixing coefficients, which in turn depend on the orbital coefficients. We need to find a minimum in the energy with respect to both types of coefficients simultaneously. It is possible to carry out a simultaneous optimisation but it is often simpler to take a step along each parameter space independently. In the following we shall denote the members of the different orbital sub-spaces as: i, j, k, l (inactive); t, u, v, w (active); a, b, c, d (virtual), p, q, r, s (arbitrary). Having obtained the wavefunction in eqn (2.305), we now define two Fock operators with matrix elements

$$F_{pq}^{\text{inactive}} = h_{pq} + \sum_{i}^{\text{inactive}} [2(pq|ii) - (pi|qi)]$$

$$F_{pq}^{\text{active}} = \sum_{tu}^{\text{active}} \gamma_{tu} \left[(pq|tu) - \frac{1}{2}(pt|qu) \right]$$

$$(2.309)$$

Given the fixed occupancies of the inactive and virtual sub-spaces, the only non-zero density matrix elements involving inactive orbitals are

$$\gamma_{ii} = 2$$

$$\Gamma_{iijj} = 2 \quad \Gamma_{ijji} = -1 \quad \Gamma_{iiii} = 1 \quad \Gamma_{iitu} = \gamma_{tu} \quad \Gamma_{itui} = -\frac{1}{2}\gamma_{tu}$$

$$(2.310)$$

Density matrix elements involving any virtual orbital indices are zero.

In terms of the Fock operators in eqn (2.309), the gradient with respect to the rotation involving an inactive orbital, i, with an inactive or virtual orbital, p, can be written as

$$\frac{dE}{dx_{ip}} = 2\left(F_{ip}^{\text{inactive}} + F_{ip}^{\text{active}} \right)$$

$$(2.311)$$

Rotations involving one active orbital, t, with an active or virtual orbital, p, can similarly be expressed as

$$\frac{dE}{dx_{tp}} = \sum_{u}^{\text{active}} \gamma_{tu} F_{pu}^{\text{inactive}} + 2 \sum_{uvw}^{\text{active}} \Gamma_{tuvw}(pu|vw)$$

$$(2.312)$$

For a complete active space, $dE/dx_{tu} = 0$.

These gradient expressions can be employed in a range of schemes to find a stationary point defining optimal orbitals.[31] Some schemes are particularly simple and in effect only require the gradient quantities in eqns (2.311) and (2.312). More elaborate schemes require a knowledge of the second derivatives of the energy with respect to the **x** parameters, $d^2E/dx_{pq}dx_{rs}$. The MCSCF equations have been written over molecular orbitals, implying that a transformation of the one- and two-electron integrals will be required. Since the inactive orbitals are doubly occupied in all determinants the contribution of the inactive orbitals can be evaluated directly over the basis function integrals, as in eqn (2.234),

$$E_{\text{inactive}} = 2 \sum_{i}^{\text{inactive}} h_{ii} + \sum_{ij}^{\text{inactive}} [2(ii|jj) - (ij|ji)]$$

$$= \sum_{\mu\nu}^{m} h_{\mu\nu} P_{\mu\nu} + \frac{1}{2} \sum_{\mu\nu\lambda\sigma}^{m} P_{\mu\nu} P_{\lambda\sigma} \left[(\mu\nu|\lambda\sigma) - \frac{1}{2}(\mu\sigma|\lambda\nu) \right]$$

(2.313)

where **P** is the closed-shell density matrix corresponding to the inactive orbitals

$$P_{\mu\nu} = 2 \sum_{i}^{\text{inactive}} c_{\mu i} c_{\nu i}$$

(2.314)

The energy of the active space electrons, in the presence of the inactive electrons, is given by

$$E_{\text{active}} = 2 \sum_{tu}^{\text{active}} h_{tu}^{C} + \sum_{tuvw}^{\text{active}} (tu|vw) \Gamma_{tuvw}$$

(2.315)

where $\hat{h}^C$ is a modified one-electron operator that incorporates the inactive–active interaction and has the form and matrix elements given below

$$\hat{h}^C = \hat{h} + \sum_{i}^{\text{inactive}} 2\hat{J}_i - \hat{K}_i$$

$$h_{pq}^{C} = h_{pq} + \sum_{i}^{\text{inactive}} [2(pq|ii) - (pi|iq)]$$

(2.316)

The density matrices γ and Γ must be stored, but they only span the active space and so are usually small enough to fit into the computer's fast memory. To evaluate the energy, $E = E_{\text{inactive}} + E_{\text{active}}$, only requires the transformation of integrals over the active space, the number of which is

$$\frac{1}{2} \frac{m_{\text{active}}(m_{\text{active}} + 1)}{2} \left[\frac{m_{\text{active}}(m_{\text{active}} + 1)}{2} + 1 \right]$$

(2.317)

where $m_{\text{active}} \ll m$. To evaluate the gradient terms, eqns (2.311) and (2.312), requires only one index to span the full range of m, with three indices restricted to m_{active}, giving only $m_{\text{active}}^3 m$ integrals of the form $(pt|uv)$ to form.

Second-order optimisation methods requiring second derivative information will need more transformed integrals, $m_{\text{active}}^2 m^2$, since two indices will now span the full orbital space. This tells us that the MCSCF methods are significantly more expensive than the single determinant Hartree–Fock method, but they are substantially less demanding than the higher order correlation methods we have discussed.

2.11.2 Electron Correlation Methods Based on a CASSCF Reference Wavefunction

Given a qualitatively correct multi-configuartional reference wavefunction, we need to look at ways of extending the correlation treatment to electrons outside the active space. We have seen that for single determinant reference wavefunctions the dynamic electron correlation can be recovered using configuration interaction, perturbation theory and coupled-cluster methods. We now need to extend these ideas to the multi-reference case.

The conceptually simplest approach is the multi-reference CI (MRCI) method, in which a truncated set of substitutions is generated from all determinants in the reference wavefunction. Typically, single and double substitutions are taken from a CASSCF reference, for example the determinants depicted in Figure 1.20. Including single and double substitutions from all four determinants effectively introduces triple and quadruple substitutions of the first, closed-shell, determinant. Hence the number of determinants in the MR-CISD wavefunction will be much larger than the corresponding single reference CISD. Having defined the required determinants, the MR-CISD procedure is essentially the same as that of the CISD method: the vector $\mathbf{Hc}$ must be formed and an eigenvalue of the hamiltonian must be found as discussed in Section 2.7. Each eigenvalue of the hamiltonian represents an approximation to the energy of an electronic state.

This type of MR-CISD calculation is only applicable to quite small molecules due to the rapid increase in the number of determinants that must be included. One way of reducing the size of the hamiltonian to be diagonalised is to use the idea of an internally contracted CI. An internally contracted CI wavefunction is generated by applying the substitution operator, $\hat{T}$, to the reference wavefunction, $|\Psi_{\text{CASSCF}}\rangle$, as a whole

$$|\Psi_{\text{CASSCF}}\rangle = \sum_M C_M |\Psi_M\rangle$$

$$\left|\Psi_{\substack{\text{internally} \\ \text{contracted}}}\right\rangle = \hat{T}|\Psi_{\text{CASSCF}}\rangle = \hat{T}\sum_M C_M |\Psi_M\rangle \tag{2.318}$$

The substitution operator generates a linear combination of substituted determinants weighted by the coefficients, C_M. The advantage of such an approach is that the number of substitutions is essentially independent of the size of the reference wavefunction, $|\Psi_{\text{CASSCF}}\rangle$. A disadvantage is that each internally contracted determinant $\hat{T}\sum_M C_M |\Psi_M\rangle$ is now a complicated mixture of terms for which matrix elements must be efficiently evaluated. Schemes for achieving this have been developed.[32] It has been found that internal contractions in which $\hat{T}$ includes single and double substitutions can show poor convergence properties. This can be circumvented by including all single substitutions of the determinants in $|\Psi_{CASSCF}\rangle$ explicitly and then limiting the internal contraction to double substitutions. Having defined the necessary matrix elements, the eigenvalues of the internally contracted hamiltonian are found using diagonalisation methods for large matrices in the usual manner. The eigenvalues obtained will be slightly higher than those of the corresponding uncontracted hamiltonian. The internal contraction restricts the variational freedom available and the energy is correspondingly higher.

Another useful idea in MRCI schemes is that of the restricted active space (RAS). Figure 2.25 shows the correspondence between RAS and CAS orbital sub-spaces. The active space in the RAS scheme is now partitioned into three. From Figure 2.25 it can be seen that the RAS2 space is equivalent to the active space of the CAS scheme. In addition, the RAS1 space is restricted by defining the number of 'holes' that can enter it, that is the number of electrons that may be substituted from the RAS1 space. The RAS3 space is defined by the number of 'particles' that can enter it, that is the number of electrons that may be substituted *into* the RAS3 space. For example, if we wish to use the RAS scheme to define the uncontracted MR-CISD calculation we discussed previously, we could do so by choosing the maximum number of holes/particles in RAS1/RAS3 to be two. Provided the orbitals of the inactive and virtual spaces were equivalent, then the RAS and MR-CISD calculations would be identical. The RAS idea can also be used to treat a much larger active space than is possible within the CAS scheme. For example, a fairly small number of orbitals and electrons could be chosen in RAS2, with RAS1/RAS3 providing a more limited multi-determinant expansion of a much larger number of electrons and orbitals. The molecular orbitals could also be optimised, using the ideas of Section 2.11.1. Such a procedure constitutes the RASSCF method.[33] With a judicious choice of the RAS spaces, the RASSCF method can be applied to medium-sized molecules.[34]

Perturbation theory methods can also be extended to the multi-reference case. Many formulations exist[35–38] and differ considerably in their theoretical and implementational details. A particularly simple scheme uses the generalised Fock operator, with elements

$$F_{pq} = h_{pq} + \sum_{ij}^{\text{inactive}+\text{active}} \gamma_{ij}\left[(pq|ij) - \frac{1}{2}(pj|iq)\right] \qquad (2.319)$$

where γ is obtained from a preliminary CASSCF calculation. The CASSCF energy is invariant to rotations *within* the inactive, active and virtual orbital sub-spaces. The generalised Fock operator is diagonalised within each sub-space independently and the resulting diagonal elements are used to define orbital eigenvalues, from which the zeroth-order energies, $E_A^{(0)}$, of the model hamiltonian, $\hat{H}_0$, can be obtained. The standard apparatus of

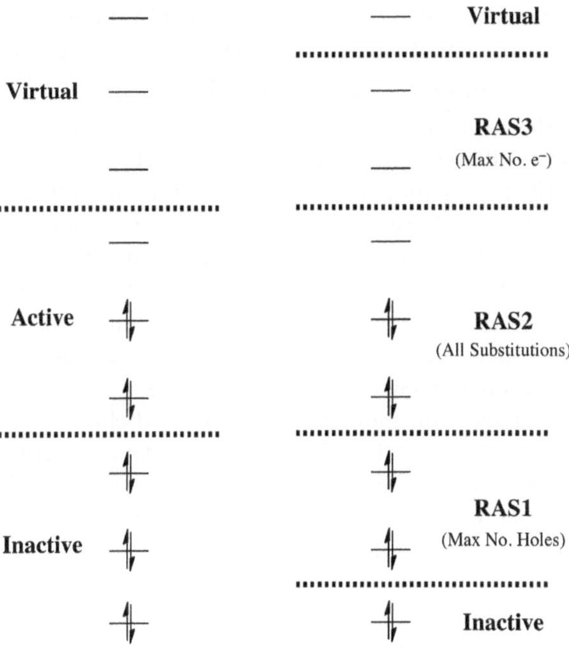

Figure 2.25 Comparison of inactive, active and virtual orbital sub-spaces in MCSCF methods with the inactive, RAS1, RAS2, RAS3 and virtual sub-spaces of the RASSCF method.

Rayleigh–Schrödinger perturbation theory then gives the second-order correction to the energy as

$$E_P^{(2)} = \sum_Q \frac{\left|\langle \Psi_P^{\mathrm{CASSCF}} | H | \Psi_Q \rangle\right|^2}{E_P^{(0)} - E_Q^{(0)}} \qquad (2.320)$$

where the label, P, refers to the specific CASSCF state of interest, and Q labels all allowed double substitutions. We can identify nine types of double substitution

- inactive–inactive $\rightarrow$ active–active   (i)
- inactive–inactive $\rightarrow$ active–virtual   (ii)
- inactive–inactive $\rightarrow$ virtual–virtual   (iii)
- inactive–active $\rightarrow$ active–active   (iv)
- inactive–active $\rightarrow$ active–virtual   (v)
- inactive–active $\rightarrow$ virtual–virtual   (vi)
- active–active $\rightarrow$ active–active   (vii)
- active–active $\rightarrow$ active–virtual   (viii)
- active–active $\rightarrow$ virtual–virtual   (ix)

Since the reference wavefunction is of the CASSCF type, the double substitutions (vii) are already included in the reference wavefunction and need not be considered further. The zeroth-order energies, $E_Q^{(0)}$, are defined analogously to the single reference case, so that the denominator in eqn (2.320) is evaluated as the difference of orbital eigenvalues. This scheme is very closely related to the MP2 formulation of Section 2.8.1, and is usually referred to as the MR-MP2 method. As we have previously noted, there are other formulations of multi-reference perturbation theory which use the Møller–Plesset type of zeroth-order hamiltonian. The energy correction in eqn (2.320) can be applied to any of the CASSCF states and is not limited to the ground state. This makes the MR-MP2 method a widely applicable and relatively efficient method for adding dynamic correlation to CASSCF-type wavefunctions.[39]

Finally we must mention multi-reference coupled-cluster theory. This is a very active area of research, which has seen many developments over the past few years. Unlike the configuration interaction and perturbation theory methods, the generalisation of coupled-cluster methods to multi-reference wavefunctions is not at all straightforward. Many subtly differing variants have been developed and their underlying equations are sufficiently complex that automated symbolic algebra tools have been developed to deal with them. In summary, multi-reference coupled-cluster methods are still highly specialised endeavours and have yet to be widely applied. This will no doubt change in the future, given the widespread activity in this area. The interested reader should consult ref. 40 for a contemporary review.

2.12 Density Functional Theory

The methods we have described so far for the calculation of electronic structure depend on knowledge of a wavefunction. The simplest wavefunction we have considered has the form of a single Slater determinant. At the opposite end of the scale, we have met the full CI wavefunction, which is exact within a given basis set. Regardless of the complexity of the wavefunction we wish to consider, they all share the common feature that they depend on the coordinates of every electron in the system. Each electron has three spatial coordinates and one spin coordinate, which gives the N-electron wavefunction a vast dimensional complexity. We have seen a manifestation of this in the heavy scaling of computational cost with size, Table 1.2.

In contrast to the N-electron wavefunction, the one-electron density depends only on three spatial coordinates regardless of the size of the system. The N-electron density is obtained from the wavefunction, $|\Psi(\mathbf{x}_1,\mathbf{x}_2,\cdots\mathbf{x}_N)\rangle$ as

$$\Gamma(\mathbf{x}_1,\mathbf{x}_2,\cdots\mathbf{x}_N) = |\Psi(\mathbf{x}_1,\mathbf{x}_2,\cdots\mathbf{x}_N)|^2 \qquad (2.321)$$

The one-electron density is obtained by integration of $\Gamma(\mathbf{x}_1,\mathbf{x}_2,\cdots\mathbf{x}_N)$ over the coordinates of $(N-1)$ electrons

$$\rho(\mathbf{r}_1) = N \int |\Psi(\mathbf{x}_1,\mathbf{x}_2,\ldots\mathbf{x}_N)|^2 d\mathbf{x}_2 d\mathbf{x}_3 \ldots d\mathbf{x}_N d\boldsymbol{\sigma}_1 \qquad (2.322)$$

The one-electron density simply depends on $\mathbf{r}_1$, assuming the integration includes the spin coordinate $\boldsymbol{\sigma}_1$. The idea of using $\rho(\mathbf{r}_1)$ in place of $|\Psi(\mathbf{x}_1,\mathbf{x}_2,\cdots\mathbf{x}_N)\rangle$ as the fundamental quantity for evaluating the electronic energy goes back to the work of L. H. Thomas and E. Fermi in the late 1920s. The Thomas–Fermi approach was very unsuccessful for chemical problems, not least because it failed to predict any binding in molecules. Nevertheless, the idea of electronic structure theory based on the one-electron density continued to percolate. In the 1950s, J. C. Slater developed the X_α method in which the exchange energy was treated using a functional of the one-electron density,

$$E^{X_\alpha} = -\frac{9}{8}\left(\frac{3}{\pi}\right)^{\frac{1}{3}}\alpha\int\rho(\mathbf{r})^{\frac{4}{3}}d\mathbf{r} \qquad (2.323)$$

Slater intended this as an approximation to the exchange interaction of the Hartree–Fock method. The simple form of eqn (2.323) (compare with eqn (2.56) in Hartree–Fock theory) is straightforwardly amenable to evaluation by numerical integration techniques. The X_α method found some utility but ultimately was not sufficiently accurate for the treatment of molecular problems since it did not include any treatment of the correlation between electrons of opposite spin. The rigorous foundation for a density-based theory of electronic structure was provided in 1964 by W. Kohn and P. Hohenberg.

The one-electron density $\rho(\mathbf{r})$ is a *function* of the positional coordinate, $\mathbf{r}$, and we show this dependence by writing $\mathbf{r}$ in parentheses after $\rho(\mathbf{r})$. Hence for a given point in space, $\mathbf{r}=(x,y,z)$, $\rho(\mathbf{r})$ is a number. A *functional* similarly produces a number but *depends on a function*. A familiar example of a functional is the energy obtained from the Schrödinger equation when a trial wavefunction, Ψ_{trial}, is used $E[\Psi_{\text{trial}}] = \langle\Psi_{\text{trial}}|\hat{H}|\Psi_{\text{trial}}\rangle / \langle\Psi_{\text{trial}}|\Psi_{\text{trial}}\rangle$. A functional is denoted by writing the function on which it depends in square brackets. For example, the exchange energy in the X_α method, eqn (2.323), should be denoted $E^{X_\alpha}[\rho]$. Finding functionals of the one-electron density that produce accurate atomic and molecular energies is the central purpose of density functional theory.

As we shall see in the following sections, it is possible to formulate density functional theory within a self-consistent field formalism that offers great computational efficiency. This coupled with accurate density functionals have made density functional theory methods the most widely used techniques in current studies of atomic and molecular problems.

2.12.1 The Hohenberg–Kohn Theorems

The electronic hamiltonian is given, in atomic units, by

$$\hat{H} = -\frac{1}{2} \sum_i^{\text{electrons}} \nabla_i^2 - \sum_A^{\text{nuclei}} \sum_i^{\text{electrons}} \frac{Z_A}{r_{iA}} + \sum_{i<j}^{\text{electrons}} \frac{1}{r_{ij}} \qquad (2.324)$$

and depends on the number of electrons, N, and the M nuclei through the second term on the right-hand side. In density functional theory (DFT), the nuclear attraction terms are referred to as the external potential and denoted as

$$\hat{V}_{\text{ext}}(\mathbf{r}_i) = - \sum_A^{\text{nuclei}} \sum_i^{\text{electrons}} \frac{Z_A}{r_{iA}} \qquad (2.325)$$

The origin of the term is that the nuclei are external to the electronic system for which we wish to solve the Schrödinger equation. For our purposes and the electronic problem we shall only consider $V_{\text{ext}}(\mathbf{r}_i)$ in the form of the nuclear attraction operator, but in general $V_{\text{ext}}(\mathbf{r}_i)$ can take other forms.

The first Hohenberg–Kohn theorem establishes that the external potential, $V_{\text{ext}}(\mathbf{r})$, is uniquely determined by the ground-state electron density, $\rho(\mathbf{r})$. The proof is straightforward and begins by considering a non-degenerate ground state with exact wavefunction, $|\Psi_{\text{Exact}}\rangle$, and exact energy, E_{Exact}. Now consider another electronic hamiltonian, with the same number of electrons but differing in $V_{\text{ext}}(\mathbf{r})$. Call this other hamiltonian $\hat{H}'$, and its exact wavefunction and energy, $|\Psi'_{\text{Exact}}\rangle$ and E'_{Exact}. Now suppose that the first Hohenberg–Kohn theorem does *not* hold. Using the variational principle we can write

$$E_{\text{Exact}} < \langle \Psi'_{\text{Exact}} | \hat{H} | \Psi'_{\text{Exact}} \rangle \qquad (2.326)$$

The hamiltonian, $\hat{H}$, can be written as

$$\hat{H} = \hat{H}' + \left(\hat{H} - \hat{H}' \right) \qquad (2.327)$$

therefore

$$E_{\text{Exact}} < \langle \Psi'_{\text{Exact}} | \hat{H}' | \Psi'_{\text{Exact}} \rangle + \langle \Psi'_{\text{Exact}} | \hat{H} - \hat{H}' | \Psi'_{\text{Exact}} \rangle \qquad (2.328)$$

Since both hamiltonians contain the same number of electrons, their difference reduces to the difference in $V_{\text{ext}}(\mathbf{r})$, so that $\hat{H} - \hat{H}' = V_{\text{ext}}(\mathbf{r}) - V'_{\text{ext}}(\mathbf{r})$ and eqn (2.328) becomes

$$E_{\text{Exact}} < E'_{\text{Exact}} + \langle \Psi'_{\text{Exact}} | V_{\text{ext}}(\mathbf{r}) - V'_{\text{ext}}(\mathbf{r}) | \Psi'_{\text{Exact}} \rangle \qquad (2.329)$$

Given that $V_{\text{ext}}(\mathbf{r}_i)$ is a one-electron operator

$$E_{\text{Exact}} < E'_{\text{Exact}} + \int \rho(\mathbf{r})[V_{\text{ext}}(\mathbf{r}) - V'_{\text{ext}}(\mathbf{r})]d\mathbf{r} \tag{2.330}$$

If we carry out the same procedure using $|\Psi_{\text{Exact}}\rangle$ and $\hat{H}'$ we obtain

$$E'_{\text{Exact}} < E_{\text{Exact}} + \int \rho(\mathbf{r})\left[V'_{\text{ext}}(\mathbf{r}) - V_{\text{ext}}(\mathbf{r})\right]d\mathbf{r} \tag{2.331}$$

Adding eqns (2.330) and (2.331) gives

$$E_{\text{Exact}} + E'_{\text{Exact}} < E'_{\text{Exact}} + E_{\text{Exact}} \tag{2.332}$$

which is a contradiction, and implies that our initial assumption that the density *does not* uniquely determine the external potential is false. The ground-state electron density *does* determine the external potential and also the number of electrons, since

$$\int \rho(\mathbf{r})d\mathbf{r} = N \tag{2.333}$$

thereby determining the electronic hamiltonian and in turn the ground-state wavefunction and energy. The ground-state energy is then a functional of the electron density

$$E_{\text{Exact}} = E_V[\rho] \tag{2.334}$$

Returning to eqn (2.324), the electronic energy is the average of the kinetic energies, the nuclear attraction energies and the electron–electron repulsion energies

$$E_{\text{Exact}} = E_V[\rho] = \langle T[\rho]\rangle + \langle V_{\text{ext}}[\rho]\rangle + \langle V_{\text{ee}}[\rho]\rangle \tag{2.335}$$

We know the form of $\langle V_{\text{ext}}[\rho]\rangle$

$$\langle V_{\text{ext}}[\rho]\rangle = \left\langle \Psi_{\text{Exact}} \left| \sum_i^N V_{\text{ext}}(\mathbf{r}_i) \right| \Psi_{\text{Exact}} \right\rangle$$
$$= \int \rho(\mathbf{r}) V_{\text{ext}}(\mathbf{r})d\mathbf{r} \tag{2.336}$$

However, the forms of the kinetic energy and electron–electron repulsion functionals are not known. We can combine these into one functional $F[\rho] = \langle T[\rho]\rangle + \langle V_{\text{ee}}[\rho]\rangle$ to obtain

$$E_{\text{Exact}} = E_V[\rho] = \int \rho(\mathbf{r}) V_{\text{ext}}(\mathbf{r}) d\mathbf{r} + F[\rho] \qquad (2.337)$$

At this stage we have a formally exact expression for the ground-state energy but cannot proceed to evaluate E_{Exact} since we do not know the form of $F[\rho]$.

The second Hohenberg–Kohn theorem establishes that any trial density, $\rho_{\text{trial}}(\mathbf{r})$, that is non-negative throughout space, $\rho_{\text{trial}}(\mathbf{r}) \geq 0$ for all $\mathbf{r}$, and integrates to the correct number of electrons, $\int \rho_{\text{trial}}(\mathbf{r}) d\mathbf{r} = N$, provides an upper bound to the exact energy when used with eqn (2.337), that is

$$E_V[\rho_{\text{trial}}] \geq E_{\text{Exact}} \qquad (2.338)$$

This applies to the exact density functional, which is unknown. There are additional requirements to those we have mentioned, such as the need for $\rho_{\text{trial}}(\mathbf{r})$ to correspond to the density obtained from a properly antisymmetric wavefunction, this is known as the "N-representability requirement". Many more details and extensions of these theorems may be found in ref. 41.

2.12.2 The Kohn–Sham Method

The discussion of the previous section provides a number of formal results, but these do not provide us with a route to practical electronic structure computations. Two difficulties are immediately apparent. First of all, we do not know the form of $F[\rho]$. So if we were presented with the exact density of a system, we still could not evaluate the energy from $E_V[\rho]$. Second, we do not have a means of calculating the electronic density without first calculating the wavefunction, and then integrating it to obtain the density from eqn (2.322). Pragmatic solutions to both these problems were suggested by W. Kohn and L. J. Sham in 1965.

In the expression for $E_V[\rho]$, one of the quantities we do not know is the form of the kinetic energy functional, $T[\rho]$. Kohn and Sham considered a model system in which the N electrons do not interact with each other. If a system is comprised of non-interacting particles, then the corresponding Schrödinger equation contains one-electron operators only. Denoting the hamiltonian for the non-interacting system as $\hat{H}_S$, we have

$$\hat{H}_S = \sum_i^N -\frac{1}{2}\nabla_i^2 + \sum_i^N V_S(\mathbf{r}_i) = \hat{T} + \hat{V}_{\text{ext}} \qquad (2.339)$$

The exact wavefunction corresponding to $\hat{H}_S$ can be written in terms of a single Slater determinant, $|\Psi_S\rangle$. To relate $\hat{H}_S$ to the real system, with interacting electrons, Kohn and Sham proposed the following hamiltonian

$$\hat{H}(\lambda) = \hat{T} + \hat{V}_{\text{ext}}(\lambda) + \lambda \hat{V}_{\text{ee}} \tag{2.340}$$

in which λ is a parameter with range $0 \leq \lambda \leq 1$. If $\lambda = 1$, the exact hamiltonian of eqn (2.324) is obtained. It is assumed that $\hat{V}_{\text{ext}}(\lambda)$ is adjusted so that the *same density* is obtained when $\lambda = 0$ as when $\lambda = 1$. That is, the external potential is altered so as to reproduce the exact ground-state density from the hamiltonian of the model, non-interacting, system.

If $|\Psi_S\rangle$ is a single Slater determinant composed of a set of orthonormal orbitals, $\left\{\phi_i^{\text{KS}}\right\}$,

$$|\Psi_S\rangle = \left|\phi_i^{\text{KS}} \phi_j^{\text{KS}} \cdots \phi_k^{\text{KS}}\right\rangle \tag{2.341}$$

then the corresponding Schrödinger equation is separable and gives

$$\hat{H}_S\left|\phi_i^{\text{KS}}\right\rangle = \varepsilon_i^{\text{KS}}\left|\phi_i^{\text{KS}}\right\rangle \quad \left(\text{for all } \phi_i^{\text{KS}}\right) \tag{2.342}$$

Each occupied orbital, $\left|\phi_i^{\text{KS}}\right\rangle$, contains a single electron and so the exact kinetic energy functional, for the non-interacting system, is given by

$$\langle T_S[\rho]\rangle = \sum_i^N \left\langle\phi_i^{\text{KS}}\left|-\frac{1}{2}\nabla_i^2\right|\phi_i^{\text{KS}}\right\rangle \tag{2.343}$$

By construction, the exact density is equivalent to the density obtained from $|\Psi_S\rangle$

$$\rho(\mathbf{r}) = \rho_S(\mathbf{r}) = \sum_i^N \left|\phi_i^{\text{KS}}(\mathbf{r})\right|^2 \tag{2.344}$$

The exact kinetic energy functional can then be written as

$$\langle T[\rho]\rangle = \langle T_S[\rho]\rangle + \langle \Delta T[\rho]\rangle \tag{2.345}$$

where $\Delta T[\rho]$ is unknown, but is assumed to be small compared to $T_S[\rho]$.

The inter-electronic term, $\langle V_{\text{ee}}[\rho]\rangle$, can also be separated as

$$\langle V_{\text{ee}}[\rho]\rangle = \frac{1}{2}\int\int\frac{\rho(\mathbf{r}_1)\rho(\mathbf{r}_2)}{r_{12}}d\mathbf{r}_1 d\mathbf{r}_2 + \langle \Delta V_{\text{ee}}[\rho]\rangle \tag{2.346}$$

The first term on the right-hand side is the classical coulombic repulsion energy arising from the density distribution, $\rho(\mathbf{r})$, of the electrons. The factor of $\frac{1}{2}$

ensures that the inter-electronic interactions are not counted twice. $\langle \Delta V_{ee}[\rho] \rangle$ contains all remaining inter-electronic interactions and is again unknown.

Returning to the expression for $E_V[\rho]$, we can now write

$$E_V[\rho] = \langle T_S[\rho] \rangle + \int \rho(\mathbf{r}) V_{ext}(\mathbf{r}) dr + \frac{1}{2} \int \int \frac{\rho(\mathbf{r}_1)\rho(\mathbf{r}_2)}{r_{12}} d\mathbf{r}_1 d\mathbf{r}_2 + \langle \Delta T_S[\rho] \rangle + \langle \Delta V_{ee}[\rho] \rangle \tag{2.347}$$

The final two terms are combined into the exchange–correlation functional

$$E_{XC}[\rho] = \langle \Delta T[\rho] \rangle + \langle \Delta V_{ee}[\rho] \rangle \tag{2.348}$$

The first three terms of eqn (2.347) are straightforward to evaluate using the density, $\rho(\mathbf{r})$, and provide the largest components of the electronic energy. The contribution from $E_{XC}[\rho]$ is relatively small but critical for the proper description of chemical problems. The energy scale of chemical properties, for example bond strengths, reaction energies and activation barriers, is such that a poor estimate of $E_{XC}[\rho]$ will lead to erroneous descriptions of such quantities.

The Kohn–Sham method uses the second Hohenberg–Kohn theorem to find the ground-state density by varying $\rho(\mathbf{r})$ so as to minimise $E_V[\rho]$. In fact, given eqn (2.344) for the density, we can carry out the minimisation by varying the orbitals $\left\{\phi_i^{KS}\right\}$ under the constraint of orthonormality of the orbitals. This constraint ensures that the density will integrate to the correct number of electrons as in eqn (2.333). The antisymmetry requirement is incorporated through the use of the Slater determinant, $|\Psi_S\rangle$. As might be imagined, the orbital optimisation problem in the Kohn–Sham method is closely analogous to that of the Hartree–Fock SCF technique. The orbital equations are

$$\left[-\frac{1}{2}\nabla_1^2 - \sum_A^{nuclei} \frac{Z_A}{r_{1A}} + \int \frac{\rho(\mathbf{r}_2)}{r_{12}} d\mathbf{r}_2 + \hat{V}_{XC}(\mathbf{r}_1) \right] |\phi_i^{KS}(\mathbf{r}_1)\rangle = \varepsilon_i^{KS} |\phi_i^{KS}(\mathbf{r}_1)\rangle \tag{2.349}$$

$$\hat{h}^{KS}(\mathbf{r}_1) |\phi_i^{KS}(\mathbf{r}_1)\rangle = \varepsilon_i^{KS} |\phi_i^{KS}(\mathbf{r}_1)\rangle$$

The $\left\{\phi_i^{KS}\right\}$ are called "Kohn–Sham orbitals" and the ε_i^{KS} are the corresponding eigenvalues. The ε_i^{KS} are the orbital energies corresponding to the non-interacting system. The term $\hat{V}_{XC}$ is the exchange–correlation potential

$$\hat{V}_{XC}(\mathbf{r}) = \frac{\delta E_{XC}[\rho(\mathbf{r})]}{\delta \rho(\mathbf{r})} \tag{2.350}$$

In practice, the Kohn–Sham equations are usually solved by using an LCAO expansion of the $\left\{\phi_i^{KS}\right\}$

$$\phi_i^{KS}(\mathbf{r}) = \sum_{\mu}^{m} c_{\mu i}^{KS}\chi(\mathbf{r}) \tag{2.351}$$

In analogy with the Roothaan–Hall procedure, the matrix Kohn–Sham equations are

$$\mathbf{h}^{KS}\mathbf{C} = \mathbf{SC}\boldsymbol{\varepsilon} \tag{2.352}$$

and can be solved by standard matrix methods. The kinetic energy and nuclear attraction integrals over basis functions are obtained exactly as in the Roothaan–Hall procedure. The Coulomb repulsion can also be obtained using analytic integrals over basis functions, equivalent to the J_{ij} in the Fock matrix. It is also possible to evaluate the Coulomb term using numerical integration and this proves to be very efficient for large molecules. The $E_{XC}[\rho]$ and $\hat{V}_{XC}[\rho]$ terms, as we shall see, are sufficiently complicated that numerical integration is often the only means available for their evaluation.

Before moving to a consideration of the remaining issue, the form of $E_{XC}[\rho]$, we must note some important properties of the Kohn–Sham method:

(i) The wavefunction $|\Psi_S\rangle$ is not equivalent to the exact wavefunction, even though it is used to construct the, in principle, exact density.

(ii) The exchange–correlation functional, $E_{XC}[\rho]$, includes a component from the kinetic energy, $\langle \Delta T[\rho]\rangle$, which is usually included by modifying the correlation functional.

(iii) Koopmans' theorem does *not* hold for the ε_i^{KS}, but ε_{HOMO}^{KS} corresponds to the negative of the exact first ionisation energy, $IE = -\varepsilon_{HOMO}^{KS}$, this is Janak's theorem. We shall consider ionisation energies further in Section 2.12.9.

(iv) $E_{XC}[\rho]$ contains a correction for the self-interaction error introduced by the classical form of the coulombic repulsion $\langle J[\rho]\rangle = \frac{1}{2}\int\int\frac{\rho(\mathbf{r}_1)\rho(\mathbf{r}_2)}{r_{12}}d\mathbf{r}_1 d\mathbf{r}_2$. The self-interaction error arises because this form allows the density arising from a given electron in the volume element $d\mathbf{r}_1$, to interact with the density of the same electron over the rest of space ($d\mathbf{r}_2$ ranges over all space). Hence $\langle J[\rho]\rangle$ allows an electron to interact with itself, which is physically incorrect.

2.12.3 The Local Density Approximation

We do not know the form of $E_{XC}[\rho]$ for a general electronic system, so it is natural to consider a simpler system that lends itself to exact treatment and so can be used to guide our understanding of $E_{XC}[\rho]$. Such a model is that of the uniform electron gas, which consists of a homogeneously distributed, electrically neutral, system of infinite volume containing an infinite number of electrons. The density, $\rho(\mathbf{r})$, is constant throughout the gas. In the local density approximation (LDA) we use the form

$$E_{XC}^{LDA}[\rho] = \int \rho(\mathbf{r})\varepsilon_{XC}[\rho]d\mathbf{r} \tag{2.353}$$

where $\varepsilon_{XC}[\rho]$ is the exchange–correlation energy per electron of a uniform electron gas with density $\rho(\mathbf{r})$. $\varepsilon_{XC}[\rho]$ can be separated into exchange and correlation components

$$\varepsilon_{XC}[\rho] = \varepsilon_X[\rho] + \varepsilon_C[\rho] \tag{2.354}$$

The form used for $\varepsilon_X[\rho]$ is due to Dirac and has the form

$$\varepsilon_X[\rho] = -\frac{3}{4}\left(\frac{3}{\pi}\right)^{\frac{1}{3}}\rho^{\frac{1}{3}} \tag{2.355}$$

For the correlation component a parameterisation exists, due to Vosko, Wilk and Nusair (VWN), of exact Quantum Monte-Carlo simulations of the uniform electron gas. It has a quite complicated form and may be found in ref. 42, we shall not reproduce it here.

Given $\varepsilon_X[\rho]$ and $\varepsilon_C[\rho]$, using eqn (2.354) it is straightforward to evaluate the exchange–correlation energy and, for the Kohn–Sham procedure, the exchange–correlation potential

$$V_{XC}^{LDA} = V_X^{LDA} + V_C^{LDA} = \frac{\delta E_X^{LDA}[\rho]}{\delta\rho} + \frac{\delta E_C^{LDA}[\rho]}{\delta\rho} \tag{2.356}$$

For systems in which the α-spin and β-spin densities are different, the LDA is modified to the local spin density approximation (LSDA), in which individual spin densities, $\rho_\alpha(\mathbf{r})$ and $\rho_\beta(\mathbf{r})$, are used. For example

$$\varepsilon_X[\rho_\alpha,\rho_\beta] = -2^{\frac{1}{3}}\frac{3}{4}\left(\frac{3}{\pi}\right)^{\frac{1}{3}}\left(\rho_\alpha^{\frac{1}{3}}+\rho_\beta^{\frac{1}{3}}\right) \tag{2.357}$$

Local density methods often predict correct chemical trends, for example bond lengths and bond angles, for covalent and ionic systems. For weakly bound non-covalent systems, they predict significant over-binding with excessively contracted bond lengths. The Dirac form of the exchange energy introduces errors of the order of 10%, which is larger than the whole of the correlation energy contribution!

2.12.4 Generalised Gradient Approximation

The local density approaches are based on the uniform electron gas, which is often a poor description of a chemical system. For example, consider the bonding between two atoms of very different electronegativity, the electron density will not be uniformly distributed. The LDA is based on a slow variation of the density with position, to improve upon it we must include non-local information or deal with a non-uniform electron gas. A way to do this is to include a dependence on the density *and* the gradient of the density

$$E_{XC}[\rho] = \int \rho(\mathbf{r})\varepsilon_{XC}[\rho]d\mathbf{r} + \int F_{XC}(\rho,\nabla\rho)d\mathbf{r} \qquad (2.358)$$

The gradient gives non-local information by providing information on how the density changes in the immediate vicinity of a point in space. This is the realm of the generalised gradient approximation (GGA). Typically, GGA exchange-correlation functionals perform much better than the LDA/LSDA for chemical properties.

An example of a GGA functional is Becke's exchange functional of 1988,[43]

$$E_X^{B88}[\rho,\nabla\rho] = E_X^{LDA}[\rho] - b\sum_{\sigma=\alpha,\beta}\int [\rho_\sigma(\mathbf{r})]^{\frac{4}{3}}\frac{[x_\sigma(\mathbf{r})]^2}{1 + 6bx_\sigma(\mathbf{r})\sinh^{-1}(x_\sigma(\mathbf{r}))}d\mathbf{r} \qquad (2.359)$$

In the second term on the right-hand side

$$x_\sigma(\mathbf{r}) = \frac{|\nabla\rho_\sigma(\mathbf{r})|}{(\rho_\sigma(\mathbf{r}))^{\frac{4}{3}}} \qquad (2.360)$$

and b is an empirical parameter with value 0.0042 au. The B88 functional is a significant improvement over the LDA and has found very wide use.

An example of a GGA correlation functional is the 1986 functional of Perdew,[44] the correlation energy per electron is given by

$$\varepsilon_C^{P86}[\rho,\nabla\rho] = \varepsilon_C^{LDA}[\rho] + e^{-\Phi}C_C(\rho)\frac{|\nabla\rho(\mathbf{r})|^2}{[\rho_\sigma(\mathbf{r})]^{\frac{4}{3}}} \qquad (2.361)$$

We simply note that this is a LDA correlation term with a correction that depends on the gradient of the density. The specific values of the constants and functions can be found in ref. 44.

These two GGA functionals are often combined into the widely used BP86 functional

$$\varepsilon_{XC}^{BP86}[\rho,\nabla\rho] = \varepsilon_X^{B88}[\rho,\nabla\rho] + \varepsilon_C^{P86}[\rho,\nabla\rho] \qquad (2.362)$$

Not all GGA functionals are derived as corrections to the LDA. Any functional depending on the density and its gradient falls under the banner of the GGA, for example the Lee–Yang–Parr (LYP) correlation functional.[45] Given the separation of the exchange and correlation functionals, eqn (2.354), any exchange functional may be combined with any exchange functional. This provides great flexibility, in principle, but also introduces a degree of arbitrariness into the choice of the exchange–correlation functional since there is no systematic way of choosing a hierarchy of functionals.

2.12.5   Meta-Generalised Gradient Approximation

To improve on the GGA the next step is to introduce the second derivative of the density, $\nabla^2\rho(\mathbf{r})$, or the kinetic energy density

$$\tau_\sigma = \frac{1}{2}\sum_{i_\sigma}^{occupied}\left|\nabla\phi_i^{\mathrm{KS}}\right|^2 \tag{2.363}$$

to give meta-GGAs of the form

$$E_{\mathrm{XC}}^{\mathrm{meta\text{-}GGA}}[\rho] = \int \rho(\mathbf{r})\varepsilon_{\mathrm{XC}}(\rho,\nabla\rho,\tau)d\mathbf{r} \tag{2.364}$$

Meta-GGA functionals often show improvement over GGA functionals but are more computationally demanding. The additional effort comes in computing the exchange–correlation potential, which now requires the evaluation of the functional derivative, $\dfrac{\delta\tau_\sigma(\mathbf{r})}{\delta\rho(\mathbf{r})}$, of the kinetic energy density. Well known meta-GGA functionals include B95,[46] BR[47] and TPSS.[48]

2.12.6   Adiabatic Connection: Hybrid Functionals

A very significant improvement in the design of exchange–correlation functionals came about in 1993 from the work of Becke,[49] who used the adiabatic connection idea of Langreth and Perdew.[50] In the adiabatic connection, the inter-electronic interaction is allowed to vary according to the coupling strength, λ,

$$\hat{V}_{\mathrm{ee}}(\lambda) = \lambda\sum_{i<j}^{N}\frac{1}{r_{ij}} \tag{2.365}$$

When $\lambda = 0$, the non-interacting electronic system is obtained and $\lambda = 1$ corresponds to the real physical system. A family of hamiltonians can be defined according to the value of λ,

$$\hat{H}(\lambda) = \sum_i^N -\frac{1}{2}\nabla_i^2 + \sum_i^N \hat{V}_{\mathrm{ext}}(\mathbf{r}_i,\lambda) + \lambda\sum_{i<j}^{N}\frac{1}{r_{ij}} \tag{2.366}$$

with corresponding Schrödinger equations

$$\hat{H}(\lambda)|\Psi(\lambda)\rangle = E(\lambda)|\Psi(\lambda)\rangle \tag{2.367}$$

If the energy is minimised by varying $|\Psi(\lambda)\rangle$ under the constraint that $\rho(\mathbf{r},\lambda) = \rho(\mathbf{r})$ for all values of λ, the exchange–correlation energy can be written as an integral over the coupling strength

$$E_{\mathrm{XC}}[\rho] = \int_0^1 \langle \Psi(\lambda)|V_{\mathrm{ee}}(\lambda)|\Psi(\lambda)\rangle - \frac{1}{2}\iint \frac{\rho(\mathbf{r}_1)\rho(\mathbf{r}_2)}{r_{12}}\,d\mathbf{r}_1 d\mathbf{r}_2 \tag{2.368}$$

Knowing the form of the exchange interaction in the two limits: $\lambda=0$ and $\lambda=1$, Becke argued that the exchange–correlation should contain some exact exchange. For a single Slater determinant, the exact exchange has the same form as in the Hartree–Fock theory,

$$E_{\mathrm{X}}^{\mathrm{H\text{-}F}} = -\frac{1}{2}\sum_{ij}^{N} \langle \phi_i\phi_j|\frac{1}{r_{12}}|\phi_j\phi_i\rangle \tag{2.369}$$

The amount of exact exchange to be included was fixed by fitting to experimental data and the B3 hybrid functional was obtained[49] as

$$E_{\mathrm{XC}}^{\mathrm{B3}} = aE_{\mathrm{X}}^{\mathrm{H\text{-}F}} + (1-a)E_{\mathrm{X}}^{\mathrm{LDA}} + b\Delta E_{\mathrm{X}}^{\mathrm{B88}} + c\Delta E_{C}^{\mathrm{GGA}} + (1-c)E_{C}^{\mathrm{LDA}} \tag{2.370}$$

with coefficients $a=0.20$, $b=0.72$, $c=0.81$. The most popular implementation of this functional is as the B3LYP hybrid functional, the details of which are specified in refs. 42, 45, 49 and 51,

$$E_{\mathrm{XC}}^{\mathrm{B3LYP}} = 0.20E_{\mathrm{X}}^{\mathrm{H\text{-}F}} + 0.80E_{\mathrm{X}}^{\mathrm{LDA}} + 0.72\Delta E_{\mathrm{X}}^{\mathrm{B88}} + 0.81E_{C}^{\mathrm{LYP}} + 0.19E_{C}^{\mathrm{VWN}} \tag{2.371}$$

The B3LYP functional has been extremely successful and has enabled the study of a wide range of chemical systems.

It is possible to combine hybrid functionals with terms involving the kinetic energy density to yield meta-hybrid exchange–correlation functionals. Things now become quite difficult to develop from first principles. This has encouraged the development of functionals in which many parameters are introduced and determined by fitting to experimental data. This kind of approach can be quite successful. The M06 family of functionals[52] is an example of highly parameterised meta-hybrid functionals that can outperform B3LYP for some molecular properties.

2.12.7 Double Hybrid Functionals

The idea of double hybrid density functionals is to mix in exact exchange and some correlation from an MP2-like calculation. The starting point is a hybrid exchange–correlation functional

$$E_{XC}^{Hybrid} = a_1 E_X^{GGA} + (1-a_1)E_X^{H\text{-}F} + a_2 E_C^{GGA} \tag{2.372}$$

which is used to obtain the Kohn–Sham orbitals. An improved energy is then calculated as

$$E_{XC}^{\substack{Double\\Hybrid}} = E_{XC}^{Hybrid} + (1-a_2)E_C^{KS\text{-}MP2} \tag{2.373}$$

where

$$E_C^{KS\text{-}MP2} = \frac{1}{4}\sum_{ij}^{occupied}\sum_{ab}^{unoccupied}\frac{\left|\left(\phi_i^{KS}\phi_a^{KS}||\phi_j^{KS}\phi_b^{KS}\right)\right|^2}{\varepsilon_i^{KS}+\varepsilon_j^{KS}-\varepsilon_a^{KS}-\varepsilon_b^{KS}} \tag{2.374}$$

in analogy with eqn (2.278). It is important to note that $E_C^{KS\text{-}MP2}$ is not the same as the MP2 energy, since the orbitals and orbital energies are those of the Kohn–Sham scheme and not the Hartree–Fock method. The parameters a_1 and a_2 are obtained by fitting to experimental data. A particular scheme, due to Grimme,[53] is called "B2PLYP" and uses

$$\begin{aligned} E_X^{GGA} &= E_X^{B88} \\ E_C^{GGA} &= E_C^{LYP} \\ a_1 &= 0.47 \\ a_2 &= 0.73 \end{aligned} \tag{2.375}$$

For a test data set of 270 enthalpies of formation, 105 ionisation energies, 63 electron affinities, 10 proton affinities and six hydrogen bond energies the mean absolute deviation from experiment is 10.5 kJ mol^{-1}.[54] In comparison, the B3LYP functional has an error of 18.4 kJ mol^{-1} for the same data set. The additional cost of the $E_C^{KS\text{-}MP2}$ term adds substantially to the cost of the calculation, but a good reduction in errors is obtained.

2.12.8 Non-Covalent Interactions

The first Hohenberg–Kohn theorem states that given the exact exchange–correlation functional, the exact ground-state energy may be obtained. Accordingly DFT should be able to address all chemical phenomena, assuming we know the exact exchange–correlation functional! Current local and gradient corrected functionals are successful for strongly bound systems involving covalent and ionic bonds. For non-covalently bound molecules, for example the water dimer or DNA base pairs, the binding comes from the dispersion interaction which has a $1/R^6$ behaviour and this is not reproduced by current density functionals. This long-range behaviour can be incorporated into density functionals through the inclusion of empirical pairwise dispersion terms, such approaches are usually known as "DFT-D methods". The energy is obtained as

$$E_{\text{DFT-D}} = E_{\text{DFT}} + E_{\text{Dispersion}} \tag{2.376}$$

where

$$E_{\text{Dispersion}} = -s_6 \sum_{A<B}^{M} \frac{C_6^{AB}}{R_{AB}^6} f(R_{AB}) \tag{2.377}$$

C_6^{AB} is a dispersion coefficient for the AB pair of atoms, R_{AB} is the inter-atomic distance. $f(R_{AB})$ is a damping function

$$f(R_{AB}) = \frac{1}{1 + e^{-d(R_{AB}/R_{\text{vdw}} - 1)}} \tag{2.378}$$

d is a parameter and R_{vdw} is the sum of the van der Waals radii for atoms A and B. s_6 is a global scaling factor that is determined by the specific exchange–correlation functional used. The dispersion coefficients can be obtained from experimental information or from high-level calculations. Details may be found in the work of Grimme[55] who has extended the parameterisation to the elements for $Z = 1$–94. DFT is now routinely applicable to non-covalent interactions.[56]

2.12.9 Ionisation Energies in Density Functional Theory

Given the exact exchange–correlation functional it can be shown that the eigenvalue of the highest occupied Kohn–Sham orbital is the *exact* first ionisation energy.

$$IE_1 = -\varepsilon_{\text{HOMO}}^{\text{KS}} \tag{2.379}$$

This should not be confused with Koopmans' theorem in Hartree–Fock theory, which provides an approximation to the ionisation energy. In practice, approximate exchange–correlation functionals are quite poor at predicting ionisation energies from the highest occupied Kohn–Sham eigenvalue. The relative positions of orbital energies can be predicted quite well but the absolute magnitudes are in error. A simple scheme for addressing this problem is to shift all the Kohn–Sham eigenvalues by a constant value. The constant can be obtained by a calculation on the cation, from which the ionisation energy can be calculated. The highest occupied orbital is then shifted to exactly reproduce this ionisation energy. The same shift is applied to all the orbital energies.

As a simple illustration of this idea, consider the methanal molecule, at its experimental geometry, a B3LYP calculation with the 6-31G(d,p) basis set gives an energy of -114.503162 au. The eight doubly occupied Kohn–Sham orbitals have the energies given in Table 2.13 (left-hand table). The first five ionisation energies obtained from experiment are also given in Table 2.13 (right-hand table). Simply using eqn (2.379) gives a value for the first ionisation energy of 7.3 eV at the B3LYP/6-31G(d,p) level, which is in error by more than 3.6 eV.

A calculation on the cation gives an energy of -114.109401 au. From this we can calculate the first ionisation energy to be 10.7 eV, which is in good agreement with the experimental result. Now we can obtain the shift that makes $-\varepsilon_{HOMO}^{KS}$ equal to the ionisation energy calculated from total energy differences of the cation and neutral molecule, that is

$$IE_1 = E_{\text{Cation}} - E_{\text{Neutral}}$$

$$= -114.109401 + 114.503162$$

$$= 0.393761 \text{ au}$$

$$-\varepsilon_{HOMO}^{KS} = 0.26779 \text{ au}$$

$$\Delta = 0.393761 - 0.26779$$

$$= 0.125971 \text{ au}$$

Table 2.13   The experimental geometry of methanal is shown. The Kohn–Sham occupied orbital energies (au) obtained at the B3LYP/6-31G(d,p) level (left-hand table). The first five experimental ionisation energies (eV) (right-hand table).

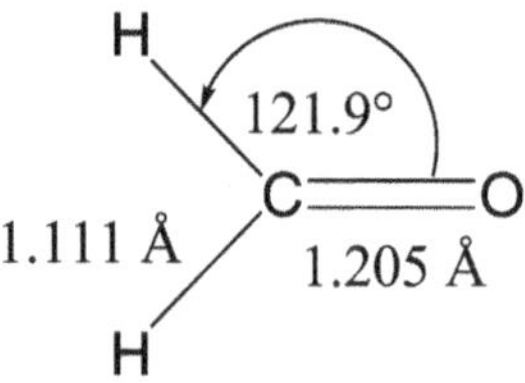

i	ε_i^{KS} /au	i	IE_i (Expt) / eV
1	-19.16939	1	10.9
2	-10.28912	2	14.5
3	-1.06159	3	16.1
4	-0.63559	4	17.0
5	-0.49556	5	21.4
6	-0.44889		
7	-0.39953		
8	-0.26779		

Applying the shift, Δ, to all the orbital energies and then calculating the ionisation energies as the negative of the orbital energy gives the results shown in Table 2.14. By construction, IE_1, is the same as obtained from $E_{\text{Cation}} - E_{\text{Neutral}}$, however the rest of the ionisation energies are now reproduced quite well with a maximum error of 0.7 eV. Given the efficiency of DFT calculations this is not an expensive correction to make and it enables the interpretation of orbital eigenvalues as ionisation energies to be retained. Of course, the theoretical basis for doing this with a hybrid density functional is somewhat muddy.

Appendix 2A   The Method of Lagrange Multipliers

Here we give a brief geometric justification for the method of Lagrange multipliers. This technique is often used to treat constrained optimisation problems, such as finding the minimum of a function of many variables, $F(x_1,x_2 \cdots)$, subject to the constraint, or set of constraints, $C(x_1,x_2 \ldots)=0$. Figure 2.26 shows an example, where the function to be minimised, $F(x_1,x_2 \cdots)$, is the surface shown in grey and the constraint surface, $C(x_1,x_2 \cdots)=0$, is shown in black. The contour levels of each surface are shown. In this example $F(x_1,x_2 \cdots)$ and $C(x_1,x_2 \cdots)=0$ intersect along a two-dimensional curve. If we follow the contours of each surface along the curve of intersection we can see that at the point where $F(x_1,x_2 \cdots)$ reaches its minimum value, the contours of $F(x_1,x_2 \cdots)$ and $C(x_1,x_2 \cdots)=0$ touch. The contour lines of $F(x_1,x_2 \cdots)$ and $C(x_1,x_2 \cdots)=0$ touch when the tangent vectors of the contour lines are parallel or, equivalently, when the gradients of $F(x_1,x_2 \cdots)$ and $C(x_1,x_2 \cdots)=0$ are parallel. At any other point along the curve of intersection, other than at the minimum value of $F(x_1,x_2 \cdots)$, the tangents of the two surfaces are no longer parallel to each other. Hence at the minimum point of $F(x_1,x_2 \cdots)$ that satisfies the requirement $C(x_1,x_2 \cdots)=0$ we can write

$$\nabla F(x_1,x_2 \cdots) = \lambda \, \nabla C(x_1,x_2 \cdots) \tag{2A.1}$$

Table 2.14   The shifted Kohn–Sham orbital energies (au), see text, and the ionisation energies obtained from the shifted orbital eigenvalues (eV) compared with the experimental ionisation energies (eV).

i	$-(\varepsilon_i^{\text{KS}}+\Delta)$/au	i	$IE_i = -(\varepsilon^{\text{KS}}+\Delta)$/eV	$IE_i(\text{Expt})$/eV
8	0.393761	1	10.7	10.9
7	0.525501	2	14.3	14.5
6	0.574861	3	15.6	16.1
5	0.621531	4	16.9	17.0
4	0.761561	5	20.7	21.4

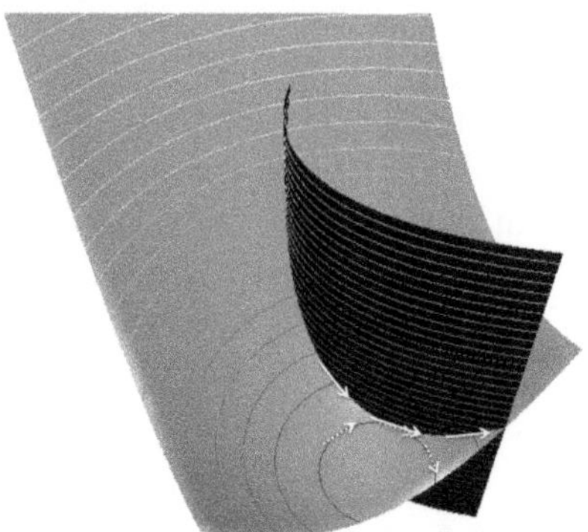

Figure 2.26   The surface to be minimised is shown in grey, with the constraint function in black. The arrows in white are the tangents to the contours of each surface and are parallel at the point at which the constrained minimum lies.

where $\nabla = \left(\dfrac{d}{dx_1}, \dfrac{d}{dx_2}, \cdots \right)$. Another way to represent the condition in eqn (2A.1) is to define a Lagrange function, $L(\mathbf{x},\lambda)$, which when minimised satisfies eqn (2.A1). We can do this by forming

$$L(\mathbf{x},\lambda) = F(\mathbf{x}) - \lambda C(\mathbf{x}) \tag{2A.2}$$

If we now minimise $L(\mathbf{x},\lambda)$ with respect to $\mathbf{x} = (x_1, x_2 \cdots)$ and λ, by requiring that $\dfrac{dL(\mathbf{x},\lambda)}{d\mathbf{x}} = 0$ and $\dfrac{dL(\mathbf{x},\lambda)}{d\lambda} = 0$, we obtain

$$\frac{dL(\mathbf{x},\lambda)}{d\mathbf{x}} = \frac{dF(\mathbf{x})}{d\mathbf{x}} - \lambda \frac{dC(\mathbf{x})}{d\mathbf{x}} = 0$$
$$\frac{dL(\mathbf{x},\lambda)}{d\lambda} = C(\mathbf{x}) = 0 \tag{2A.3}$$

The first equation is equivalent to the requirement in eqn (2A.1) and the second equation ensures that the constraint is satisfied. Hence our constrained variation of $F(x_1, x_2 \cdots)$ subject to $C(x_1, x_2 \cdots) = 0$ can be achieved by the free variation of the Lagrange function in eqn (2A.2). We can use the form of the Lagrange function in eqn (2A.2) for any number of equality constraints, with each constraint, $C_i(x_1, x_2 \cdots) = 0$, being assigned a multiplier, λ_i.

Appendix 2B   Orthogonalisation Methods

Consider the three vectors shown in Figure 2.27, we wish to generate three orthonormal vectors from these. We shall discuss two ways of accomplishing this. The first method we shall use is known as the "Gram–Schmidt process". In this technique we move through each vector normalising it and then projecting it out of each remaining vector. This means that the first vector we start with remains unchanged but each subsequent vector is changed to ensure orthogonality with all vectors that precede it.

Some care must be taken as to how this is done to ensure numerical stability. The procedure applied to n vectors is as follows:

For $i=1,2,\cdots n$, normalise $\mathbf{v}_i$ by forming $\mathbf{v}'_i = \dfrac{\mathbf{v}_i}{\mathbf{v}_i \cdot \mathbf{v}_i}$

For $j=i+1,2,\cdots n$, remove the component of $\mathbf{v}'_i$ from $\mathbf{v}_j$ by forming

$$\mathbf{v}_j = \mathbf{v}_j - \frac{\mathbf{v}'_i \cdot \mathbf{v}_j}{\mathbf{v}'_i \cdot \mathbf{v}'_i}\mathbf{v}'_i$$

next j

next i

Applying this process to the vectors in Figure 2.27 produces the following:

(i) $\mathbf{v}_1$ is already normalised and so remains unchanged: $\mathbf{v}_1 = (1,0,0)$

(ii) Subtract $\mathbf{v}_1$ from $\mathbf{v}_2$: $\mathbf{v}_2 = \mathbf{v}_2 - \mathbf{v}_1 = (0,1,0)$
      Subtract $\mathbf{v}_1$ from $\mathbf{v}_3$: $\mathbf{v}_3 = \mathbf{v}_3 - \mathbf{v}_1 = (0,1,1)$

(iii) $\mathbf{v}_2$ is already normalised and so remains unchanged.

(iv) Subtract $\mathbf{v}_2$ from $\mathbf{v}_3$: $\mathbf{v}_3 = \mathbf{v}_3 - \mathbf{v}_2 = (0,0,1)$

(v) $\mathbf{v}_3$ is already normalised and so remains unchanged.

The vectors produced are: $\mathbf{v}_1 = (1,0,0)$, $\mathbf{v}_2 = (0,1,0)$ and $\mathbf{v}_3 = (0,0,1)$ which are coincident with the cartesian axes and so are orthogonal, they are also clearly normalised.

The other orthogonalisation technique we shall describe is known as "Löwdin" or "symmetric" orthogonalisation. Here we must form $\mathbf{S}^{-\frac{1}{2}}$, where $\mathbf{S}$ is the overlap matrix. The procedure is as follows:

$$\text{(i)} \quad \text{Form the overlap matrix: } \mathbf{S} = \begin{pmatrix} \mathbf{v}_1 \cdot \mathbf{v}_1 & \mathbf{v}_1 \cdot \mathbf{v}_2 & \mathbf{v}_1 \cdot \mathbf{v}_3 & \cdots \\ \mathbf{v}_2 \cdot \mathbf{v}_1 & \mathbf{v}_2 \cdot \mathbf{v}_2 & \mathbf{v}_2 \cdot \mathbf{v}_3 & \cdots \\ \mathbf{v}_3 \cdot \mathbf{v}_1 & \mathbf{v}_3 \cdot \mathbf{v}_2 & \mathbf{v}_3 \cdot \mathbf{v}_3 & \cdots \\ \vdots & \vdots & \vdots & \ddots \end{pmatrix}$$

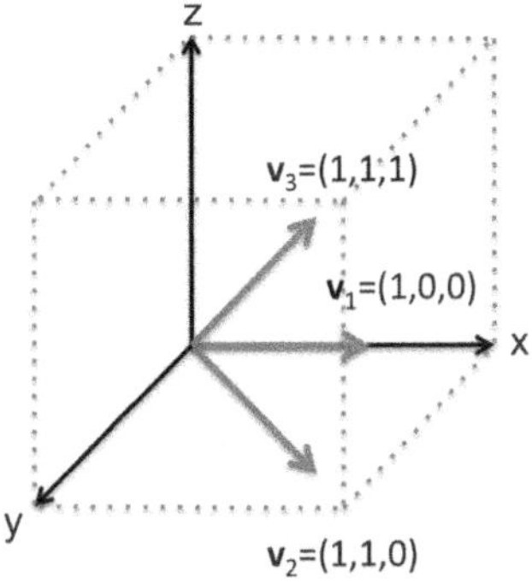

Figure 2.27   Three non-orthogonal and un-normalised cartesian vectors.

(ii)   Diagonalise the overlap matrix: $\mathbf{U}^{\dagger}\mathbf{S}\mathbf{U}=\sigma=\begin{pmatrix} \sigma_1 & 0 & 0 & \cdots \\ 0 & \sigma_2 & 0 & \cdots \\ 0 & 0 & \sigma_3 & \cdots \\ \vdots & \vdots & \vdots & \ddots \end{pmatrix}$

(iii)   Form $\mathbf{S}^{-\frac{1}{2}}$: $\sigma^{-\frac{1}{2}}=\begin{pmatrix} \dfrac{1}{\sqrt{\sigma_1}} & 0 & 0 & \cdots \\ 0 & \dfrac{1}{\sqrt{\sigma_2}} & 0 & \cdots \\ 0 & 0 & \dfrac{1}{\sqrt{\sigma_3}} & \cdots \\ \vdots & \vdots & \vdots & \ddots \end{pmatrix}$ $\mathbf{S}^{-\frac{1}{2}}=\mathbf{U}\sigma^{-\frac{1}{2}}\mathbf{U}^{+}$

(iv)   Transform the vectors by $\mathbf{S}^{-\frac{1}{2}}$: $\mathbf{V}\mathbf{S}^{-\frac{1}{2}}$, where $\mathbf{V}$ contains the vectors arranged by columns.

Applying this process to the vectors in Figure 2.27 gives:

(i)

$$\mathbf{S}=\begin{pmatrix} 1 & 1 & 1 \\ 1 & 2 & 2 \\ 1 & 2 & 3 \end{pmatrix}$$

(ii)

$$\sigma=\begin{pmatrix} 0.307979 & 0 & 0 \\ 0 & 0.643104 & 0 \\ 0 & 0 & 5.04892 \end{pmatrix}$$

$$\mathbf{U}=\begin{pmatrix} 0.591009 & 0.736976 & 0.327985 \\ -0.736976 & 0.327985 & 0.591009 \\ 0.327985 & -0.591009 & 0.736976 \end{pmatrix}$$

(iii)

$$\mathbf{S}^{-\frac{1}{2}}=\begin{pmatrix} 1.35455 & -0.397166 & -0.0862683 \\ -0.397166 & 1.26829 & -0.483434 \\ -0.0862683 & -0.483434 & 0.871119 \end{pmatrix}$$

(iv)

$$\mathbf{VS}^{-\frac{1}{2}} = \begin{pmatrix} 0.871119 & 0.387685 & 0.301416 \\ -0.483434 & 0.784851 & 0.387685 \\ -0.0862683 & -0.483434 & 0.871119 \end{pmatrix}$$

The three new vectors are:

$$\mathbf{v}_1 = (\,0.871119 \quad -0.483434 \quad -0.0862683\,)$$

$$\mathbf{v}_2 = (\,0.387685 \quad 0.784851 \quad -0.483434\,)$$

$$\mathbf{v}_3 = (\,0.301416 \quad 0.387685 \quad 0.871119\,)$$

It is easily verified that these vectors are orthonormal. So we have generated another set of three orthonormal vectors. These are quite distinct from those generated by the Gram–Schmidt procedure, but span the same three-dimensional space. Both the Gram–Schmidt and the symmetric orthogonalisation methods are frequently used in quantum chemical programs.

Appendix 2C Computing Eigenvalues and Eigenvectors of Large Matrices

In Appendix 1C we discussed the complete solution of a symmetric eigenvalue problem using the Jacobi algorithm. When the dimension of the matrix becomes large it is not feasible, or desirable, to compute all eigenvalues and eigenvectors. Additionally, when the matrix for which we wish to find the eigenvalues and eigenvectors is the electronic hamiltonian, $\mathbf{H}$, represented in a basis of Slater determinants, there is much sparsity in the matrix that is not easily exploited with full matrix methods.

Our purpose is to find a few of the lowest eigenvalues and eigenvectors of a large matrix. The basic method we shall describe is known as the "Davidson method".[57] The key feature that makes this method, and its many variants, applicable to large matrices is that the matrix is never explicitly assembled. Rather, only matrix-vector products, $\mathbf{Hx}$, are formed. A projection of $\mathbf{H}$ into a small sub-space of vectors is built and this sub-space problem is solved exactly, using standard small matrix techniques. The sub-space is then expanded and the process repeated until the eigenvalues of the sub-space problem become equivalent to that of the full matrix. The basic procedure for finding the lowest eigenvalue is as follows:

(i) Decide on the minimum and maximum sub-space dimensions, k_{min} and k_{max}, respectively. k_{min} must greater than the number of eigenvalues sought. Let N be the dimension of the matrix.

(ii) Choose k_{min} orthonormal starting vectors of dimension N, $\mathbf{b}_i\ (i=1,2,\cdots k_{min})$.

(iii) Form and save the products $\mathbf{Hb}_i\ (i=1,2,\cdots k_{min})$.

(iv) Form the projected matrix $\tilde{\mathbf{H}}$ with elements $\tilde{H}_{ij} = \mathbf{b}_i \mathbf{H} \mathbf{b}_j$ and compute all the eigenvalues $(\tilde{E}_i)$ and eigenvectors $(\tilde{\mathbf{c}}_i)$ of $\tilde{\mathbf{H}}$. $\tilde{E}_1$ is an approximation to E_1 and the corresponding approximate eigenvector is $\mathbf{c}_1 = \sum_i^{k_{\min}} \tilde{c}_{i1} \mathbf{b}_i$.

(v) Form the residual vector $\mathbf{r} = (\mathbf{H} - \tilde{E}_1 \mathbf{I}) \mathbf{c}_1 = \sum_i^{k_{\min}} \tilde{c}_{i1} \mathbf{H} \mathbf{b}_i - \sum_i^{k_{\min}} \tilde{c}_{i1} \tilde{E}_1 \mathbf{b}_i$ and calculate its norm $\|\mathbf{r}\|$.

(vi) If $\|\mathbf{r}\| < \tau$, where τ is a chosen threshold, then $\tilde{E}_1 = E_1$ and the process stops. Otherwise, a new vector $\mathbf{b}_{k_{\min}+1}$ must be formed.

(vii) Form the vector $\mathbf{d}$ with elements $d_M = \dfrac{r_M}{\tilde{E}_1 - H_{MM}}$ $(M = 1...N)$ and orthogonalise it to all $\mathbf{b}_i$ $(i = 1,2, \cdots k_{\min})$ using the Gram–Schmidt procedure and form $\mathbf{b}_{k_{\min}+1} = \dfrac{\mathbf{d}}{\|\mathbf{d}\|}$.

(viii) Expand the projection sub-space from $k_{\min}$ to $k_{\min}+1$ and go to step (iii).

If the dimension of the projection sub-space exceeds $k_{\max}$, then the process can be restarted with the initial $\mathbf{b}_i$ $(i = 1,2, \cdots k_{\min})$ chosen as $\mathbf{b}'_i = \sum_k^{k_{\min}} \tilde{c}_{ki} \mathbf{b}_k$

using the current eigenvectors of $\tilde{\mathbf{H}}$. If several, say p, eigenvalues are wanted, then in step (v) the residual vector should be formed for each eigenvalue sought, providing p residual vectors, from which p new $\mathbf{d}$ and $\mathbf{b}$ vectors can be formed in step (vii). The projection sub-space is then expanded by $k_{\min}+p$ in each iteration.

References

1. C. C. J. Roothaan, *Rev. Mod. Phys.*, 1951, **23**, 69.
2. G. G. Hall, *Proc. R. Soc. London*, 1951, **A205**, 541.
3. H. B. Schlegel and J. J. W. McDouall, in *Computational Advances in Organic Chemistry: Molecular Structure and Reactivity*, ed. C. Ögretir and I. G. Csizmadia, NATO ASI Series, 1991, vol. **330**, pp. 167–185.
4. I. H. Hillier and V. R. Saunders, *Int. J. Quantum Chem.*, 1970, **4**, 503; *Proc. R. Soc. London*, 1970, **A320**, 161.
5. P. Pulay, *Chem. Phys. Lett.*, 1980, **73**, 393; *J. Comput. Chem.*, 1982, **3**, 556.
6. J. Almlöf, in *Modern Electronic Structure Theory*, ed. D. R. Yarkony, World Scientific, Singapore, 1995, ch. **1**, pp. 110–151.
7. T. Helgaker, P. Jorgensen and J. Olsen, *Modern Electronic Structure Theory*, Wiley, Chichester, 2000, ch. 9.
8. T. Helgaker and P. R. Taylor, in *Modern Electronic Structure Theory*, ed. D. R. Yarkony, World Scientific, Singapore, 1995, ch. **2**, pp. 725–856.
9. H. B. Schlegel and M. J. Frisch, *Int. J. Quantum Chem.*, 1995, **54**, 83.

10. S. Huzinaga, *J. Chem. Phys.*, 1965, **42**, 1293.

11. T. H. Dunning Jr., *J. Chem. Phys.*, 1989, **90**, 1007.

12. R. A. Kendall, T. H. Dunning Jr. and R. J. Harrison, *J. Chem. Phys.*, 1992, **96**, 6796.

13. F. Weigend and R. Ahlrichs, *Phys. Chem. Chem. Phys.* 2005, **7**, 3297.

14. B. E. Rocher-Casterline, L. C. Ch'ng, A. K. Mollner and H. Reisler, *J. Chem. Phys.*, 2011, **115**, 6903.

15. N. C. Handy, *Chem. Phys. Lett.*, 1980, **74**, 280.

16. W. Duch and G. H. F Diercksen, *J. Chem. Phys.*, 1994, **101**, 3018.

17. D. Maurice and M. Head-Gordon, *Int. J. Quantum Chem. Symp.*, 1995, **29**, 361.

18. I. Shavitt and R. J. Bartlett, *Many-Body Methods in Chemistry and Physics*, Cambridge University Press, Cambridge, 2009.

19. J. A. Pople, J. S. Binkley and R. Seeger, *Int. J. Quantum Chem.*, 1976, **10**, 1.

20. R. J. Bartlett and D. M. Silver, *Int. J. Quantum Chem. Symp.*, 1974, **8**, 271.

21. H.-J. Werner, F. R. Manby and P. J. Knowles, *J. Chem. Phys.*, 2003, **118**, 8149.

22. S. Grimme, L. Goerigk and R. F. Fink, *Wiley Interdiscip. Rev.: Comput. Mol. Sci.*, 2012, **2**, 886.

23. F. Neese, T. Schwabe, S. Kossmann, B. Schirmer and S. Grimme, *J. Chem. Theory Comput.*, 2009, **5**, 3060.

24. G. E. Scuseria and P. Y. Ayala, *J. Chem. Phys.*, 1999, **111**, 8330.

25. M. Kállay and P. R. Surján, *J. Chem. Phys.*, 2001, **115**, 2945.

26. S. Saebo and P. Pulay, *Annu. Rev. Phys. Chem.*, 1993, **44**, 213.

27. C. Hampel and H.-J. Werner, *J. Chem. Phys.*, 1996, **104**, 6286.

28. J. E. Subotnik, A. Sodt and M. Head-Gordon, *J. Chem. Phys.*, 2006, **125**, 74116.

29. J. W. Boughton and P. Pulay, *J. Comput. Chem.*, 1993, **14**, 736.

30. R. A. Mata and H.-J. Werner, *J. Chem. Phys.*, 2006, **125**, 184110.

31. B. O. Roos, *Adv. Chem. Phys.*, 1987, **69**, 399.

32. H.-J. Werner and P. J. Knowles, *J. Chem. Phys.*, 1988, **89**, 5803.

33. J. Olsen, B. O. Roos, P. Jørgensen and H. J. A. Jensen, *J. Chem. Phys.*, 1988, **89**, 2185.

34. M. Rubio, B. O. Roos, L. Serrano-Andrés and M. Merchán, *J. Chem. Phys.*, 1999, **110**, 7202.

35. K. Hirao, *Chem. Phys. Lett.*, 1992, **190**, 374.

36. K. Andersson, P.-A. Malmqvist, B. O. Roos, A. J. Sadlej and K. Wolinski, *J. Phys. Chem.*, 1990, **94**, 5483.

37. R. B. Murphy and R.P. Messmer, *Chem. Phys. Lett.*, 1991, **183**, 443.

38. C. Angeli, R. Cimiraglia, S. Evangelisti, T. Leininger and J.-P. Malrieu, *J. Chem. Phys.*, 2001, **114**, 10252.

39. J. J. W. McDouall and D. Robinson, *AIP Conf. Proc.*, 2007, **963**, 268.

40. D. I. Lyakh, M. Musiaz, V. F. Lotrich and R. J. Bartlett, *Chem. Rev.*, 2012, **112**, 182.

41. R. G. Parr and W. Yang, *Density-Functional Theory of Atoms and Molecules*, Oxford University Press, Oxford, 1995.

42. S. H. Vosko, L. Wilk and M. Nusair, *Can. J. Phys.*, 1980, **58**, 1200.

43. A. D. Becke, *Phys. Rev. A*, 1988, **38**, 3098.

44. J. P. Perdew, *Phys. Rev. B*, 1986, **33**, 8822.
45. B. Miehlich, A. Savin, H. Stoll and H. Preuss, *Chem. Phys. Lett.*, 1989, **157**, 200.
46. A. D. Becke, *J. Chem. Phys.*, 1996, **104**, 1040.
47. A. D. Becke and M. R. Roussel, *Phys. Rev. A*, 1989, **39**, 3761.
48. J. M. Tao, J. P. Perdew, V. N. Staroverov and G. E. Scuseria, *Phys. Rev. Lett.*, 2003, **91**, 146401.
49. A. D. Becke, *J. Chem. Phys.*, 1993, **98**, 5648.
50. D. C. Langreth and J. P. Perdew, *Phys. Rev. B*, 1977, **15**, 2884.
51. P. J. Stephens, F. J. Devlin, C. F. Chabalowski and M. J. Frisch, *J. Phys. Chem.*, 1994, **98**, 11623.
52. Y. Zhao and D. G. Truhlar, *Theor. Chem. Account.*, 2008, **120**, 215.
53. S. Grimme, *J. Chem. Phys.*, 2006, **124**, 34108.
54. T. Schwabe and S. Grimme, *Phys. Chem. Chem. Phys.*, 2006, **8**, 4398.
55. S. Grimme, *J. Comput. Chem.*, 2004, **25**, 1463.
56. J. Černý, P. Jurečka, P. Hobza and H. Valdés, *J. Phys. Chem. A*, 2007, **111**, 1146.
57. E. R. Davidson, *J. Comput. Phys.*, 1975, **17**, 87.

CHAPTER 3
The Computation of Molecular Properties

3.1 Molecular Properties as Derivatives of the Potential Energy

Much of chemistry is governed by consideration of relative energies, corresponding to different points on the Born–Oppenheimer potential energy surface. In addition there are many spectroscopic, and other experimentally derived quantities, that require more than just a knowledge of the energy and wavefunction. The number and variety of molecular properties that are of interest to chemists is vast. We cannot hope to cover more than a small selection of key properties here. We can distinguish two types of molecular property: those that depend on a single electronic state and those that involve transitions between electronic states. Typical examples of the former type of molecular property include: geometries of molecules at equilibrium; geometries of transition structures; infrared spectra; multi-pole moments (dipole, quadrupole, octapole,...); and solvation energies. Properties that depend on transitions between electronic states include: absorption and emission spectra; fluorescence and phosphorescence lifetimes.

We shall begin with a discussion of properties for which we can confine our attention to a single electronic state. When a molecule is exposed to a perturbation, for example a geometrical distortion or if the molecule is placed in an electric field, the energy is altered by the perturbation. Denoting a general perturbation by λ, the energy can be expanded in a Taylor series around $\lambda = 0$ as

RSC Theoretical and Computational Chemistry Series No. 5
Computational Quantum Chemistry: Molecular Structure and Properties *in Silico*
By Joseph J W McDouall
© Joseph J W McDouall 2013
Published by the Royal Society of Chemistry, www.rsc.org

$$E(\lambda) = E(0) + \frac{dE(0)}{d\lambda}\lambda + \frac{1}{2}\frac{d^2E(0)}{d\lambda^2}\lambda^2 + \frac{1}{6}\frac{d^3E(0)}{d\lambda^3}\lambda^3 + \dots \qquad (3.1)$$

The derivatives describe the response of the energy to the perturbation and can be identified with molecular properties. For example, if $\lambda = X$, where X is a geometric variable then we can make the following associations:

$$-\frac{dE(0)}{dX}$$ the force acting along coordinate X;

$$\frac{d^2E(0)}{dX^2}$$ the harmonic force constant with respect to coordinate X;

$$\frac{d^3E(0)}{dX^3}$$ the cubic force constant with respect to coordinate X.

If $\lambda = \varepsilon = \{\varepsilon_x, \varepsilon_y, \varepsilon_z\}$ where ε is an electric field with components along the cartesian axes, then

$$\frac{dE(0)}{d\varepsilon_y}$$ the component of the electric dipole moment along the y-coordinate;

$$\frac{d^2E(0)}{d\varepsilon_x d\varepsilon_y}$$ the xy-component of the polarisability;

$$\frac{d^3E(0)}{d\varepsilon_x d\varepsilon_y d\varepsilon_z}$$ the xyz-component of the first hyperpolarisability.

We can also consider a 'mixed' derivative such as

$$\frac{d^2E(0)}{dX\, d\varepsilon_x}$$

which is the *dipole derivative* that determines the intensity of infrared bands in the harmonic approximation. Other chemically important molecular properties and the corresponding energy derivatives include the coupling of geometric variables with external fields (electric and magnetic) and internal magnetic moments (nuclear spins). For example denoting, the cartesian components as $i, j = x, y, z$, a magnetic field as $\mathbf{B}$, and the electronic and nuclear spin moments as $\mathbf{S}$ and $\mathbf{I}^K$, respectively, we have

$$\frac{d^2E(0)}{dB_i dB_j}$$ the i, j component of the magnetisability;

$$\frac{\mathrm{d}^2 E(0)}{\mathrm{d}S_i \mathrm{d}B_j}$$ the i, j component of the electronic **g** matrix;

$$\frac{\mathrm{d}^2 E(0)}{\mathrm{d}I_i^K \mathrm{d}B_j}$$ the i, j component of the NMR shielding tensor for nucleus K.

There are, or course, a great many other molecular properties that we shall not consider.

We saw in Section 2.8, that perturbations could also be incorporated by modifying the hamiltonian. This is closely related to the discussion above and can be seen by writing a hamiltonian that depends on λ

$$\hat{H}(\lambda) = \hat{H}^{(0)} + \lambda\hat{H}^{(1)} + \lambda^2\hat{H}^{(2)} + \ldots \tag{3.2}$$

where $\hat{H}^{(0)} = \hat{H}(0)$ is the unperturbed hamiltonian. Comparing with a Taylor expansion

$$\hat{H}(\lambda) = \hat{H}(0) + \frac{\mathrm{d}\hat{H}(0)}{\mathrm{d}\lambda}\lambda + \frac{1}{2}\frac{\mathrm{d}^2\hat{H}(0)}{\mathrm{d}\lambda^2}\lambda^2 + \ldots \tag{3.3}$$

allows us to identify

$$\hat{H}^{(n)} = \frac{1}{n!}\frac{\mathrm{d}^n\hat{H}(0)}{\mathrm{d}\lambda^n} \tag{3.4}$$

The corresponding energies can also be treated similarly

$$E(\lambda) = E(0) + \frac{\mathrm{d}E(0)}{\mathrm{d}\lambda}\lambda + \frac{1}{2}\frac{\mathrm{d}^2 E(0)}{\mathrm{d}\lambda^2}\lambda^2 + \ldots$$
$$= E^{(0)} + \lambda E^{(1)} + \lambda^2 E^{(2)} + \ldots \tag{3.5}$$

We can obtain expressions for the energy at various orders using the apparatus of perturbation theory that we encountered in Section 2.8. The Hellmann–Feynman theorem states that the derivative of the total energy with respect to a perturbation is equivalent to the expectation value of the derivative of the hamiltonian with respect to that same perturbation, that is

$$\frac{\mathrm{d}E(0)}{\mathrm{d}\lambda} = \langle\Psi|\frac{\partial H}{\partial\lambda}|\Psi\rangle \tag{3.6}$$

The perturbation expressions are strictly valid only for exact wavefunctions, but as we shall see in Section 3.2, approximate variational wavefunctions will obey a generalised form of the Hellmann–Feynman theorem. We shall continue our treatment using derivative techniques.

Our concern is the evaluation of energy derivatives and it may be imagined that this can be done simply by numerical differentiation of the energy. In principle this is true. However numerical differentiation schemes are prone to numerical errors and require a great number of evaluations of the energy, since the outcome of each perturbation must be evaluated individually. In the case of geometrical derivatives, there are $3N_{atoms}$ nuclear coordinates and the application of a central difference, Δ, for each coordinate

$$\frac{\mathrm{d}E(0)}{\mathrm{d}x} = \frac{E(\Delta) - E(-\Delta)}{2\Delta} \qquad (3.7)$$

would require $6N_{atoms}$ energy evaluations. This makes numerical differentiation a very inefficient strategy for derivatives with respect to nuclear coordinates.

3.2   Analytic Differentiation of the Energy Expression

The energy obtained from the various approximations discussed in Chapter 2 can be said to depend on parameters that are external to the electronic system, such as the molecular geometry or the presence of an electric field. We shall denote these collectively as $\mathbf{X}$. The energy also depends on the parameters that determine the wavefunction. These may be molecular orbital coefficients or the configuration mixing coefficients of the CI method. We shall denote all such wavefunction parameters as $\mathbf{C}$. To proceed we note that the molecular energy depends on both $\mathbf{X}$ and $\mathbf{C}$. Additionally $\mathbf{C} = \mathbf{C}(\mathbf{X})$, that is $\mathbf{C}$ depends implicitly on $\mathbf{X}$, since the wavefunction parameters will be specific to $\mathbf{X}$. For example, the molecular orbital coefficients ($\mathbf{C}$) will differ according to geometry, $\mathbf{X}$. Hence we write $\mathbf{C}(\mathbf{X})$ and

$$E(\mathbf{X},\mathbf{C}(\mathbf{X})) \qquad (3.8)$$

is the quantity we must deal with. Additionally we can use the equations that determine the wavefunction parameters, writing them as $f(\mathbf{X},\mathbf{C}(\mathbf{X})) = 0$. For example, in the Hartree–Fock method

$$f(\mathbf{X},\mathbf{C}(\mathbf{X})) = (\mathbf{F} - \mathbf{S}E)\mathbf{C}^{MO} = \mathbf{0} \qquad (3.9)$$

while in the CI method, based on Hartree–Fock orbitals, we must add to $f(\mathbf{X},\mathbf{C}(\mathbf{X}))$ the conditions

$$f(\mathbf{X},\mathbf{C}(\mathbf{X})) = (\mathbf{H} - \mathbf{I}E)\mathbf{C}^{CI} = 0 \qquad (3.10)$$

Now let us consider the derivative of eqn (3.8). The total differential is

$$dE(\mathbf{X},\mathbf{C}(\mathbf{X})) = \frac{\partial E(\mathbf{X},\mathbf{C}(\mathbf{X}))}{\partial \mathbf{X}} d\mathbf{X} + \frac{\partial E(\mathbf{X},\mathbf{C}(\mathbf{X}))}{\partial \mathbf{C}(\mathbf{X})} d\mathbf{C}(\mathbf{X}) \qquad (3.11)$$

and so

$$\frac{dE(\mathbf{X},\mathbf{C}(\mathbf{X}))}{d\mathbf{X}} = \frac{\partial E(\mathbf{X},\mathbf{C}(\mathbf{X}))}{\partial \mathbf{X}} + \frac{\partial E(\mathbf{X},\mathbf{C}(\mathbf{X}))}{\partial \mathbf{C}(\mathbf{X})} \frac{d\mathbf{C}(\mathbf{X})}{d\mathbf{X}} \qquad (3.12)$$

The first term on the right-hand side corresponds to the explicit dependence of the energy on the external parameters $\mathbf{X}$. This term is relatively straightforward to calculate as it involves the derivatives of the molecular integrals with respect to $\mathbf{X}$. The second term is complicated by the presence of the *response* of the wavefunction parameters to the perturbation, $\dfrac{d\mathbf{C}(\mathbf{X})}{d\mathbf{X}}$. At this stage it is useful to distinguish between wavefunction parameters that are variationally determined and those that are not.

3.2.1   Variational Wavefunctions

For variational wavefunctions, the term, $\dfrac{d\mathbf{C}(\mathbf{X})}{d\mathbf{X}}$, in eqn (3.12) presents no difficulty since

$$\frac{\partial E(\mathbf{X},\mathbf{C}(\mathbf{X}))}{\partial \mathbf{C}(\mathbf{X})} = 0 \qquad (3.13)$$

This eliminates the need to evaluate the second term in eqn (3.12) and

$$\frac{dE(\mathbf{X},\mathbf{C}(\mathbf{X}))}{d\mathbf{X}} = \frac{\partial E(\mathbf{X},\mathbf{C}(\mathbf{X}))}{\partial \mathbf{X}} = \langle \Psi | \frac{\partial H}{\partial X} | \Psi \rangle \qquad (3.14)$$

which is equivalent to the expression obtained from the Hellmann–Feynman theorem, eqn (3.6). Moving on to the case of the second derivative, we can use eqn (3.12) to obtain an expression for the derivative operator as

$$\frac{d}{d\mathbf{X}} = \frac{\partial}{\partial \mathbf{X}} + \frac{\partial}{\partial \mathbf{C}(\mathbf{X})} \frac{d\mathbf{C}(\mathbf{X})}{d\mathbf{X}} \qquad (3.15)$$

Applying this operator to the first derivative of the energy gives

$$\frac{\mathrm{d}^2 E(\mathbf{X},\mathbf{C}(\mathbf{X}))}{\mathrm{d}\mathbf{X}^2} = \frac{\mathrm{d}}{\mathrm{d}\mathbf{X}}\left(\frac{\partial E(\mathbf{X},\mathbf{C}(\mathbf{X}))}{\partial \mathbf{X}}\right)$$

$$= \frac{\partial^2 E(\mathbf{X},\mathbf{C}(\mathbf{X}))}{\partial \mathbf{X}^2} + \frac{\partial^2 E(\mathbf{X},\mathbf{C}(\mathbf{X}))}{\partial \mathbf{X}\partial \mathbf{C}(\mathbf{X})}\frac{\mathrm{d}\mathbf{C}(\mathbf{X})}{\mathrm{d}\mathbf{X}} \quad (3.16)$$

We can no longer avoid the calculation of the wavefunction response, $\dfrac{\mathrm{d}\mathbf{C}(\mathbf{X})}{\mathrm{d}\mathbf{X}}$. To evaluate the response we can make use of the stationary condition, eqn (3.13), which holds for all values of the external perturbation $\mathbf{X}$. Differentiating eqn (3.13) gives

$$\frac{\mathrm{d}}{\mathrm{d}\mathbf{X}}\frac{\partial E(\mathbf{X},\mathbf{C}(\mathbf{X}))}{\partial \mathbf{C}(\mathbf{X})} = \frac{\partial^2 E(\mathbf{X},\mathbf{C}(\mathbf{X}))}{\partial \mathbf{X}\partial \mathbf{C}(\mathbf{X})} + \frac{\partial^2 E(\mathbf{X},\mathbf{C}(\mathbf{X}))}{\partial \mathbf{C}(\mathbf{X})^2}\frac{\mathrm{d}\mathbf{C}(\mathbf{X})}{\mathrm{d}\mathbf{X}} \quad (3.17)$$

$$= 0$$

from which the first-order response equations can be obtained as

$$\frac{\partial^2 E(\mathbf{X},\mathbf{C}(\mathbf{X}))}{\partial \mathbf{C}(\mathbf{X})^2}\frac{\mathrm{d}\mathbf{C}(\mathbf{X})}{\mathrm{d}\mathbf{X}} = -\frac{\partial^2 E(\mathbf{X},\mathbf{C}(\mathbf{X}))}{\partial \mathbf{X}\partial \mathbf{C}(\mathbf{X})} \quad (3.18)$$

Defining the matrices $\mathbf{A}$, $\mathbf{U}^a$ and $\mathbf{b}^a$ with elements

$$A_{ij} = \left(\frac{\partial^2 E(\mathbf{X},\mathbf{C}(\mathbf{X}))}{\partial C_i(\mathbf{X})\partial C_j(\mathbf{X})}\right)$$

$$U_i^a = \left(\frac{\mathrm{d}C_i(\mathbf{X})}{\mathrm{d}X_a}\right) \quad (3.19)$$

$$b_i^a = -\left(\frac{\partial^2 E(\mathbf{X},\mathbf{C}(\mathbf{X}))}{\partial X_a\partial C_i(\mathbf{X})}\right)$$

we can write eqn (3.18) in matrix form and obtain the wavefunction response by inverting the matrix $\mathbf{A}$,

$$\mathbf{A}\mathbf{U}^a = \mathbf{b}^a$$

$$\mathbf{U}^a = \mathbf{A}^{-1}\mathbf{b}^a \quad (3.20)$$

Note that $\dfrac{\partial^2 E(\mathbf{X},\mathbf{C}(\mathbf{X}))}{\partial \mathbf{C}(\mathbf{X})^2}$ does not explicitly depend on the external perturbation $\mathbf{X}$. Hence the single quantity, $\mathbf{A}^{-1}$, suffices to determine $\mathbf{U}^a = \dfrac{\mathrm{d}\mathbf{C}(\mathbf{X})}{\mathrm{d}X_a}$ for all X_a.

This has very significant consequences for the efficient evaluation of derivatives since $\mathbf{A}$ is typically of large dimension and its inverse must be obtained using iterative techniques.

We have now seen that for variational wavefunctions we need the first-order response of the wavefunction in order to evaluate the second-derivative (Hessian) matrix but we do not require it for the first-derivative (gradient) vector. A general rule exists which states that for variational wavefunctions the derivatives of the wavefunction parameters to order n are sufficient to determine the energy to order $2n + 1$. This is similar to Wigner's rule that we encountered in Section 2.8. This was first exploited in the context of derivative calculations by Handy and Schaefer[1] and developed further by Helgaker and Jorgensen.[2-3]

3.2.2   Non-Variational Wavefunctions

The situation is more complicated for non-variationally determined wavefunction parameters. We have met the idea of Lagrange multipliers and their use in constrained variational problems. To treat the derivatives for non-variational wavefunctions we introduce a Lagrange multiplier, μ, and construct a lagrangian function

$$L(\mathbf{X},\mathbf{C}(\mathbf{X}),\mu(\mathbf{X})) = E(\mathbf{X},\mathbf{C}(\mathbf{X})) + \mu f(\mathbf{X},\mathbf{C}(\mathbf{X})) \tag{3.21}$$

Provided that the constraint equations are met,

$$f(\mathbf{X},\mathbf{C}(\mathbf{X})) = 0 \tag{3.22}$$

then

$$L(\mathbf{X},\mathbf{C}(\mathbf{X}),\mu(\mathbf{X})) = E(\mathbf{X},\mathbf{C}(\mathbf{X})) \tag{3.23}$$

We add the conditions

$$\frac{\mathrm{d}L(\mathbf{X},\mathbf{C}(\mathbf{X}),\mu(\mathbf{X}))}{\mathrm{d}\mu(\mathbf{X})} = 0 \tag{3.24}$$

$$\frac{\mathrm{d}L(\mathbf{X},\mathbf{C}(\mathbf{X}),\mu(\mathbf{X}))}{\mathrm{d}\mathbf{C}(\mathbf{X})} = 0 \tag{3.25}$$

The first of these conditions is equivalent to enforcing eqn (3.22) and the second provides a means for evaluating the Lagrange multipliers. Since eqn (3.23) establishes the equivalence of $E(\mathbf{X},\mathbf{C}(\mathbf{X}))$ and $L(\mathbf{X},\mathbf{C}(\mathbf{X}),\mu(\mathbf{X}))$, the derivative of $L(\mathbf{X},\mathbf{C}(\mathbf{X}),\mu(\mathbf{X}))$ is also the required derivative of $E(\mathbf{X},\mathbf{C}(\mathbf{X}))$,

$$\frac{\mathrm{d}E(\mathbf{X},\mathbf{C}(\mathbf{X}))}{\mathrm{d}\mathbf{X}} = \frac{\mathrm{d}L(\mathbf{X},\mathbf{C}(\mathbf{X}),\mu(\mathbf{X}))}{\mathrm{d}\mathbf{X}}$$

$$= \frac{\partial E(\mathbf{X},\mathbf{C}(\mathbf{X}))}{\partial \mathbf{X}} + \mu(\mathbf{X})\frac{\partial f(\mathbf{X},\mathbf{C}(\mathbf{X}))}{\partial \mathbf{X}} + \frac{\partial \mu(\mathbf{X})}{\partial \mathbf{X}}f(\mathbf{X},\mathbf{C}(\mathbf{X})) \quad (3.26)$$

$$= \frac{\partial E(\mathbf{X},\mathbf{C}(\mathbf{X}))}{\partial \mathbf{X}} + \mu(\mathbf{X})\frac{\partial f(\mathbf{X},\mathbf{C}(\mathbf{X}))}{\partial \mathbf{X}}$$

The last line is obtained by using eqn (3.22). Equivalently, we could have directly invoked the $2n + 1$ rule for the lagrangian, which is variational, and concluded the same. Eqn (3.25) imposes a condition on the lagrangian from which we can obtain the multipliers

$$\mu(\mathbf{X})\frac{\partial f(\mathbf{X},\mathbf{C}(\mathbf{X}))}{\partial \mathbf{C}(\mathbf{X})} = -\frac{\partial E(\mathbf{X},\mathbf{C}(\mathbf{X}))}{\partial \mathbf{C}(\mathbf{X})} \quad (3.27)$$

The Lagrange multipliers are part of the wavefunction parameters since they depend implicitly on the external perturbation, $\mathbf{X}$. However, in addition to the $2n + 1$ rule, it has been established[3] that for non-variational wavefunctions, provided the energy is expressed as a fully variational lagrangian, the derivatives of the Lagrange multipliers to order n, $\dfrac{\partial^n \mu(\mathbf{X})}{\partial \mathbf{X}^n}$, are sufficient to determine the energy to order $2n + 2$. Hence the quantity $\dfrac{\partial \mu(\mathbf{X})}{\partial \mathbf{X}}$ need not be considered until we require the third derivative of the energy.

3.3 First Derivative with Respect to Geometric Coordinates: Variational Case

We begin by applying the findings of Section 3.2.1 to the case of the first derivatives of the Hartree–Fock energy with respect to geometric coordinates. The Hartree–Fock energy is variational and the energy is stationary with respect to the molecular orbital coefficients, subject to the constraint of orthonormality. In terms of integrals over m basis functions, the Hartree–Fock energy can be written as

$$E_{\mathrm{HF}} = \sum_{\mu\nu}^{m} P_{\mu\nu}h_{\mu\nu} + \frac{1}{2}\sum_{\mu\nu\lambda\sigma}^{m} P_{\mu\nu}P_{\lambda\sigma}[(\mu\nu|\lambda\sigma) - (\mu\sigma|\lambda\nu)] + V_{AB} \quad (3.28)$$

In this expression the density matrix $\mathbf{P}$ is defined over spin–orbitals (in contrast to eqn (2.130)) as

$$P_{\mu\nu} = \sum_{i}^{N} c_{\mu i}c_{\nu i} \quad (3.29)$$

The orthonormality condition is

$$\sum_{\mu\nu}^{m} c_{\mu p} S_{\mu\nu} c_{\nu q} = \delta_{pq} \tag{3.30}$$

where p, q label arbitrary molecular orbitals. The inclusion of this constraint leads to the lagrangian function

$$L = \sum_{\mu\nu}^{m} P_{\mu\nu} h_{\mu\nu} + \frac{1}{2} \sum_{\mu\nu\lambda\sigma}^{m} P_{\mu\nu} P_{\lambda\sigma} [(\mu\nu|\lambda\sigma) - (\mu\sigma|\lambda\nu)] \\ + V_{AB} - \sum_{pq}^{m} \mu_{pq} \left(\sum_{\mu\nu}^{m} c_{\mu p} S_{\mu\nu} c_{\nu q} - \delta_{pq} \right) \tag{3.31}$$

Note that the Lagrange multipliers, μ_{pq}, are the same as those we met in Section 2.2.1 and they form a symmetric matrix which can be made diagonal. Now applying eqn (3.26) to this lagrangian yields

$$\frac{\mathrm{d}L}{\mathrm{d}X} = \frac{\mathrm{d}E_{\mathrm{HF}}}{\mathrm{d}X} \\ = \sum_{\mu\nu}^{m} P_{\mu\nu} \frac{\partial h_{\mu\nu}}{\partial X} + \frac{1}{2} \sum_{\mu\nu\lambda\sigma}^{m} P_{\mu\nu} P_{\lambda\sigma} \left[\frac{\partial(\mu\nu|\lambda\sigma)}{\partial X} - \frac{\partial(\mu\sigma|\lambda\nu)}{\partial X} \right] + \frac{\partial V_{AB}}{\partial X} - \sum_{\mu\nu}^{m} W_{\mu\nu} \frac{\partial S_{\mu\nu}}{\partial X} \tag{3.32}$$

The matrix $\mathbf{W}$ is called the "energy-weighted density matrix" and its elements are given by

$$W_{\mu\nu} = \sum_{i=1}^{N} c_{\mu i} \, \varepsilon_i \, c_{\nu i} \tag{3.33}$$

To obtain eqn (3.32) we have ignored the wavefunction response of the orthonormality constraint, in accordance with eqn (3.26). The only term to be differentiated is the overlap integral. The orthonormality constraint ensures that only the diagonal terms of the Lagrange multiplier matrix survive and so the last term in eqn (3.32) is obtained.

The explicit dependence on the geometric coordinate, X, arises from the basis functions which are centred on the nuclei and so must move with the nuclei. Additionally the nuclear repulsion energy, V_{AB}, has an explicit dependence on X. The derivative of V_{AB} with respect to, say X_B, is easily evaluated since

$$V_{AB} = \sum_{A<B}^{M} \frac{Z_A Z_B}{R_{AB}} = Z_A Z_B \overline{R}_{AB}^{-1/2}$$

$$R_{AB} = \sqrt{(X_B - X_A)^2 + (Y_B - Y_A)^2 + (Z_B - Z_A)^2}$$

$$\overline{R}_{AB} = R_{AB}^2$$

$$\frac{\partial V_{AB}}{\partial X_B} = \frac{\partial V_{AB}}{\partial \overline{R}_{AB}} \frac{\partial \overline{R}_{AB}}{\partial X_B} \tag{3.34}$$

$$\frac{\partial V_{AB}}{\partial \overline{R}_{AB}} = -\frac{1}{2} \sum_{A<B}^{M} Z_A Z_B \overline{R}_{AB}^{-3/2}$$

$$\frac{\partial \overline{R}_{AB}}{\partial X_B} = 2(X_B - X_A)$$

so

$$\frac{\partial V_{AB}}{\partial X_B} = -\sum_{A(\neq B)} \frac{Z_A Z_B (X_B - X_A)}{R_{AB}^3} \tag{3.35}$$

To evaluate the derivatives of the one- and two-electron integrals, the basis functions must be differentiated and the integral formed. For a normalised primitive gaussian centred on atom B

$$\chi(\alpha,i,j,k,\mathbf{r}-\mathbf{R}_B) = N(X_B-x)^i(Y_B-y)^j(Z_B-z)^k e^{-\alpha[(X_B-x)^2+(Y_B-y)^2+(Z_B-z)^2]} \tag{3.36}$$

differentiation with respect to X_B gives

$$\frac{\partial \chi(\alpha,i,j,k,\mathbf{r}-\mathbf{R}_B)}{\partial X_B} = N(X_B-x)^{i-1}(Y_B-y)^j(Z_B-z)^k e^{-\alpha[(X_B-x)^2+(Y_B-y)^2+(Z_B-z)^2]}$$
$$-2N\alpha(X_B-x)^{i+1}(Y_B-y)^j(Z_B-z)^k e^{-\alpha[(X_B-x)^2+(Y_B-y)^2+(Z_B-z)^2]} \tag{3.37}$$

So the derivative of a gaussian basis function is a linear combination of two basis functions with the same exponent but with the angular momentum decreased in one function and increased in the other. For example, the derivative integral of a p-type basis function will be a linear combination of an s-type integral and a d-type integral. Efficient procedures for the evaluation of derivative integrals, and property integrals in general, are discussed in ref. 4.

3.4 Second Derivative with Respect to Geometric Coordinates: Variational Case

Differentiation of the first-derivative expression for the Hartree–Fock method, eqn (3.32), with respect to a second geometric coordinate, X', gives

$$\frac{\mathrm{d}^2 E_{\mathrm{HF}}}{\mathrm{d}X'\mathrm{d}X} = \sum_{\mu\nu} P_{\mu\nu} \frac{\partial^2 h_{\mu\nu}}{\partial X'\partial X} + \frac{1}{2} \sum_{\mu\nu\lambda\sigma} P_{\mu\nu} P_{\lambda\sigma} \frac{\partial^2}{\partial X'\partial X} [(\mu\nu|\lambda\sigma) - (\mu\sigma|\lambda\nu)] + \frac{\partial^2 V_{AB}}{\partial X'\partial X}$$

$$- \sum_{\mu\nu} W_{\mu\nu} \frac{\partial^2 S_{\mu\nu}}{\partial X'\partial X} + \sum_{\mu\nu} \frac{\partial P_{\mu\nu}}{\partial X'} \frac{\partial h_{\mu\nu}}{\partial X} + \sum_{\mu\nu\lambda\sigma} \frac{\partial P_{\mu\nu}}{\partial X'} P_{\lambda\sigma} \frac{\partial}{\partial X} [(\mu\nu|\lambda\sigma) \quad (3.38)$$

$$- (\mu\sigma|\lambda\nu)] - \sum_{\mu\nu} \frac{\partial W_{\mu\nu}}{\partial X'} \frac{\partial S_{\mu\nu}}{\partial X}$$

in which all indices are summed over the m basis functions. We now need to evaluate expressions involving derivatives of the molecular orbital coefficients, $\frac{\partial P_{\mu\nu}}{\partial X}$ and $\frac{\partial W_{\mu\nu}}{\partial X}$. This is achieved through the coupled-perturbed Hartree–Fock equations. Recalling the orthonormality condition, eqn (3.30), and the Roothaan–Hall equations

$$\sum_{\nu}^{m} \left(F_{\mu\nu} - \varepsilon_p S_{\mu\nu} \right) c_{\nu p} = 0 \tag{3.39}$$

we differentiate both equations with respect to the external perturbation X. The derivative of eqn (3.30) gives

$$\sum_{\mu\nu}^{m} \left[\frac{\partial c_{\mu p}}{\partial X} S_{\mu\nu} c_{\nu q} + c_{\mu p} \frac{\partial S_{\mu\nu}}{\partial X} c_{\nu q} + c_{\mu p} S_{\mu\nu} \frac{\partial c_{\nu q}}{\partial X} \right] = 0 \tag{3.40}$$

and that of eqn (3.39)

$$\sum_{\nu}^{m} \left(\frac{\partial F_{\mu\nu}}{\partial X} - \frac{\partial \varepsilon_q}{\partial X} S_{\mu\nu} - \varepsilon_q \frac{\partial S_{\mu\nu}}{\partial X} \right) c_{\nu q} + \sum_{\nu}^{m} \left(F_{\mu\nu} - \varepsilon_q S_{\mu\nu} \right) \frac{\partial c_{\nu q}}{\partial X} = 0 \tag{3.41}$$

It is possible to write the molecular orbital coefficient derivatives as[5]

$$\frac{\partial c_{\mu p}}{\partial X} = \sum_{r}^{\mathrm{MOs}} c_{\mu r} U_{rp} \tag{3.42}$$

Applying this to eqn (3.40) we obtain

$$\sum_{\mu\nu}^{m}\left[\sum_{r}^{\text{MOs}}c_{\mu p}U_{rp}S_{\mu\nu}c_{\nu q}+c_{\mu p}\frac{\partial S_{\mu\nu}}{\partial X}c_{\nu q}+\sum_{r}^{\text{MOs}}c_{\mu p}S_{\mu\nu}c_{\nu r}U_{rq}\right]=0 \qquad (3.43)$$

Since $\sum_{\mu\nu}^{m}c_{\mu p}S_{\mu\nu}c_{\nu q}=\delta_{pq}$, performing the summations over the basis functions, eqn (3.43) becomes

$$U_{qp}+S_{pq}^{(X)}+U_{pq}=0 \qquad (3.44)$$

where

$$S_{pq}^{(X)}=\sum_{\mu\nu}^{m}c_{\mu p}\frac{\partial S_{\mu\nu}}{\partial X}c_{\nu q} \qquad (3.45)$$

Applying the same substitutions to eqn (3.41) and pre-multiplying by $c_{\mu p}$ gives

$$\sum_{\mu\nu}^{m}c_{\mu p}\left(\frac{\partial F_{\mu\nu}}{\partial X}-\frac{\partial \varepsilon_q}{\partial X}S_{\mu\nu}-\varepsilon_q\frac{\partial S_{\mu\nu}}{\partial X}\right)c_{\nu q}+\sum_{\mu\nu}^{m}\sum_{r}^{\text{MOs}}c_{\mu p}\left(F_{\mu\nu}-\varepsilon_q S_{\mu\nu}\right)c_{\nu r}U_{rq}=0 \qquad (3.46)$$

Using superscript (X) to denote derivative integrals transformed by the reference molecular orbital coefficients we obtain

$$F_{pq}^{(X)}-\delta_{pq}\frac{\partial \varepsilon_q}{\partial X}-\varepsilon_q S_{pq}^{(X)}+\left(\varepsilon_p-\varepsilon_q\right)U_{pq}=0 \qquad (3.47)$$

The term $\mathbf{F}^{(X)}$ involves the expression for the Fock matrix evaluated with derivative integrals transformed by the molecular orbital coefficients. In analogy with eqn (3.45), the one-electron term, $\mathbf{h}^{(X)}$, is simply

$$h_{pq}^{(X)}=\sum_{\mu\nu}^{m}c_{\mu p}\frac{\partial h_{\mu\nu}}{\partial X}c_{\nu q} \qquad (3.48)$$

The two-electron terms depend on the molecular orbital coefficients through the density matrix

$$\begin{aligned}\frac{\partial P_{\mu\nu}}{\partial X}&=\sum_{i}^{\text{occupied}}\left[\frac{\partial c_{\mu i}}{\partial X}c_{\nu i}+c_{\mu i}\frac{\partial c_{\nu i}}{\partial X}\right]\\[6pt]&=\sum_{i}^{\text{occupied}}\sum_{r}^{\text{MOs}}\left[c_{\mu r}U_{ri}c_{\nu i}+c_{\mu i}c_{\nu r}U_{ri}\right]\end{aligned} \qquad (3.49)$$

The two-electron terms are

$$G_{pq}^{(X)} = \sum_i^{\text{occupied}} \sum_r^{\text{MOs}} U_{ri}[(pq|ri) - (pi|rq) + (pq|ir) - (pr|iq)]$$

$$+ \sum_{\mu\nu\lambda\sigma}^{m} c_{\mu p} c_{\nu q} P_{\lambda\sigma} \frac{\partial}{\partial X} \{(\mu\nu|\lambda\sigma) - (\mu\sigma|\lambda\nu)$$

(3.50)

and with these quantities we have

$$F_{pq}^{(X)} = h_{pq}^{(X)} + G_{pq}^{(X)}$$

(3.51)

Using the diagonal elements of eqn (3.47) we can obtain

$$\frac{\partial \varepsilon_p}{\partial X} = F_{pp}^{(X)} - \varepsilon_p S_{pp}^{(X)}$$

(3.52)

and the off-diagonal elements give

$$U_{pq} = \frac{F_{pq}^{(X)} - \varepsilon_q S_{pq}^{(X)}}{\left(\varepsilon_q - \varepsilon_p\right)}$$

(3.53)

We need only solve for U_{pq} where p, q refer to occupied-virtual components. The occupied part can be obtained from the differentiated orthonormality condition, eqn (3.44) as

$$U_{ji} + U_{ij} = -S_{ij}^{(X)}$$

(3.54)

For the occupied-virtual block, after some manipulation of the equations, the following expression may be obtained

$$(\varepsilon_i - \varepsilon_a) U_{ai} - \sum_j^{\text{occupied}} \sum_b^{\text{unoccupied}} A_{ai,bj} U_{bj} = b_{ai}^X$$

(3.55)

where

$$A_{ai,bj} = 2(ai|bj) - (aj|bi) - (ab|ji)$$

$$b_{ai}^X = h_{ai}^{(X)} - S_{ai}^{(X)} \varepsilon_i - \sum_{kl}^{\text{occupied}} S_{ai}^{(X)}[(ai|kl) - (al|ki)]$$

(3.56)

$$+ \sum_{\mu\nu\lambda\sigma}^{m} c_{\mu a} c_{\nu i} P_{\lambda\sigma} \frac{\partial}{\partial X} \{(\mu\nu|\lambda\sigma) - (\mu\sigma|\lambda\nu)\}$$

These equations can be solved by iterative techniques to yield U_{ai}.

Our discussion of the treatment of first and second derivatives of the energy with respect to geometric coordinates has illustrated some of the concepts and techniques involved. Computational implementation of these methods requires a much more in-depth discussion of the details and for this the reader should consult the appropriate research literature, for example refs. 3 and 6–8. These matters are advanced and complex in their theoretical foundations and also in their computational implementation. The material in this section should provide a starting background for tackling the advanced literature on this subject.

3.5 Application of Energy Derivatives with Respect to Geometric Coordinates: Geometry Optimisation

In the preceding sections we have discussed some of the principles involved in the evaluation of molecular gradients and Hessians. In Section 1.3 we saw the importance of these quantities in locating and characterising molecular structures. Knowledge of molecular geometry is a key interest in many types of chemical investigation. So given a molecular gradient vector and possibly the Hessian matrix, how can we use these quantities to locate chemical equilibrium structures? A first consideration is the type of coordinate system in which we express the gradient vector and Hessian matrix. Molecular integrals and their derivatives are evaluated over cartesian coordinates. Consequently the gradient vector is initially evaluated in cartesian coordinates. We can proceed to locate stationary points on the surface in cartesian coordinates but there may be advantages in using other types of coordinate. In quantum chemical calculations the cost of energy and gradient evaluations can be very high in terms of computational resources. So an optimal choice of coordinate system, such that the number of energy and gradient evaluations is minimised, is an important consideration. A familiar alternative to cartesian coordinates are the internal coordinates (bond distances, valence angles and dihedral angles). A well-chosen set of internal coordinates is often very successful for non-cyclic molecules. For cyclic molecules there is some evidence that cartesian coordinates are to be preferred,[9] but it has been argued that a well-chosen set of internal coordinates is probably preferable.[10] It is also possible to use a redundant set of internal coordinates[11] such that there are more than $3N_{atoms} - 6$ variables, and these have become widely employed with great success. In what follows we shall not refer to any particular coordinate system, but note that cartesian and internal coordinates may be inter-converted using the Wilson **B** matrix. Denoting internal coordinates by q and cartesian coordinates by X,

$$B_{ij} = \frac{dq_j}{dX_i} \qquad \text{and} \qquad \mathbf{q} = \mathbf{BX} \qquad (3.57)$$

B is a rectangular matrix, since the j index in eqn (3.57) ranges over the internal coordinate set and index i over the $3N_{atoms}$ cartesian coordinates. The inversion of **B** requires the use of a generalised inverse and amounts to an iterative transformation. The details of this transformation, and that required for gradient vectors and Hessian matrices, are discussed in ref. 12. In addition to the choice of coordinate system, the rate of convergence of any method will depend on well-chosen starting geometries and, for reasons that will become clear, good initial Hessian matrices.

The geometry optimisation methods we shall briefly describe are based on a local quadratic representation of the potential energy surface. Denoting the coordinates at the current geometry by $\mathbf{X_0}$, we can expand the energy in a Taylor series to second-order as

$$E(\mathbf{X_0}+\mathbf{\Delta}) = E(\mathbf{X_0}+\mathbf{\Delta}) = \mathbf{g}^\mathrm{T}\cdot\mathbf{\Delta} + \frac{1}{2}\mathbf{\Delta}^\mathrm{T}\mathbf{H}\mathbf{\Delta} \tag{3.58}$$

where the gradient vector, $\mathbf{g}$, and the Hessian matrix, $\mathbf{H}$, are evaluated at $\mathbf{X_0}$

$$\begin{aligned}
g_i &= \frac{\mathrm{d}E(\mathbf{X_0})}{\mathrm{d}X_i} \\
H_{ij} &= \frac{\mathrm{d}^2 E(\mathbf{X_0})}{\mathrm{d}X_i \mathrm{d}X_j}
\end{aligned} \tag{3.59}$$

Requiring the energy to be stationary with respect to the displacements $\mathbf{\Delta}$,

$$\Delta_i = (X_i - X_{0i}) \tag{3.60}$$

gives the Newton–Raphson equation

$$\mathbf{g} + \mathbf{H}\mathbf{\Delta} = 0 \tag{3.61}$$

from which the optimal displacement, assuming a quadratic surface, may be obtained

$$\mathbf{\Delta} = -\mathbf{H}^{-1}\mathbf{g} \tag{3.62}$$

The straightforward application of this equation poses a number of problems. First of all, the evaluation of $\mathbf{H}$ may be sufficiently demanding that it is not available! If $\mathbf{H}$ is available but the starting geometry, $\mathbf{X_0}$, is far from the stationary point then the quadratic representation of the energy surface will not be valid and simple application of eqn (3.62) may produce wild and oscillatory step vectors that can drive the optimisation process away from the stationary point. We can address this by limiting the size of the step taken with a step control parameter, s. The displacement is evaluated as

$$\mathbf{\Delta} = -s\mathbf{H}^{-1}\mathbf{g} \tag{3.63}$$

and s is chosen such that a minimum in the energy is obtained in the search direction $-\mathbf{H}^{-1}\mathbf{g}$. This is known as a "line search" and it can be carried out accurately, often requiring several energy evaluations, or it can be done approximately so as to ensure that the energy is reduced relative to the previous point, but is not necessarily a minimum in the search direction. An approximate line search is often adequate and leads to a stable minimisation process.

We have assumed the availability of the Hessian matrix $\mathbf{H}$. If $\mathbf{H}$ is expensive to calculate there is little incentive to evaluate it at a geometry, $\mathbf{X_0}$, which may be very far from a stationary point. Consequently it is often preferable to approximate $\mathbf{H}$ and allow it to evolve as information is gained from the energy and gradient vector at several points on the potential energy surface. An approximate $\mathbf{H}$ can be obtained from calculations using a lower level computational method or by using empirical force fields[13] or empirical rules.[14] Given an initial approximate Hessian $\bar{\mathbf{H}}$, or its inverse $\bar{\mathbf{H}}^{-1}$, an update can be defined by requiring that the quadratic representation of the surface reproduces the gradient at the current and any previous points, namely that

$$\boldsymbol{\Delta}^{(i)} = \bar{\mathbf{H}}^{-1}\left(\mathbf{g}^{(n)} - \mathbf{g}^{(n-1)}\right) \tag{3.64}$$

where the superscript (n) denotes the points on the potential energy surface that have been traversed prior to the current point, (i), so that $(n) < (i)$. This is known as the "quasi-Newton condition". There are a variety of update methods and we shall only mention the one due to Broyden, Fletcher, Goldfarb and Shanno (BFGS).[15–18] The update can be applied directly to the inverse Hessian, $\bar{\mathbf{H}}^{-1}$, and takes the form

$$\bar{\mathbf{H}}^{-1(i)} = \bar{\mathbf{H}}^{-1(i-1)} + \frac{\boldsymbol{\Delta}^{(i)}\boldsymbol{\Delta}^{\mathrm{T}(i)}}{\boldsymbol{\Delta}^{\mathrm{T}(i)}\boldsymbol{\Delta}^{(i)}} - \frac{\bar{\mathbf{H}}^{-1(i-1)}\boldsymbol{\Delta}\mathbf{g}^{(i)}\boldsymbol{\Delta}\mathbf{g}^{\mathrm{T}(i)}\bar{\mathbf{H}}^{-1(i-1)}}{\boldsymbol{\Delta}\mathbf{g}^{\mathrm{T}(i)}\bar{\mathbf{H}}^{-1(i-1)}\boldsymbol{\Delta}\mathbf{g}^{(i)}} +$$

$$\left(\boldsymbol{\Delta}\mathbf{g}^{\mathrm{T}(i)}\bar{\mathbf{H}}^{-1(i-1)}\boldsymbol{\Delta}\mathbf{g}^{(i)}\right)\mathbf{w}\mathbf{w}^{\mathrm{T}} \tag{3.65}$$

$$\boldsymbol{\Delta}\mathbf{g}^{(i)} = \mathbf{g}^{(i)} - \mathbf{g}^{(i-1)}$$

$$\mathbf{w} = \frac{\boldsymbol{\Delta}^{(i)}}{\boldsymbol{\Delta}^{\mathrm{T}(i)}\boldsymbol{\Delta}\mathbf{g}^{(i)}} - \frac{\bar{\mathbf{H}}^{-1(i-1)}\boldsymbol{\Delta}\mathbf{g}^{(i)}}{\boldsymbol{\Delta}\mathbf{g}^{\mathrm{T}(i)}\bar{\mathbf{H}}^{-1(i-1)}\boldsymbol{\Delta}\mathbf{g}^{(i)}}$$

The BFGS update scheme is particularly successful in minimisation problems as it ensures that the eigenvalues of $\bar{\mathbf{H}}^{-1}$ remain positive, provided that they are so at the start of the process.

In Section 1.3.2 we also discussed the optimisation of transition structures, which correspond to saddle points on the potential energy surface. Saddle points pose a much bigger optimisation challenge than do minima. For the latter, any step that lowers the energy corresponds to an improvement in the geometry. For saddle points it is necessary to ascend along one unique direction and descend along all other directions. The choice of the direction of ascent defines the reaction path and there are potentially many reaction paths leading to products or conformations that are of no chemical interest. To successfully locate a transition state, using methods similar to the ones we have discussed for minimisation, it is necessary to have a starting geometry that is within the quadratic region of the saddle point and is characterised by a Hessian with one negative eigenvalue. The eigenvector (normal mode) of $\mathbf{H}$ corresponding to this negative eigenvalue describes the reaction coordinate at that point. Choosing a geometry that is close to a saddle point with a Hessian of the correct signature often requires chemical insight, experience and skill. In the absence of these, there exist a variety of semi-automatic procedures for obtaining an approximate saddle point geometry. For example, if the geometries of the reactants and products have been obtained by minimisation then an approximate reaction path can be obtained by a linear interpolation between the two minima. The maximum along this approximate path is then an approximation to the transition structure. This is an easy procedure to apply but has the drawback that the Hessian corresponding to the maximum along the interpolated path may have more than a single negative eigenvalue. In such cases chemical knowledge can be used to decide which direction corresponds to the required path. Provided this can be done, the full optimisation of the saddle point structure can be carried out using quasi-Newton techniques. For saddle points, the BFGS update is not usually appropriate. A better choice is the Powell update method.[19] The control of step size and direction are also modified such that the search direction corresponds to ascent along the lowest eigenvector of the Hessian and descent along all other eigenvectors.

Having located a saddle point structure, it is sometimes useful to follow the reaction path. A convenient definition of the reaction reaction path is the path of steepest descent, the direction $-\mathbf{g}$, from a transition structure down to the two minima (reactant and product) that are connected by the transition structure. The reaction path is often further specified as the path of steepest descent in mass-weight cartesian coordinates. At the simplest level the gradient can be followed in very small steps, starting at the transition structure and moving

$$\mathbf{\Delta} = -s\frac{\mathbf{g}}{|\mathbf{g}|} \tag{3.66}$$

where s is a step size. This method can work if s is chosen sufficiently small. Far more robust methods exist and a discussion of these and many other details associated with geometry optimisation and reaction path following can be found in the review by Schlegel.[20]

3.6   Electric and Magnetic Field Perturbations

The interactions of molecules with external electric and magnetic fields provide a wealth of chemical information. These interactions are probed in very familiar experiments. For example, electric dipole moments can be determined from measurements of dielectric constants, while chemical shifts and molecular electronic **g** matrices are investigated in magnetic resonance experiments. The theoretical treatment of such quantities is rather complex and here we shall provide only an overview to make the connection with the derivative theory we have discussed so far in this chapter.

3.6.1   External Electric Fields

The treatment of an external electric field perturbation is quite straightforward. Consider an electric field, ε, with components $(\varepsilon_x, \varepsilon_y, \varepsilon_z)$. Provided the field is a small perturbation, with respect to the total molecular energy, then we can expand around the field free solution $E(0) = E(\varepsilon = 0)$

$$E(\varepsilon) = E(0) + \frac{\mathrm{d}E(0)}{\mathrm{d}\varepsilon}\varepsilon + \frac{1}{2}\frac{\mathrm{d}^2 E(0)}{\mathrm{d}\varepsilon^2}\varepsilon^2 + \ldots \tag{3.67}$$

From a consideration of the physical situation we know that a molecule with a permanent dipole moment, $\boldsymbol{\mu}$, placed in an electric field ε will interact with the electric field through its dipole moment

$$\hat{H}^{(1)} = -\boldsymbol{\mu} \cdot \boldsymbol{\varepsilon} \tag{3.68}$$

Let us assume a variational wavefunction, $|\Psi_0\rangle$, and invoke the Hellmann–Feynman theorem, eqn (3.6),

$$\frac{\mathrm{d}E}{\mathrm{d}\varepsilon} = \langle \Psi_0 | \frac{\partial H^{(1)}}{\partial \varepsilon} | \Psi_0 \rangle = -\boldsymbol{\mu} \tag{3.69}$$

The electric field will induce a dipole, $\boldsymbol{\mu}^*$, which depends on the polarisability, $\boldsymbol{\alpha}$,

$$\boldsymbol{\mu}^* = \boldsymbol{\alpha}\varepsilon \tag{3.70}$$

The change in the energy will be

$$\Delta E = -\int_0^\varepsilon \boldsymbol{\mu}^* \mathrm{d}\varepsilon = -\int_0^\varepsilon \boldsymbol{\alpha} \cdot \boldsymbol{\varepsilon} \, \mathrm{d}\varepsilon = -\frac{1}{2}\boldsymbol{\alpha} \cdot \boldsymbol{\varepsilon}^2 \tag{3.71}$$

Comparing eqns (3.69) and (3.71) with eqn (3.67) allows us to identify

$$-\frac{\mathrm{d}E(0)}{\mathrm{d}\varepsilon} = \mu$$

$$-\frac{\mathrm{d}^2 E(0)}{\mathrm{d}\varepsilon^2} = \alpha \tag{3.72}$$

We can continue to the first hyperpolarisability, β,

$$-\frac{\mathrm{d}^3 E(0)}{\mathrm{d}\varepsilon^3} = \beta \tag{3.73}$$

and so on. We can, therefore, obtain expressions for these quantities, μ, α, β, as derivatives of the molecular energy with respect to the components of an electric field.

3.6.2   External Magnetic Fields and Internal Magnetic Moments

The description of the interaction of a molecule with a magnetic field is somewhat more complicated than that of an electric field. The required forms of the perturbations that enter the hamiltonian can be derived through consideration of the relativistic Dirac equation, see Chapter 5, and approximations to it. We shall not pursue that approach but rather attempt to connect the common observables of magnetic resonance spectroscopies to energy derivatives with respect to magnetic fields and electronic and nuclear spin moments.

Magnetic resonance experiments are generally interpreted with the aid of a phenomenological spin hamiltonian, built of spin-operators and applied fields and a set of numerical parameters that we can associate with experimentally determined molecular constants. Denoting the nuclear spin-operator of nucleus, A, as $\mathbf{I}^A$, the total electronic spin as $\mathbf{S}$, and the external magnetic field as $\mathbf{B}$, the spin hamiltonian can be written as

$$\hat{H}_{\mathrm{spin}} = \mathbf{S}\cdot\mathbf{g}\cdot\mathbf{B} + \sum_A^{\mathrm{nuclei}} \mathbf{S}\cdot\mathbf{A}^A\cdot\mathbf{I}^A + \mathbf{S}\cdot\mathbf{D}\cdot\mathbf{S} + \sum_A^{\mathrm{nuclei}} \mathbf{I}^A\cdot(\mathbf{I}-\sigma_A)\cdot\mathbf{B}$$

$$+ \sum_{AB}^{\mathrm{nuclei}} \mathbf{I}^A\cdot(\mathbf{D}_{AB}+\mathbf{K}_{AB})\cdot\mathbf{I}^B \tag{3.74}$$

The first three terms require a non-zero total spin and so are only of relevance in paramagnetic molecules. Experimentally, this is the realm of electron paramagnetic resonance (EPR) spectroscopy. The first term in eqn (3.74) is the electronic Zeeman term. The matrix $\mathbf{g}$ parameterises the coupling of the total electronic spin with the applied magnetic field. The second term describes the interaction of the electronic spin with the nuclear magnetic moments and is the hyperfine interaction term. When a molecule contains more than one unpaired electron, $S > \frac{1}{2}$, the dipolar interaction between the unpaired electronic spins causes a perturbation of the electronic energy levels, even in the absence of an external magnetic field. This is the zero-field splitting, which is described by the parameters in **D**. The last two terms in eqn (3.74) parameterise the nuclear Zeeman term and the nuclear spin–spin coupling, respectively. σ_A is the magnetic shielding which modifies the local field experienced by the nucleus, A, due to the shielding effect of the electrons. The spin–spin coupling is given by the direct coupling term, $\mathbf{D}_{AB}$, which is a dipolar interaction between nuclear spins. There is also an indirect coupling of nuclear spins mediated through interaction with the electronic distribution, $\mathbf{K}_{AB}$. σ_A, $\mathbf{D}_{AB}$ and $\mathbf{K}_{AB}$ are determined experimentally by nuclear magnetic resonance (NMR) experiments.

Inspection of eqn (3.74) suggests that the experimental observables will correspond to second derivatives of the total energy with respect to an external magnetic field, or the electronic spin moment, or the nuclear spin moments, or combinations thereof. For example, the first term in eqn (3.74) refers to the electronic **g** matrix and its cartesian components $(i,j = x,y,z)$ are

$$g_{ij} = \frac{1}{\mu_B} \left(\frac{\mathrm{d}^2 E}{\mathrm{d} B_i \mathrm{d} S_j} \right)_{\substack{\mathbf{B}=0 \\ \mathbf{S}=0}} \tag{3.75}$$

evaluated at zero field, $\mathbf{B} = 0$, and zero electronic spin, $\mathbf{S} = 0$. Similarly the components of the **A** tensor are

$$A_{ij}^A = \left(\frac{\mathrm{d}^2 E}{\mathrm{d} I_i^A \mathrm{d} S_j} \right)_{\substack{I^A=0 \\ \mathbf{S}=0}} \tag{3.76}$$

where the superscript reminds us that this refers specifically to nucleus A.

To illustrate the evaluation of the second derivatives we shall again concentrate on the variational Hartree–Fock method. For a one-electron perturbation, λ, we can use eqns (3.14) and (3.16), and recall that the Hellman–Feynman theorem is obeyed and

$$\frac{\mathrm{d} E}{\mathrm{d} \lambda} = \frac{\partial E}{\partial \lambda} \tag{3.77}$$

Since λ is a one-electron perturbation

$$\frac{\mathrm{d} E}{\mathrm{d} \lambda} = \left\langle \Psi_0 \left| \frac{\partial h}{\partial \lambda} \right| \Psi_0 \right\rangle \tag{3.78}$$

which in terms of the Hartree–Fock density matrix over basis functions, eqn (3.29), is simply

$$\frac{dE}{d\lambda} = \sum_{\mu\nu}^{m} P_{\mu\nu} \langle \chi_\mu | \frac{\partial h}{\partial \lambda} | \chi_\nu \rangle \tag{3.79}$$

Provided that the basis functions do *not* depend on the perturbation, in contrast to the case of geometrical derivatives, we can obtain the second derivative by direct differentiation of eqn (3.79) with respect to a second perturbation, κ, to obtain

$$\frac{d^2 E}{d\kappa\, d\lambda} = \sum_{\mu\nu}^{m} P_{\mu\nu} \langle \chi_\mu | \frac{\partial^2 h}{\partial\kappa\, \partial\lambda} | \chi_\nu \rangle + \sum_{\mu\nu}^{m} \frac{\partial P_{\mu\nu}}{\partial\kappa} \langle \chi_\mu | \frac{\partial h}{\partial \lambda} | \chi_\nu \rangle \tag{3.80}$$

The integrals over basis functions can be obtained relatively straightforwardly, but we must evaluate the response in the form of the change in the density matrix with respect to the perturbation. Introducing the perturbed molecular orbital coefficients as in Section 3.4,

$$\frac{\partial c_{\mu p}}{\partial\kappa} = \sum_{r}^{\text{MOs}} c_{\mu r} U_{rp} \tag{3.81}$$

we have

$$\frac{\partial P_{\mu\nu}}{\partial\kappa} = \sum_{i}^{N} \left[\frac{\partial c_{\mu i}}{\partial\kappa} c_{\nu i} + c_{\mu i} \frac{\partial c_{\nu i}}{\partial\kappa} \right]$$

$$= \sum_{i}^{N} \sum_{r}^{\text{MOs}} \left[c_{\mu r} U_{ri} c_{\nu i} + c_{\mu i} c_{\nu r} U_{ri} \right] \tag{3.82}$$

Our remaining task is to evaluate perturbed molecular orbital coefficients. We proceed as in the case of geometrical derivatives in setting up the coupled perturbed Hartree–Fock equations. Given the assumption of one-electron perturbations and perturbation-independent basis functions the outcome is simpler than for geometrical perturbations. The CPHF equations

$$\mathbf{AU} = -\mathbf{b} \tag{3.83}$$

which must be solved for $\mathbf{U}$, now have a simple form which depends only on unperturbed quantities. The elements of $\mathbf{A}$ are

$$A_{ia,jb} = \delta_{ij}\delta_{ab}(\varepsilon_a - \varepsilon_i) - [(ij|ab) - (ib|aj)] \tag{3.84}$$

The vector $\mathbf{b}$ includes derivative integrals transformed to the basis of the unperturbed molecular orbitals, for example

$$b_{ia} = \langle \phi_i | \frac{\partial h}{\partial \lambda} | \phi_a \rangle \tag{3.85}$$

The index "*ia*" in eqns (3.84) and (3.85) is a compound index over the occupied (*i*) and virtual (*a*) molecular orbitals. Hence, eqn (3.83) is a linear equation and can be solved by standard methods for large matrices.

We must now consider the form of the hamiltonian in the presence of an external magnetic field and the internal magnetic moments provided by the nuclei. The theory of electromagnetism tell us that a particle of charge q moving with velocity, $\mathbf{v}$, in an electromagnetic field experiences a force, the Lorentz force,

$$\mathbf{F} = q(\boldsymbol{\varepsilon} + \mathbf{v} \times \mathbf{B}) \tag{3.86}$$

in which $\boldsymbol{\varepsilon}$ and $\mathbf{B}$ are the electric and magnetic components of the field. $\boldsymbol{\varepsilon}$ and $\mathbf{B}$ must obey Maxwell's equations. For our purposes we simply state that the introduction of a vector potential, $\mathbf{A}$, allows the magnetic field, $\mathbf{B}$, to be defined as

$$\mathbf{B} = \nabla \times \mathbf{A} = \begin{vmatrix} i & j & k \\ \partial/\partial x & \partial/\partial y & \partial/\partial z \\ A_x & A_y & A_z \end{vmatrix} \tag{3.87}$$

and satisfies the Maxwell equation $\nabla \cdot \mathbf{B} = 0$. Similarly the introduction of a scalar potential, φ, allows the electric field to be determined

$$\boldsymbol{\varepsilon} = -\nabla \varphi - \frac{\partial \mathbf{A}}{\partial t} \tag{3.88}$$

Here we shall concern ourselves with the magnetic field alone, and specifically the case of a uniform, static magnetic field.

The vector potential is not unique. The reasons for this need not divert us at this point. We shall simply state that the form of $\mathbf{A}$ is usually chosen to be

$$\mathbf{A}_O(\mathbf{r}_i) = \frac{1}{2} \mathbf{B} \times \mathbf{r}_{iO} \tag{3.89}$$

in which $\mathbf{r}_{iO} = \mathbf{r}_i - \mathbf{R}_O$ and $\mathbf{R}_O$ is the origin of the coordinate system used. $\mathbf{R}_O$ is known as the "gauge origin". It happens that we can choose $\mathbf{R}_O$ arbitrarily and still satisfy the requirement in eqn (3.87). Consequently, properties which depend on $\mathbf{B}$ will also depend on the choice of $\mathbf{R}_O$. Calculations involving a complete basis set will not show this *gauge dependence*, but we can seldom work at that level for general molecular systems and steps must be taken to minimise or eliminate the gauge dependence. The most reliable approach at this time is the use of gauge-invariant atomic orbitals[21] (GIAO), sometimes called "London orbitals" after Fritz London who first applied the idea to simple Hückel calculations. London orbitals eliminate the gauge dependence by including a complex phase factor, $e^{-i\mathbf{A}\cdot\mathbf{r}}$, which pre-multiplies each basis function. When integrals are evaluated, the expressions only contain the difference in the vector potentials and consequently the gauge origin is eliminated.

The form of the vector potential in eqn (3.89) is known as the "Coulomb gauge" and has the useful property that $\nabla\cdot\mathbf{A}=0$. The magnetic moments provided by the nuclei are treated as magnetic dipoles, since contributions of higher moments, for example the magnetic quadrupole, are very much smaller. The vector potential associated with a nucleus, C, is

$$\mathbf{A}_C(\mathbf{r}_i)=\alpha^2 g_C \mu_N \frac{\mathbf{I}^C \times \mathbf{r}_{iC}}{r_{iC}^3} \tag{3.90}$$

here α is the fine-structure constant, g_C is the g-value of nucleus C and μ_N is the nuclear magneton. Another factor we must consider is the intrinsic spin of the electron. Since the Born–Oppenheimer hamiltonian contains no reference to spin, we introduce it here in the operator

$$\sum_i^N \mathbf{s}_i\cdot\mathbf{B}(\mathbf{r}_i) \tag{3.91}$$

The proper introduction of spin appears in Dirac's relativistic treatment of the electron. However, spin should not be viewed as a purely relativistic property, since it persists when the Dirac theory is considered in the non-relativistic limit.

Returning to the non-relativistic Born–Oppenheimer hamiltonian, the one-electron part can be written in terms of the momentum $\mathbf{p}_i=-i\nabla_i$ as

$$\hat{h}=\sum_i^N \frac{\mathbf{p}_i^2}{2}-\sum_A^{\text{nuclei}}\sum_i^N \frac{Z_A}{r_{iA}} \tag{3.92}$$

In the presence of a magnetic field the kinetic momentum

$$\pi_i=\mathbf{p}_i-q\mathbf{A}(\mathbf{r}_i) \tag{3.93}$$

replaces $\mathbf{p}_i$. Since q refers to the charge of the electron, $-e$, π_i becomes

$$\pi_i=\mathbf{p}_i+\mathbf{A}_O(\mathbf{r}_i)=\mathbf{p}_i+\frac{1}{2}\mathbf{B}(\mathbf{r}_i)\times\mathbf{r}_{iO} \tag{3.94}$$

To this must be added the effect of nuclear magnetic moments, each having the form of eqn (3.90). The total vector potential then becomes

$$\mathbf{A}^{\text{Total}}(\mathbf{r}_i) = \mathbf{A}_O(\mathbf{r}_i) + \sum_C^{\text{nuclei}} \mathbf{A}_C(\mathbf{r}_i) \tag{3.95}$$

and

$$\boldsymbol{\pi}_i = \mathbf{p}_i + \mathbf{A}_O(\mathbf{r}_i) + \sum_C^{\text{nuclei}} \mathbf{A}_C(\mathbf{r}_i) \tag{3.96}$$

Substituting $\boldsymbol{\pi}_i$ and the interaction of the electron spin with the magnetic field into eqn (3.92) yields the required one-electron hamiltonian as

$$\hat{h} = \frac{1}{2}\sum_i^N \boldsymbol{\pi}_i^2 + \sum_i^N \mathbf{s}_i \cdot \mathbf{B}(\mathbf{r}_i) - \sum_A^{\text{nuclei}} \sum_i^N \frac{Z_A}{r_{iA}} \tag{3.97}$$

Since the potential term is unchanged, the effect of the magnetic field is incorporated in the first two terms. Expanding $\boldsymbol{\pi}_i^2$ gives

$$\boldsymbol{\pi}_i^2 = \left[\mathbf{p}_i + \mathbf{A}_O(\mathbf{r}_i) + \sum_C^{\text{nuclei}} \mathbf{A}_C(\mathbf{r}_i) \right] \left[\mathbf{p}_i + \mathbf{A}_O(\mathbf{r}_i) + \sum_C^{\text{nuclei}} \mathbf{A}_C(\mathbf{r}_i) \right] \tag{3.98}$$

$$= \mathbf{p}_i^2 + \ldots\ldots$$

The first term is equivalent to the kinetic energy in the absence of the magnetic interactions. All other terms now depend on the external magnetic field, *via* eqn (3.89), and/or the nuclear moments, *via* eqn (3.90). We now have, in principle, a definition of the hamiltonian from which we can evaluate $\dfrac{\partial h}{\partial \lambda}$ and $\dfrac{\partial^2 h}{\partial \kappa\, \partial \lambda}$, where $\kappa, \lambda = \mathbf{S}, \mathbf{I}^C, \mathbf{B}$. From eqn (3.98) we shall find terms of the form $\mathbf{A}_O(\mathbf{r}_i) \cdot \mathbf{p}_i$ and $\mathbf{A}_C(\mathbf{r}_i) \cdot \mathbf{p}_i$. The term $\mathbf{A}_O(\mathbf{r}_i) \cdot \mathbf{p}_i$ represents the interaction of the magnetic field with the electronic orbital angular momentum

$$\mathbf{A}_O(\mathbf{r}_i) \cdot \mathbf{p}_i = \frac{1}{2}\mathbf{B}(\mathbf{r}_i) \times \mathbf{r}_{iO} \cdot \mathbf{p}_i$$

$$= \frac{1}{2}\mathbf{B}(\mathbf{r}_i) \cdot \boldsymbol{l}_{iO} \tag{3.99}$$

and $\mathbf{A}_C(\mathbf{r}_i) \cdot \mathbf{p}_i$ represents the hyperfine interaction between the electronic and nuclear spins

$$\mathbf{A}_C(\mathbf{r}_i) \cdot \mathbf{p}_i = \alpha^2 g_C \mu_N \mathbf{I}^C \frac{\boldsymbol{l}_{iC}}{r_{iC}^3} \tag{3.100}$$

The orbital angular momentum operator is used in the two preceding equations and is defined as

$$l_{iO} = \mathbf{r}_{iO} \times \mathbf{p}_i$$
$$= -i\,\mathbf{r}_{iO} \times \nabla_i \tag{3.101}$$

For example, consider the $\mathbf{s}\cdot\mathbf{B}$ term in eqn (3.97) where the $\mathbf{B}$ term refers to a nuclear magnetic moment. Taking the curl of the vector potential for nucleus C, using eqn (3.87) gives

$$\mathbf{B} = \nabla \times \mathbf{A}_C(\mathbf{r}_i)$$
$$= \frac{8\pi\alpha^2}{3} g_C \mu_N \mathbf{I}^C \delta(\mathbf{r}_{iC}) + \alpha^2 g_C \mu_N \mathbf{I}^C \frac{\mathbf{r}_{iC}(\mathbf{r}_{iC})^{\mathrm{T}} - r_{iC}^2 \mathbf{I}_3}{r_{iC}^5} \tag{3.102}$$

The first term contains the delta function and is zero everywhere except at the position of the nucleus, C, this is the Fermi contact term of the hyperfine interaction. The second term represents the spin (nucleus)–spin (electron) dipole interaction. So for $S \neq 0$ these terms give rise to the hyperfine splitting observed in EPR experiments. There remains the interaction of the nuclear magnetic moment with the orbital angular momentum of the electrons, this is a spin (nucleus)–orbit (electron) interaction. This arises from the total magnetic moment operator of the electron, which in atomic units is

$$\hat{\mu} = -\frac{1}{2}\left(\hat{L} + g_e \hat{S}\right) \tag{3.103}$$

where the Bohr magneton, is taken as $\mu_B = \frac{1}{2}$au, and g_e is the magnetogyric ratio of the electron, $g_e = 2.0023192$. This spin-orbit component of the hyperfine interaction thus contributes

$$\frac{1}{2}\mathbf{B}(\mathbf{r}_i)\cdot l_{iO} \tag{3.104}$$

to the hamiltonian.

There is also the spin (electron)–orbit (electron) interaction that arises in several terms in eqn (3.74). This electronic spin orbit operator contains one- and two-electron contributions, which in the Breit–Pauli approximation, are given as

$$\hat{H}^{\mathrm{SOC}} = \frac{\alpha^2}{2} \sum_A^{\mathrm{nuclei}} \sum_i^{N} Z_A \frac{l_{iA}\mathbf{s}_i}{r_{iA}^3} - \frac{\alpha^2}{2} \sum_{i \neq j}^{\mathrm{electrons}} \frac{l_{ij}(\mathbf{s}_i + 2\mathbf{s}_{ij})}{r_{ij}^3} \tag{3.105}$$

Approximations to this full operator are motivated by the idea that the two-electron contribution essentially provides a screening effect to the one-electron contribution. Given the complexity of the two-electron term and the large number of integrals that must be dealt with, it is important to have accurate approximations. The simplest scheme is to include only the one-electron terms but now replace the full nuclear charge with a screened effective charge, Z^{eff}, giving

$$\hat{H}^{\text{SOC}} \approx \hat{h}^{\text{SOC}} = \frac{\alpha^2}{2} \sum_{A}^{\text{nuclei}} \sum_{i}^{N} Z_A^{\text{eff}} \frac{l_{iA} \mathbf{s}_i}{r_{iA}^3} \tag{3.106}$$

Z^{eff} has been parameterised for a large part of the periodic table[22–24] and has proven to be useful for lighter elements leading to an efficient and reasonably accurate spin orbit operator. An alternative form of one-electron approximation developed by Hess[25] uses a mean-field form, akin to the way the Fock operator incorporates the two-electron repulsions in an averaged manner. This form does require the evaluation of two-electron spin–orbit integrals but they are summed into the spin–orbit mean-field operator, $\hat{h}^{\text{SOMF}}$, leaving an effective one-electron operator with matrix elements

$$\hat{h}^{\text{SOMF}} = \sum_{i}^{N} \mathbf{z}^{\text{SOMF}}(i)\mathbf{s}(i)$$

$$\langle \chi_\mu | z_x^{\text{SOMF}} | \chi_v \rangle = \langle \chi_\mu | h_x^{\text{SO}} | \chi_v \rangle$$

$$+ \sum_{\lambda\sigma}^{m} \left[\langle \chi_\mu \chi_v | g_x^{\text{SO}} | \chi_\lambda \chi_\sigma \rangle - \frac{3}{2} \langle \chi_\mu \chi_\lambda | g_x^{\text{SO}} | \chi_\sigma \chi_v \rangle \right.$$

$$\left. - \frac{3}{2} \langle \chi_\sigma \chi_v | g_x^{\text{SO}} | \chi_\mu \chi_\lambda \rangle \right] \tag{3.107}$$

In this expression h_x^{SO} and g_x^{SO} correspond to

$$h_x^{\text{SO}} = \frac{\alpha^2}{2} \sum_{A}^{\text{nuclei}} \sum_{i}^{N} Z_A \frac{l_{iA}^x}{r_{iA}^3}$$

$$g_x^{\text{SO}} = -\frac{\alpha^2}{2} \frac{l_{ij}^x}{r_{ij}^3} \tag{3.108}$$

Integrals over these spatial operators are relatively straightforward to evaluate. Some care must be taken over the permutational symmetries of the integrals because of the angular momentum operator being, $l_{iA} = -i\,\mathbf{r}_{iA} \times \nabla_i$. If real basis functions are used, as is the usual situation, then matrix elements will be pure imaginary. The term $i = \sqrt{-1}$ can be factored out, provided that we now treat the resultant matrices as antisymmetric for imaginary operators, for example

$$h_{\nu\mu}^{\text{real}} = h_{\mu\nu}^{\text{real}}$$
$$h_{\nu\mu}^{\text{imaginary}} = -h_{\mu\nu}^{\text{imaginary}} \tag{3.109}$$

Using the ideas we have introduced here it is possible to obtain expressions for all terms in the spin hamiltonian of eqn (3.74). We shall not go through this arduous task. In our discussion of the evaluation of magnetic field dependent properties we have only tried to sketch an overview of the necessary components. Detailed monographs on the subject are available in refs. 26–28.

3.7 Time-Dependent Linear Response Methods: Excited States

The study of excited-state potential energy surfaces provides considerable challenges to computational quantum chemistry. We have already discussed schemes such as the CIS, CASSCF, MRCI and MRMP2 methods, which in principle provide routes to such studies. The CIS method can be formulated in an efficient manner and can be applied to quite large molecular problems. However, the accuracy of the CIS method limits its usefulness. At the other extreme, we have the MRCI methods that are capable of very high accuracy but are sufficiently demanding that they can only be applied to relatively small molecules. Hence, there is a need to consider methods which retain the simplicity of the CIS approach, based on a single reference determinant, but provide more accurate treatments of excited-state energies and properties. We shall mention two such methods here, the time-dependent Hartree–Fock (TDHF) and time-dependent density functional (TDDFT) theories. They are very closely related. In practice the TDHF method is not very widely used since it appears to offer little advantage over the CIS method in terms of accuracy, yet requires additional computational effort. By contrast, the TDDFT method is very widely used for the study of excited-state properties, such as electronic absorption and emission spectra. Both TDHF and TDDFT rely on obtaining the linear response of the ground-state density matrix of a system under the influence of a time-dependent perturbation, such as the frequency-dependent oscillations of an electromagnetic field.

The solutions of the Schrödinger equation we have dealt with so far have come from the time-independent form

$$\hat{H}|\Psi(\mathbf{r},\mathbf{R})\rangle = E|\Psi(\mathbf{r},\mathbf{R})\rangle \tag{3.110}$$

In the time-dependent formulation we replace the energy with the operator $i\frac{\partial}{\partial t}$ and the hamiltonian is modified to contain a one-electron time-dependent perturbation, $\hat{H}(t)$. The wavefunction is now time-dependent as well, giving

$$\hat{H}(\mathbf{r},\mathbf{R},t)|\Psi(\mathbf{r},\mathbf{R},t)\rangle = i\frac{\partial|\Psi(\mathbf{r},\mathbf{R},t)\rangle}{\partial t}$$

$$\hat{H}(\mathbf{r},\mathbf{R},t) = \hat{H}(\mathbf{r},\mathbf{R}) + \hat{H}(t) \tag{3.111}$$

The hamiltonian $\hat{H}(\mathbf{r},\mathbf{R})$ is the familiar Born–Oppenheimer molecular hamiltonian. The wavefunction in the TDHF case is a single Slater determinant of time-dependent spin–orbitals

$$|\Psi(\mathbf{r},\mathbf{R},t)\rangle = |\phi_i(\mathbf{r},\mathbf{R},t)\phi_j(\mathbf{r},\mathbf{R},t)\ldots\ldots\phi_N(\mathbf{r},\mathbf{R},t)\rangle \tag{3.112}$$

We shall not provide the details since there are many paths to the outcome we seek,[29] but simply note that application of time-dependent variation theory leads to the time-dependent Hartree–Fock equation

$$\hat{F}(\mathbf{r},\mathbf{R},t)|\Psi(\mathbf{r},\mathbf{R},t)\rangle = i\frac{\partial|\Psi(\mathbf{r},\mathbf{R},t)\rangle}{\partial t} \tag{3.113}$$

The time-dependent Fock operator includes $\hat{H}(t)$, and the Coulomb and exchange operators depend on time through the spin–orbitals on which they depend. Applying the familiar basis set expansion and imposing various orthonormality constraints leads to the matrix form of the TDHF equations

$$\begin{bmatrix} \mathbf{A} & \mathbf{B} \\ \mathbf{B}^* & \mathbf{A}^* \end{bmatrix} \begin{bmatrix} \mathbf{X} \\ \mathbf{Y} \end{bmatrix} = \omega \begin{bmatrix} \mathbf{I} & \mathbf{0} \\ \mathbf{0} & -\mathbf{I} \end{bmatrix} \begin{bmatrix} \mathbf{X} \\ \mathbf{Y} \end{bmatrix} \tag{3.114}$$

This matrix equation is non-Hermitian. The eigenvalues, ω, represent the excitation energies, $\omega = E - E_0$. The dimension of this equation system is $2(n_{\mathrm{occ}}\, n_{\mathrm{virt}})$, the vector $\mathbf{X}$ represents a set of excitation amplitudes (occupied $\rightarrow$ unoccupied) and $\mathbf{Y}$ a set of de-excitation amplitudes (unoccupied $\rightarrow$ occupied). The matrix elements of the block, $\mathbf{A}$, are

$$A_{ia,jb} = \delta_{ij}\delta_{ab}(\varepsilon_a - \varepsilon_i) + (ia|jb) - (ij|ba) \tag{3.115}$$

where ia is a compound index corresponding to the single substitution, $\phi_i \rightarrow \phi_a$, in which occupied orbitals are denoted i, j, k, l and unoccupied orbitals as a, b, c, d. The $\mathbf{B}$ block has elements

$$B_{ia,jb} = (ia|jb) - (ib|ja) \tag{3.116}$$

The form of the **A** block can be compared to the hamiltonian matrix elements encountered in the CIS method, in Section 2.7.2. In fact omission of the **B** block and the **Y** vector yields the CIS equations (excluding the ground-state term). This is sometimes referred to as the Tamm–Dancoff approximation. For historical reasons, the full system of TDHF equation is also known as the "Random Phase Approximation" (RPA).

To move to the TDDFT involves a great many considerations,[30] such as a time-dependent form of the Kohn–Sham equations, but a similar system of equations may be formulated. The matrix elements of the **A** and **B** blocks are modified and require the evaluation of the second functional derivative of the exchange–correlation energy, f_{XC},

$$f_{XC} = \frac{\delta^2 E_{XC}[\rho]}{\delta\rho(\mathbf{r})\delta\rho(\mathbf{r}')} \tag{3.117}$$

They must also accommodate the inclusion of Hartree–Fock exchange to allow the use of hybrid exchange–correlation functionals. For example the popular B3LYP functional includes 20% Hartree–Fock exchange. Denoting the amount of Hartree–Fock exchange in general, as C_{HF}, the TDDFT matrix elements are

$$A_{ia,jb} = \delta_{ij}\delta_{ab}\left(\varepsilon_a^{KS} - \varepsilon_i^{KS}\right) + (ia|jb) - C_{HF}(ij|ab) + (1 - C_{HF})(ia|f_{XC}|jb)$$
$$B_{ia,jb} = (ia|jb) - C_{HF}(ib|aj) + (1 - C_{HF})(ia|f_{XC}|bj) \tag{3.118}$$

where $(ia|f_{XC}|bj)$ is the integral

$$(ia|f_{XC}|bj) = \iint \phi_i^{KS}(\mathbf{r}_1)\phi_a^{KS}(\mathbf{r}_1) \frac{\delta^2 E_{XC}[\rho]}{\delta\rho(\mathbf{r}_1)\delta\rho(\mathbf{r}_2)} \phi_j^{KS}(\mathbf{r}_2)\phi_b^{KS}(\mathbf{r}_2)\mathrm{d}\mathbf{r}_1\mathrm{d}\mathbf{r}_2 \tag{3.119}$$

It is easily seen that when $C_{HF} = 0$ and $f_{XC} = 1$, the TDHF equations are obtained.

The TDDFT method provides accurate energies and properties for many types of valence-excited electronic states. The TDDFT scheme when based on some of the commonly used exchange–correlation functionals, does produce significant errors for charge-transfer-type excited states. Charge transfer states can involve the excitation of an electron from the occupied orbitals of one molecular fragment, or unit in the case of van der Waals bound systems, to the virtual orbitals of a neighbouring fragment or unit. The commonly used form of exchange–correlation functionals typically underestimates the energies of such excitations. The origin of this problem can be understood by considering the expressions for the matrix elements in eqn (3.115) with orbitals i and j being located on one fragment and the orbitals a and b on a distant fragment. For a pure density functional, $C_{HF} = 0$, and the only term surviving in the **A** matrix element is the difference in orbital energies. Since we are *not* able to associate the Koopmans' type correspondence between ionisation energies and electron affinities with $-\varepsilon_i^{KS}$ and $-\varepsilon_a^{KS}$, respectively, then $\left(\varepsilon_a^{KS} - \varepsilon_i^{KS}\right)$ is a poor estimate of the excitation energy of the charge-transfer state. In density functional theory the occupied and virtual orbitals are both calculated for the N-electron system. This is in contrast to the Hartree–Fock method in which the occupied orbitals are calculated for the N-electron system and the virtual orbitals for the $(N + 1)$-electron system. In the pure DFT case this amounts to a self-interaction error, since the charge-transfer state corresponds to a 'hole' on one fragment and an electron on the other fragment. In the DFT approximation the electron on the distant fragment effectively still interacts with itself since the Coulomb repulsion present in ε_a^{KS} includes the repulsion of the electron in ϕ_a^{KS} with all the occupied orbitals of the ground state. Yet in the charge-transfer state, the orbital, ϕ_i^{KS}, is no longer occupied. The electron in ϕ_a^{KS} is still interacting with the electrostatic potential corresponding to it still being in ϕ_i^{KS}, hence the self-interaction error. Charge-transfer states are better described by exchange–correlation functionals that include some exact Hartree–Fock exchange, $C_{HF} > 0$. In the Hartree–Fock scheme, $(\varepsilon_a - \varepsilon_i)$ is an approximation to the difference between the ionisation energy and the electron affinity, which is a reasonable approximation to the excitation energy of the charge-transfer state. Hartree–Fock exchange improves the situation but significant errors still remain. An approach for dealing with this problem is to use a *range-separated exchange–correlation functional*. In these functionals the Coulomb operator is split into short-range and long-range parts

$$\frac{1}{r_{12}} = \frac{1 - erf(\mu r_{12})}{r_{12}} + \frac{erf(\mu r_{12})}{r_{12}} \tag{3.120}$$

where *erf* is the error function and μ is a parameter. The first term, the short-range part, is evaluated using the exchange–correlation potential from a density functional method of choice, while the long-range part is evaluated with exact Hartree–Fock exchange. In the widely used CAM-B3LYP functional,[31] the B3LYP form is used at short range with $C_{HF} = 0.2$, and an increasing amount of Hartree–Fock exchange is introduced at long range with $C_{HF} = 0.6$. CAM-B3LYP performs very much better than B3LYP for charge-transfer excited states.

Having obtained the transition energy, $\omega_i = E_i - E_0$, we can assess the strength or intensity of the transition by calculating the oscillator strength associated with the transition. The oscillator strength is given as

$$f_i = \frac{2}{3}\omega_i |\langle \Psi_i | \boldsymbol{\mu} | \Psi_0 \rangle|^2 \tag{3.121}$$

where the transition energy ω_i is in atomic units and $\langle \Psi_i | \boldsymbol{\mu} | \Psi_0 \rangle$ is the

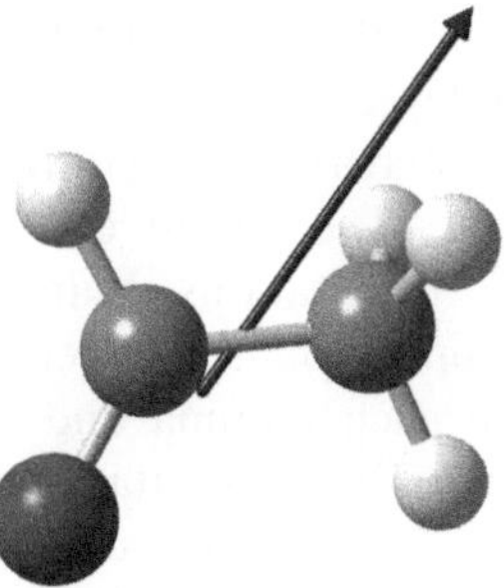

Figure 3.1   The dipole moment vector in ethanal, CH_3CHO.

transition dipole moment, with $\hat{\mu}$ being the electric dipole moment operator. A simulated absorption spectrum can be constructed using the transition dipole moment for each transition and assuming a gaussian lineshape.

3.8   Continuum Methods of Solvation

The computational methods we have discussed provide us with the energy, and possibly the molecular properties, of a single molecule in the gas phase. It is often important to be able to study the effects of solvation on a molecule of interest. For example, we may wish to know the influence of different solvents on the electronic absorption spectrum, or the change of a given vibrational normal mode when a molecule enters solution. If we consider a molecule of ethanal, CH_3CHO, we know that the isolated molecule will have a dipole moment with the negative charge accumulating around the oxygen atom, see Figure 3.1. If the molecule is now solvated by water molecules, we can imagine that the polar water molecules will orientate their dipoles so as to minimise the energy of the solute and solvent system. The approach of two dipoles will cause the induction of an additional dipole moment in both the solute and solvent. The effect of this will be to enhance the dipole moment on the solute. The same effect will be reflected on each neighbouring solvent molecule. The outcome will be a polarisation of the bulk solvent around each solute molecule. The electric field created by the polarised solvent around the solute is called the "reaction field". In terms of the electronic charge distributions of the solute and solvent, each polarises the other and, in turn, is polarised further. Consequently, many electronic properties will be different in solution than in the gas phase.

To deal with the effects of solvation, one might imagine trying to include a large number of explicitly defined solvent molecules around the solute molecule. This approach presents at least two obvious difficulties. First of all, the addition of explicit molecules of solvent incurs a significant computational cost, since the number of atoms being dealt with in the calculation increases. In practice this means that only a few solvent molecules can be dealt with at the same level of theory as is used to treat the solute. The other problem arises from the increase in the conformational space available to the atoms in the calculation. The solute can be surrounded by solvent molecules in many geometric orientations. Many of the conformations may have similar energies and appear on the potential energy surface as local minima. The mapping of all such minima and evaluating their Boltzmann populations presents a considerable computational challenge. A widely used approach that avoids these problems is the use of continuum solvation models. In these models, the geometric complexity of the solvent molecule is removed by treating the solvent as an infinitely large dielectric medium that surrounds a cavity containing the solute, see Figure 3.2. The dielectric medium is characterised by its dielectric constant, or relative permittivity, ε_r. The interaction of the dielectric medium with the solute is incorporated by modifying the hamiltonian to include the reaction field of the solvent. The electronic distribution of the solute is now changed by the presence of the reaction field. In turn the reaction field is modified by the solute's charge distribution. The interaction between solute and solvent is solved for in a self-consistent manner, yielding the self-consistent reaction field (SCRF).

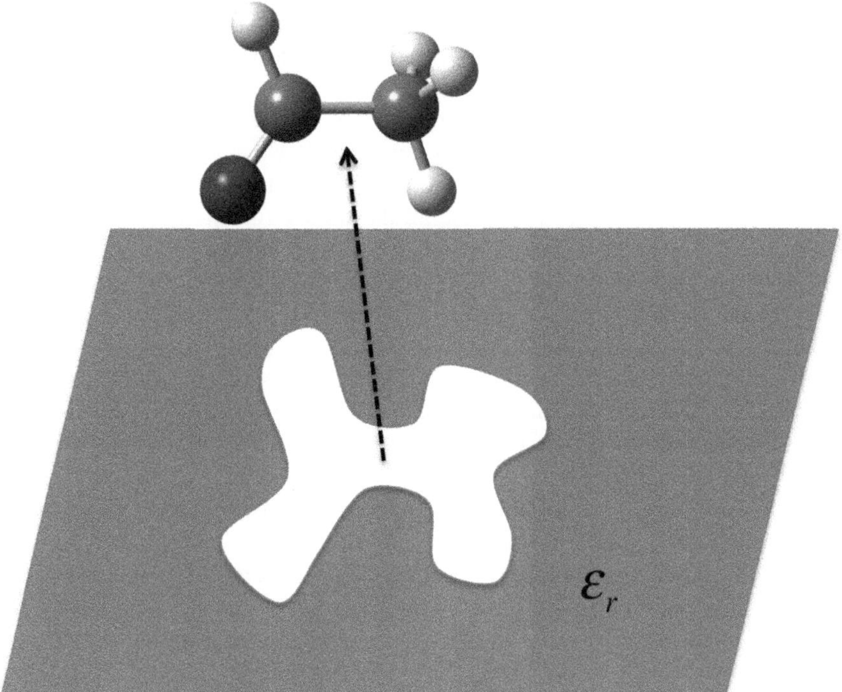

Figure 3.2 Continuum solvation model in which the bulk solvent is represented as an infinitely large medium of dielectric constant ε_r. The solute is embedded in a cavity within the dielectric medium.

The simplest SCRF scheme is the Onsager model that treats the solute as a dipole in a spherical cavity of radius, a, see Figure 3.3. The reaction field in the spherical cavity due to the solvent is given by

$$\mathbf{E}_{\mathrm{RF}} = \frac{2(\varepsilon_r - 1)}{(2\varepsilon_r + 1)a^3}\,\boldsymbol{\mu} \tag{3.122}$$

The interaction between the solute dipole and the reaction field is

$$\hat{V}_{int} = -\boldsymbol{\mu}\cdot\mathbf{E}_{\mathrm{RF}} \tag{3.123}$$

where the dipole moment is given by

$$\boldsymbol{\mu} = \sum_i^N \mathbf{r}_i + \sum_A^{\mathrm{nuclei}} Z_A \mathbf{R}_A \tag{3.124}$$

The SCRF procedure involves the following steps:

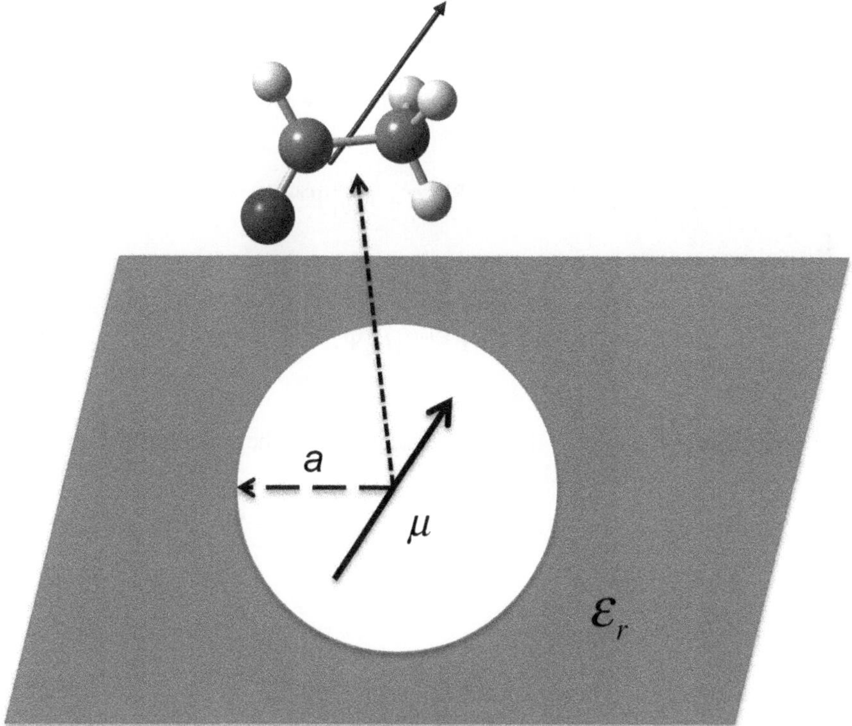

Figure 3.3   Onsager continuum model, which treats the solute as a dipole in a spherical cavity of radius a.

(i)   Initialise ρ, $\boldsymbol{\mu}$ and $\mathbf{E}_{\mathrm{RF}}$.

(ii)   Add $\hat{V}_{int} = -\boldsymbol{\mu}\cdot\mathbf{E}_{\mathrm{RF}}$ to $\hat{H}$.

(iii) Calculate the density of the solute, ρ.

(iv) Calculate the dipole moment $\boldsymbol{\mu} = -\int \rho(\mathbf{r})\mathrm{d}\mathbf{r} + \sum_{A}^{\mathrm{nuclei}} Z_A R_A$

(v)   Calculate the reaction field $\mathbf{E}_{\mathrm{RF}} = \dfrac{2(\varepsilon_r - 1)}{(2\varepsilon_r + 1)a^3}\mu$

(vi) Obtain $\Delta\rho$, $\Delta\boldsymbol{\mu}$ and $\Delta\mathbf{E}_{\mathrm{RF}}$, if all are below a chosen threshold then exit, otherwise go to step (ii).

A number of important points present themselves:

(i)   Within the Born–Oppenheimer approximation, the nuclear contribution to the dipole moment is a constant for a given geometry.

(ii)   A value must be assumed for the cavity radius, a. There is no unique way to do this. A sensible choice might be to use the van der Waals radii of the atoms of the solute to define the maximum size of the cavity. In practice, given the steep dependence of $\mathbf{E}_{\mathrm{RF}}$ on a, parameterised or scaled radii may have to be used to provide an accurate reaction field.

(iii) A non-polar molecule, such as benzene, will not yield any reaction field within this simple model, unless higher multipole moments are included in the definition of $\mathbf{E}_{\mathrm{RF}}$.

The energy of solvation is calculated from the wavefunction obtained from the SCRF procedure as

$$E^{\mathrm{SCRF}} = \langle \Psi^{\mathrm{SCRF}}|H + V_{int}|\Psi^{\mathrm{SCRF}}\rangle \tag{3.125}$$

The gas-phase solute has energy

$$E^{\mathrm{Solute}} = \langle \Psi^{\mathrm{Solute}}|H|\Psi^{\mathrm{Solute}}\rangle \tag{3.126}$$

E^{SCRF} must be corrected for the change in energy of the solvent through being polarised by the solute which is

$$E^{\mathrm{Pol}} = -\frac{1}{2}\langle \Psi^{\mathrm{SCRF}}|V_{int}|\Psi^{\mathrm{SCRF}}\rangle = \frac{(\varepsilon_r - 1)}{(2\varepsilon_r + 1)a^3}|\boldsymbol{\mu}|^2 \tag{3.127}$$

The free energy of solvation is

$$\Delta G_{\mathrm{solv}} = E^{\mathrm{SCRF}} - E^{\mathrm{Solute}} + E^{\mathrm{Pol}} \tag{3.128}$$

If the solute carries a net charge, q, as for any ionic species, the energy must be modified by the constant term

$$\Delta G_{\text{solv}}^{\text{ion}} = -\frac{q^2}{2a}\left(1 - \frac{1}{\varepsilon_r}\right) \tag{3.129}$$

The basic Onsager reaction field method is quick and simple from a computational point of view but is too crude for many purposes. Many improvements are possible, especially with regard to the assumption of the spherical cavity shape. In the polarisable continuum model (PCM) developed by Tomasi and coworkers[32] this assumption is replaced by assigning to each atom of the solute a sphere of radius $1.2\,r_{\text{vdw}}$, where r_{vdw} is the van der Waals radius of the atom. This provides a system of overlapping spheres that define a surface. The surface is a complex shape and discretised by dividing it into small surface elements of area, A_i. A charge, Q_i, is placed on each element and the electrostatic potential due to these charges is

$$\varphi(\mathbf{r}) = \sum_i \frac{Q_i}{|\mathbf{r} - \mathbf{r}_i|} \tag{3.130}$$

The electrostatic potential and the charges are treated in a self-consistent manner. Once these are obtained V_{int}^{PCM} is evaluated as

$$V_{int}^{\text{PCM}} = -\sum_i \varphi(\mathbf{r}_i) + \sum_A^M Z_A \varphi(\mathbf{R}_A) \tag{3.131}$$

and included in the hamiltonian to produce an improved density and in turn an improved electrostatic potential. The process is repeated until no changes in the charges and the electrostatic potential are observed. There are many details that we have omitted and the interested reader may consult ref. 32.

Our attention has rested on the evaluation of the electronic energy and the polarisation of the electronic system. In addition to the electrostatic contribution, the free energy of solvation must include the cavitation contribution, which is the work done in creating 'pockets' in the bulk solvent which the solute molecules occupy, see Figure 3.2. The London forces that attract solvent and solute molecules must be accounted for in the dispersion contribution and also the repulsive interactions that keep the solute and solvent molecules apart.[32]

Many variants of continuum solvation models exist. In particular the conductor-like screening model of Klamt and coworkers has proved very useful. A review may be found in ref. 33.

References

1. N. C. Handy and H. F. Schaefer III, *J. Chem. Phys.*, 1984, **81**, 5031.
2. T. Helgaker and P. Jørgensen, *Adv. Quantum Chem.*, 1988, **19**, 183.

3. T. Helgaker, S. Coriani, P. Jørgensen, K. Kristensen, J. Olsen and K. Ruud, *Chem. Rev.*, 2012, **112**, 543.
4. T. Helgaker and P. R. Taylor, in *Modern Electronic Structure Theory*, ed. D. R. Yarkony, World Scientific, Singapore, 1995, ch. **2**, pp. 725–856.
5. J. Gerratt, I. M. Mills, *J. Chem. Phys.*, 1968, **49**, 1719.
6. N. C. Handy, D. J. Tozer, C. W. Murray, G. J. Laming and R. D. Amos, *Isr. J. Chem.*, 1993, **33**, 331.
7. B. G. Johnson, M. J. Frisch, *J. Chem. Phys.*, 1994, **100**, 7429.
8. Y. Yamaguchi, Y. Osamura, J. D. Goddard and H. F. Schaefer III, in *Ab Initio Molecular Electronic Structure Theory*, Oxford University Press, New York, 1994.
9. J. Baker and W. J. Hehre, *J. Comp. Chem.*, 1991, **12**, 606.
10. H. B. Schlegel, *Int. J. Quantum Chem. Symp.*, 1992, **26**, 243.
11. P. Pulay and G. Fogarasi, *J. Chem. Phys.*, 1992, **96**, 2856.
12. P. Pulay, G. Fogarasi, F. Pang and J. E. Boggs, *J. Am. Chem. Soc.*, 1979, **101**, 2550.
13. H. B. Schlegel, *Theor. Chim. Acta*, 1984, **66**, 333.
14. T. H. Fischer and J. Almlof, *J. Phys. Chem.*, 1992, **96**, 9768.
15. C. G. Broyden, *J. Inst. Math. Appl.*, 1970, **6**, 76.
16. R. Fletcher, *Comput. J.*, 1970, **13**, 317.
17. D. Goldfarb, *Math. Comput.*, 1970, **24**, 23.
18. D. F. Shanno, *Math. Comput.*, 1970, **24**, 647.
19. J. E. Dennis and R. B. Schnabel, *Numerical Methods for Un-constrained Optimization and Nonlinear Equations*, Prentice-Hall, Englewood Cliffs, 1983.
20. H. B. Schlegel, *Wiley Interdiscip. Rev.: Comput. Mol. Sci.*, 2011, **1**, 790.
21. J. R. Cheeseman, G. W. Trucks, T. A. Keith and M. J. Frisch, *J. Chem. Phys.*, 1996, **104**, 5497.
22. S. Koseki, M. W. Schmidt and M. S. Gordon. *J. Phys. Chem.*, 1992, **96**, 10768.
23. S. Koseki, M. S. Gordon, M. W. Schmidt and N. Matsunaga, *J. Phys. Chem.*, 1995, **99**, 12764.
24. S. Koseki, M. W. Schmidt, and M. S. Gordon, *J. Phys. Chem. A*, 1998, **102**, 10430.
25. B. A. Hess, C. M. Marian, U. Wahlgren and O. Gropen, *Chem. Phys. Lett.*, 1996, **251**, 365.
26. *Calculation of NMR and EPR Parameters*, ed. Martin Kaupp, Michael Bühl and V. G. Malkin, Wiley-VCH, Weinheim, 2004.
27. J. E. Harriman, *Theoretical Foundations of Electron Spin Resonance*, Academic Press, New York, 1978.
28. F. Neese, in *Magnetism: Molecules to Materials IV*, ed. J. S. Miller and M. Drillon, Wiley-VCH, Weinheim, 2003, pp. 345–466.
29. A. Dreuw and M. Head-Gordon, *Chem. Rev.*, 2005, **105**, 4009.
30. M. E. Casida and M. Huix-Rotllant, *Annu. Rev. Phys. Chem.*, 2012, **63**, 287.

31. T. Yanai, D. Tew and N. C. Handy, *Chem. Phys. Lett.*, 2004, **393**, 51.
32. J. Tomasi, B. Mennucci and R. Cammi, *Chem. Rev.*, 2005, **105**, 2999.
33. A. Klamt, *Wiley Interdiscip. Rev.: Comput. Mol. Sci.*, 2011, **1**, 699.

CHAPTER 4

Understanding Molecular Wavefunctions, Orbitals and Densities

4.1 Isosurface Representations

Molecular wavefunctions, obtained by any of the methods we have discussed, depend on the $4N$ coordinates of the electronic system, as well as the nuclear geometry. This may seem a daunting quantity to deal with, but we can use a variety of techniques to unravel this complexity and pare it down to a few essential ideas. Often, when trying to understand chemical trends and principles, we look for the overriding factors that dominate. We do not, cannot, dwell on all details to the same degree. The fact that we are able to draw general principles from quantum chemical studies attests to the validity of this approach.

The first quantity that we must address is the set of canonical molecular orbitals obtained from any type of SCF calculation. The molecular orbital is a one-electron function and as such is simply a mathematical form that is used to build the N-electron wavefunction. Molecular orbitals cannot be observed experimentally and do not exist! Yet if we consider the form of the molecular electronic density, $\rho(\mathbf{r})$, expressed over m atomic basis functions

$$\rho(\mathbf{r}) = \sum_{i}^{\text{occupied}} |\phi_i(\mathbf{r})|^2 = \sum_{\mu\nu}^{m} \sum_{i}^{\text{occupied}} c_{\mu i} c_{\nu i} \chi_\mu(\mathbf{r}) \chi_\nu(\mathbf{r}) \qquad (4.1)$$

RSC Theoretical and Computational Chemistry Series No. 5
Computational Quantum Chemistry: Molecular Structure and Properties *in Silico*
By Joseph J W McDouall
© Joseph J W McDouall 2013
Published by the Royal Society of Chemistry, www.rsc.org

which can be observed in crystallographic experiments, we immediately see the key role played by the occupied molecular orbitals. Another way of writing the density is

$$\rho(\mathbf{r}) = \sum_{\mu v}^{m} P_{\mu v}\, \chi_\mu(\mathbf{r})\, \chi_v(\mathbf{r}) \tag{4.2}$$

where the density matrix, for a single determinant wavefunction, can be written in terms of the occupied spin–orbitals as

$$P_{\mu v} = \sum_{i}^{\text{occupied}} c_{\mu i} c_{v i} \tag{4.3}$$

Thus it is tempting, and has proven useful, to try to understand chemical phenomena by analysing molecular orbitals and how they distort and combine during chemical reactions. General principles have been arrived at that can often be used to explain and predict chemical trends. This must be done with some care since molecular orbitals are not unique. For example, the same variationally optimum Hartree–Fock energy can be obtained by an essentially infinite number of equivalent sets of orbitals. We can argue a particular case for the canonical molecular orbitals since they span the irreducible representations of the molecular point group. We can also argue on physical grounds that certain sets of orbitals should be more useful in couching chemical principles than others. We shall now illustrate some of these ideas using the example of 5,5-dimethyl-1-pyrroline-*N*-oxide (DMPO) and its reaction with a simple radical, OH. DMPO is used as a spin trap in EPR experiments. When reacted with a free-radical, the closed-shell DMPO molecule traps the radical forming a stable nitroxy radical that can be detected using EPR spectroscopy, see Figure 4.1. The molecular orbitals are dependent on the three cartesian coordinates, $\mathbf{r} = (x, y, z)$,

$$\phi_i(\mathbf{r}) = \sum_{\mu}^{m} c_{\mu i}\, \chi_\mu(\mathbf{r}) \tag{4.4}$$

To simplify the display of the orbital it is customary to plot a surface of constant value, called an "isosurface", such that

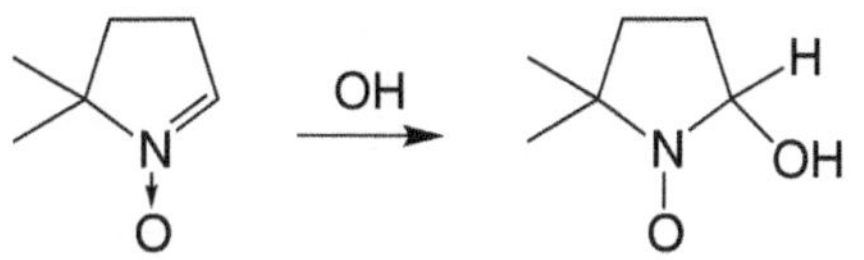

Figure 4.1 Reaction of DMPO with hydroxyl radical. The open-shell nitroxy radical formed is detectable by EPR experiments.

$$\phi(\mathbf{r}) = constant \tag{4.5}$$

This is a necessary simplification since the three cartesian coordinates and the value of the molecular orbital would otherwise require a four-dimensional plot. When considering any surface plotted as an isosurface, it is important that the constant value chosen for the isosurface be stated. The detailed shape of the surface will vary with the isosurface value. Furthermore, the visual comparison of different surfaces is only valid if they are all plotted at the same isosurface value. Often in presenting isosurface plots, the axes are referred to "arbitrary units". In atomic units, the unit for the isosurface of a molecular orbital is $electron^{1/2}a_0^{-3/2}$. This is the square root of the unit of the electron density, $electron\ a_0^{-3}$.

4.2 Canonical Orbitals, Density Matrices and Natural Orbitals

Figure 4.2 shows the highest occupied (HOMO) and lowest unoccupied (LUMO) molecular orbitals of DMPO obtained from a Hartree–Fock calculation using the 6-31G(d,p) basis set. The surfaces are plotted at an isosurface value of 0.04 au. The HOMO orbital is mainly composed of the π bonding component of the double bond and also the oxygen $2p$ orbital that is parallel to, and participates in, the π system of the $-N(\rightarrow O)=C$ unit. The LUMO is the antibonding π^* orbital and also contains a component of the $2p$ orbital on oxygen. By contrast the electron density, which is positive everywhere, serves to define the shape and size of the molecule. The nuclei alone are insufficient to define the size of the molecule since the electron cloud that is attached to the nuclear frame determines the spatial electronic extent of the molecule. The situation is a little more involved when dealing with open-shell species. The nitroxy radical formed by the reaction of the hydroxyl radical with DMPO, Figure 4.1, is described in the spin-unrestricted Hartree–Fock formalism (UHF) by two sets of orbitals, one set of α-spin orbitals and one set

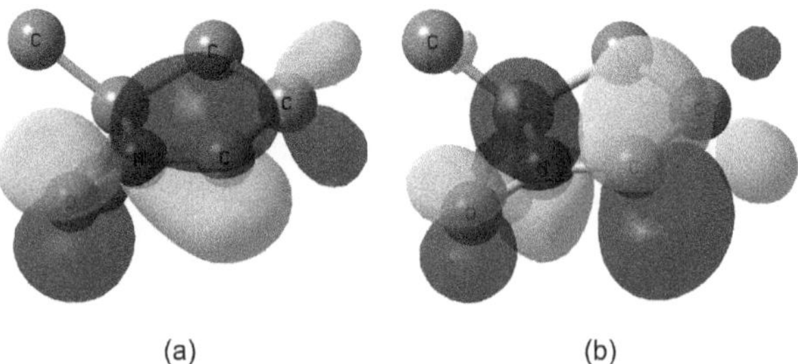

(a)   (b)

Figure 4.2   (a) HOMO and (b) LUMO of DMPO, plotted at an isosurface value of 0.04 au. Hydrogen atoms have been omitted for clarity.

of β-spin orbitals. The first question that suggests itself is *where is the unpaired spin density located?* To address this we must calculate the spin density

$$\rho^{\text{spin}}(\mathbf{r}) = \rho^\alpha(\mathbf{r}) - \rho^\beta(\mathbf{r}) \tag{4.6}$$

where

$$\rho^\alpha(\mathbf{r}) = \sum_{\mu\nu}^{m} \sum_{i}^{N_\alpha} c_{\mu i}^\alpha c_{\nu i}^\alpha \, \chi_\mu(\mathbf{r}) \, \chi_\nu(\mathbf{r})$$

$$\rho^\beta(\mathbf{r}) = \sum_{\mu\nu}^{m} \sum_{i}^{N_\beta} c_{\mu i}^\beta c_{\nu i}^\beta \, \chi_\mu(\mathbf{r}) \, \chi_\nu(\mathbf{r}) \tag{4.7}$$

or in terms of α-spin and β-spin density matrices

$$P_{\mu\nu}^\sigma = \sum_{i}^{N_\sigma} c_{\mu i}^\sigma c_{\nu i}^\sigma$$

$$\rho^{\text{spin}}(\mathbf{r}) = \sum_{\mu\nu}^{m} (P_{\mu\nu}^\alpha - P_{\mu\nu}^\beta) \chi_\mu(\mathbf{r}) \chi_\nu(\mathbf{r}) \tag{4.8}$$

We note that the total density is, as before, the sum of the α-spin and β-spin densities

$$\rho(\mathbf{r}) = \rho^\alpha(\mathbf{r}) + \rho^\beta(\mathbf{r})$$

$$= \sum_{\mu\nu}^{m} (P_{\mu\nu}^\alpha + P_{\mu\nu}^\beta) \, \chi_\mu(\mathbf{r}) \, \chi_\nu(\mathbf{r}) \tag{4.9}$$

Figure 4.3 shows the total and spin densities of the nitroxy radical. The total density defines the shape and size of the radical. The spin density shows the unpaired electron spin to be localised on the –N–O unit with very little intensity elsewhere within the radical. The HOMO and LUMO of each set of spin–orbitals are now quite different in form and energy, as shown in Figure 4.4. The interpretation of these two sets of molecular orbitals presents a difficulty. It is much easier to think about electronic structure, and its rearrangement, if there is only one set of orbitals to deal with. In Section 2.7.1 we discussed the natural orbitals that can be obtained by diagonalising the one-electron density matrix

$$(\mathbf{S}^{\frac{1}{2}}\mathbf{P}\mathbf{S}^{\frac{1}{2}})(\mathbf{S}^{\frac{1}{2}}\mathbf{C}^{\text{NO}}) = (\mathbf{S}^{\frac{1}{2}}\mathbf{C}^{\text{NO}})\mathbf{n} \tag{4.10}$$

The eigenvectors, $\mathbf{C}^{\text{NO}}$, are the natural orbitals and the corresponding

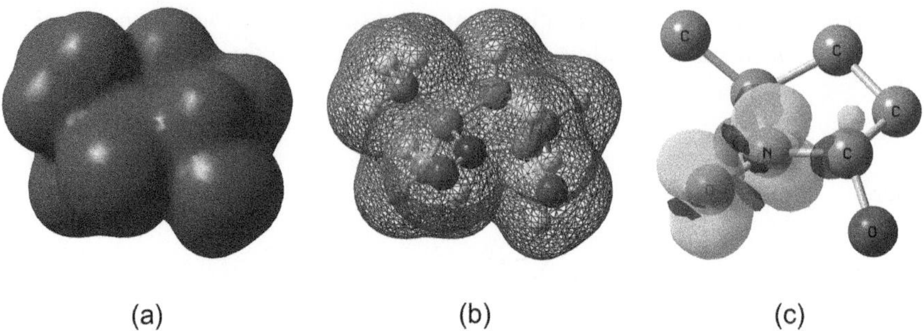

(a)   (b)   (c)

Figure 4.3   (a) Total density of nitroxy radical plotted as a solid surface, (b) the density in (a) plotted as a mesh and (c) the spin density, eqn (4.8). All surfaces are plotted at an isosurface value of 0.004 au.

eigenvalues are orbital occupations that range between 2 and 0 provided that the total density matrix, $\mathbf{P}^\alpha + \mathbf{P}^\beta$, is used. Consequently, we have only one set of natural orbitals to consider. Three of the UHF natural orbitals (UNO) of the nitroxy radical, ordered by their occupation numbers, are shown in Figure 4.5. The natural orbital with occupation number 1.000 is seen to be a π^* orbital on the >N–O unit. The square of this orbital yields the spin density. All other orbitals are equally occupied by α and β electrons and, by eqn (4.8), do not contribute to the spin density. The natural orbital below the π^* orbital is the

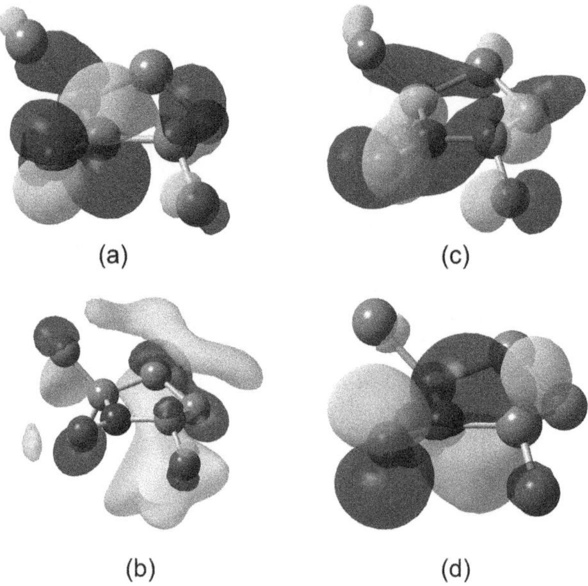

(a)   (c)

(b)   (d)

Figure 4.4   (a) HOMO of the set of α-spin orbitals (b) LUMO of the set of α-spin orbitals (c) HOMO of the set of β-spin orbitals (d) LUMO of the set of β-spin orbitals. All orbitals are plotted at an isosurface value of 0.04 au.

>N–O π bonding orbital with an occupancy of 1.995. There is a very small amount of de-population of this orbital into orbital 37, but we can essentially conclude that the >N–O π orbital is doubly occupied and the π^* orbital is singly occupied. Hence a useful chemical picture has been regained by transforming to natural orbitals. It is important to note that this transformation does not affect the variational condition on the energy. The orbitals depicted in Figures 4.4 and 4.5 will give the same electronic energy. While the natural orbitals are quite different in their spatial forms from the canonical orbitals, the total densities remain the same. An important difference is that we have the orbitals ordered by occupation number rather than orbital energy.

If we insist on a single set of orbitals and a single determinant wavefunction, then our recourse is to the spin-restricted open-shell Hartree–Fock scheme. The variational energy of the ROHF solution will be higher than that of the UHF solution. There also exists a scheme for transforming a UHF solution into a quasi-restricted set of orbitals.[1] This is very useful as it can approximate the ROHF state quite well, provided that the degree of spin-contamination in the UHF solution is small. ROHF-type solutions are not as successful in describing spin-dependent properties as the UHF scheme. This is due to the lack of a proper description of spin-polarisation, see Section 2.3.4. Hence the spin-restricted, single determinant formalism may be easier to interpret but the results obtained from it are less reliable.

4.3 Natural Bond Orbitals

The delocalised canonical molecular orbitals are sometimes difficult to interpret in terms of familiar chemical notions such as lone pairs, bonding/antibonding orbitals and core orbitals. A very successful tool for achieving such analyses has been the *natural bond orbital* idea. We have seen that diagonalisation of the total density matrix, eqn (4.10), produces natural orbitals and their corresponding occupation numbers. In the natural bond orbital (NBO) scheme, the same idea is applied to atomic and diatomic blocks

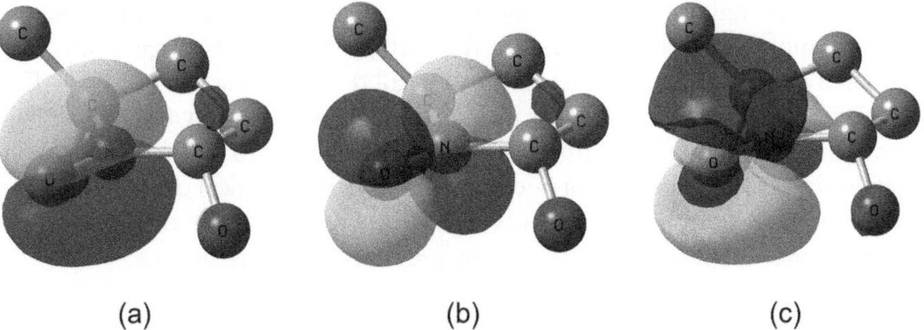

(a)   (b)   (c)

Figure 4.5 Some natural orbitals of the nitroxy radical of Figure 4.1. Occupation numbers: (a) 1.995, (b) 1.000 and (c) 0.005. All orbitals are plotted at an isosurface value of 0.04 au.

of the total density matrix. Since basis functions are usually located on the atomic centres, the partitioning of the density matrix into atomic and diatomic blocks is easily achieved. For a system composed of atoms *A, B, C, D,*

$$\mathbf{P} = \begin{pmatrix} \mathbf{P}^{AA} & \mathbf{P}^{AB} & \mathbf{P}^{AC} & \mathbf{P}^{AD} \\ \mathbf{P}^{BA} & \mathbf{P}^{BB} & \mathbf{P}^{BC} & \mathbf{P}^{BD} \\ \mathbf{P}^{CA} & \mathbf{P}^{CB} & \mathbf{P}^{CC} & \mathbf{P}^{CD} \\ \mathbf{P}^{DA} & \mathbf{P}^{DB} & \mathbf{P}^{DC} & \mathbf{P}^{DD} \end{pmatrix} \tag{4.11}$$

Each of the diagonal, atomic, blocks is diagonalised to produce a set of non-orthogonal natural atomic orbitals (NAO). The strongly occupied NAOs are orthogonalised to all other strongly occupied NAOs. Similarly the weakly occupied NAOs are first orthogonalised to the strongly occupied NAOs on the same atomic centre and then to all other weakly occupied NAOs.

From these NAOs, those with occupancies >1.999 are identified as core orbitals and their contribution is subtracted from the density matrix. NAOs with occupancies >1.9 are identified as lone pairs and their contribution is subtracted from the density matrix. The diatomic blocks of remaining density matrix are then diagonalised to define the NBOs as the eigenvectors with eigenvalue >1.9. The procedure is widely used and details may be found in ref. 2. Some NBOs of DMPO are shown in Figure 4.6 and can be clearly identified as lone pair, bonding or antibonding orbitals.

A particularly useful type of analysis is to express the canonical molecular orbitals as linear combinations of the NBO.

4.4 Localised Molecular Orbitals

The NBOs provide localised orbitals because of the atomic/diatomic blocks of the density matrix being diagonalised. It is sometimes useful to obtain orbitals localised over atoms or bonds by other means. The earliest method for localising orbitals was proposed by S. F. Boys. It is still widely used and we shall illustrate the ideas of orbital localisation using the Boys scheme. We start with the set of canonical molecular orbitals, obtained for example from the Hartree–Fock SCF method. To make the connections clear we shall denote this set as $\left\{\phi^{\text{CMO}}\right\}$. The occupied orbitals can be rotated, or mixed, amongst themselves without changing the variationally optimum nature of the set of orbitals. In orbital localisation schemes we exploit this freedom in combination with a physical or mathematical criterion to form a set of localised orbitals, $\left\{\phi^{\text{LMO}}\right\}$. To maintain orthonormality of the orbitals, the mixing is described by a unitary transformation

$$\Phi^{\text{LMO}} = \Phi^{\text{CMO}}\mathbf{U} \tag{4.12}$$

and $\mathbf{U}^{\dagger}\mathbf{U} = \mathbf{I}$. In terms of individual orbital rotations, we form a modified pair of orbitals

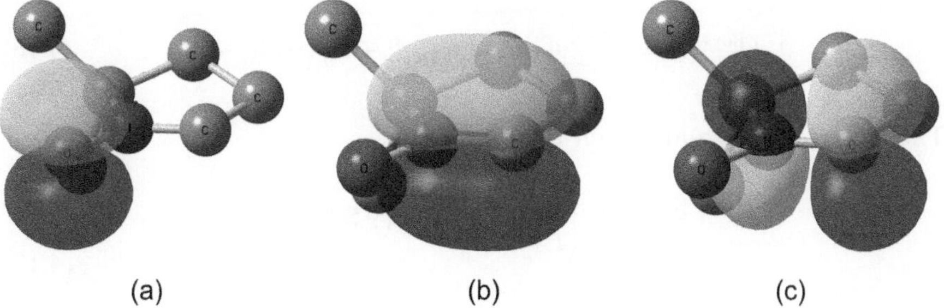

(a)   (b)   (c)

Figure 4.6 Some natural bond orbitals of DMPO corresponding to (a) an oxygen lone pair, (b) the N=C bonding π orbital and (c) the N=C antibonding π^* orbital. All orbitals are plotted at an isosurface value of 0.04 au.

$$|\phi_i'\rangle = \cos\theta_{ij}|\phi_i\rangle + \sin\theta_{ij}|\phi_j\rangle$$
$$|\phi_j'\rangle = -\sin\theta_{ij}|\phi_i\rangle + \cos\theta_{ij}|\phi_j\rangle \tag{4.13}$$

by carrying out all pairwise rotations of the orbitals, $i, j, \ldots$, to be localised. The process is repeated until all pairs meet a convergence criterion that defines the localised set of orbitals, $\left\{\phi^{\text{LMO}}\right\}$.

In the Boys localisation scheme the spatial extent of each orbital is minimised. This idea can be usefully expressed as the *maximisation* of the function

$$L^{\text{Boys}}[\Phi] = \sum_{i>j}^{\text{occupied}} \left[\langle\phi_i|\mathbf{r}|\phi_i\rangle - \langle\phi_j|\mathbf{r}|\phi_j\rangle\right]^2 \tag{4.14}$$

The operator, $\mathbf{r} = \mathbf{x} + \mathbf{y} + \mathbf{z}$, is the electric dipole moment operator, $\hat{\boldsymbol{\mu}}$, and the necessary integrals, $\langle\phi_i|\mathbf{x}|\phi_i\rangle$, $\langle\phi_i|\mathbf{y}|\phi_i\rangle$, $\langle\phi_i|\mathbf{z}|\phi_i\rangle$, are readily available. The further apart the orbitals ϕ_i and ϕ_j are, the greater the value of the $L^{\text{Boys}}[\Phi]$ functional will be. An equivalent formulation of the problem, which leads to a more economical computational implementation, is the maximisation of the distance of the orbital centroids from the origin of the coordinate system

$$L^{\text{Boys}}[\Phi] = \sum_{i}^{\text{occupied}} |\langle\phi_i|\mathbf{r}|\phi_i\rangle|^2 \tag{4.15}$$

The change in the value of the functional $L^{\text{Boys}}[\Phi]$ due to the rotation in eqn (4.13) is

$$L^{\text{Boys}}[\Phi'] = L^{\text{Boys}}[\Phi] + A_{ij} + \sqrt{A_{ij}^2 + B_{ij}^2} \cos 4(\gamma - \alpha) \qquad (4.16)$$

The 'angles' γ and α are given by

$$\cos 4\alpha = \frac{-A_{ij}}{\sqrt{A_{ij}^2 + B_{ij}^2}} \qquad \left(0 \le \alpha \le \frac{\pi}{2}\right)$$

$$\gamma = \alpha + \frac{k\pi}{2} \qquad (k = \text{integer}) \qquad (4.17)$$

and the A_{ij} and B_{ij} integrals are

$$A_{ij} = \left[\langle \phi_i | \mathbf{r} | \phi_j \rangle\right]^2 - \frac{1}{4}\left[\langle \phi_i | \mathbf{r} | \phi_i \rangle - \langle \phi_j | \mathbf{r} | \phi_j \rangle\right]^2$$

$$B_{ij} = \langle \phi_i | \mathbf{r} | \phi_j \rangle \left[\langle \phi_i | \mathbf{r} | \phi_i \rangle - \langle \phi_j | \mathbf{r} | \phi_j \rangle\right] \qquad (4.18)$$

An alternative criterion for localisation, which is very widely used, is due to Pipek and Mezey[3] and is based on the maximisation of the population of an orbital. Specifically the Mulliken population, which we shall discuss in Section 4.6. For each orbital, ϕ_i, the population is defined as

$$d_i = \left[\sum_A^M \left(\sum_{\mu \in A}^m \sum_v^m c_{\mu i} S_{\mu v} c_{v i}\right)^2\right]$$

$$= \sum_A^M Q_i^A \qquad (4.19)$$

in which the index A is summed over the nuclei of the system, and $S_{\mu v}$ is an element of the basis function overlap matrix. The localisation functional to be maximised is defined as

$$L^{\text{P-M}}[\Phi] = \frac{1}{N} \sum_{i=1}^N d_i \qquad (4.20)$$

The mixing angle, eqn (4.13), can be evaluated as

$$\theta_{ij} = sign(B_{ij}) \frac{1}{4} \cos^{-1} \left(\frac{-A_{ij}}{\sqrt{A_{ij}^2 + B_{ij}^2}} \right) \qquad \left(-\frac{\pi}{4} \leq \theta_{ij} \leq +\frac{\pi}{4} \right) \qquad (4.21)$$

where

$$A_{ij} = \sum_A^M \left[\left(Q_{ij}^A \right)^2 - \frac{1}{4} \left[Q_i^A - Q_j^A \right]^2 \right]$$

$$B_{ij} = \sum_A^M Q_{ij}^A \left[Q_i^A - Q_j^A \right] \qquad\qquad (4.22)$$

$$Q_{ij}^A = \frac{1}{2} \sum_{\mu \in A}^m \sum_{\nu}^m \left[c_{\nu i} c_{\mu j} + c_{\mu i} c_{\nu j} \right] S_{\mu\nu}$$

Other schemes for orbital localisation exist but are now less widely used than the two methods we have described. A key difference between the Boys and Pipek–Mezey localisation methods is the treatment of π bonds. The Pipek–Mezey criterion preserves the separation of the σ and π type molecular orbitals, whereas the Boys schemes allows π and σ orbitals to mix, yielding "τ" bonds or, more figuratively, 'banana' bonds. Figure 4.7 shows the localised π bond of DMPO as obtained by the Boys scheme.

By contrast, the corresponding localised orbitals obtained from the Pipek–Mezey scheme show a clear σ or π shape, and look similar to the orbitals in Figure 4.6(b) and (c). The application of these localisation schemes is straightforward for the set of occupied molecular orbitals. The virtual orbitals are usually very difficult, if not impossible, to localise successfully. When localised virtual orbitals are required, as in the local correlation methods of Section 2.10, they are often obtained by projection onto atoms and so are typically non-orthogonal.

4.5 Natural Transition Orbitals

In Section 3.7 we considered the calculation of excited states using time-dependent Hartree–Fock and density functional methods. The outcome of such calculations is a set of excitation amplitudes, eqn (3.121), based on excitations out of the set of orbitals optimised for the ground state. This means that each excited state is a linear combination of determinants. If the excited state is composed principally of one or two determinants then the interpretation of the electronic process is fairly straightforward. However, in many cases, excited states are linear combinations of a large number of determinants, each with a low amplitude. Interpreting what is happening

(a)   (b)

Figure 4.7 Boys localised orbitals. (a) τ bonding orbital between N and C, (b) τ antibonding orbital between N and C. All orbitals are plotted at an isosurface value of 0.04 au.

from an electronic point of view is very difficult in such situations. A very useful tool in this regard is the idea of the natural transition orbital (NTO).[4] An excitation from the ground state to an excited state is governed by the transition dipole moment, $\langle \Psi_i|\boldsymbol{\mu}|\Psi_0\rangle$, where $\hat{\boldsymbol{\mu}}$ is the electric dipole moment operator. To evaluate the transition dipole moment we must form the transition density matrix, $\mathbf{T}$, which in the spin–orbital basis, see eqn (3.121), has elements

$$T_{ia} = X_{ia}^{\alpha} + X_{ia}^{\beta} + Y_{ia}^{\alpha} + Y_{ia}^{\beta} \tag{4.23}$$

$\mathbf{T}$ is a rectangular matrix, since i refers to occupied orbitals and a to unoccupied orbitals of the ground state. We know from our previous discussion that diagonalising a density matrix gives us natural orbitals, which have some very useful properties. Here we have a *transition density* matrix, which is non-square. The set of molecular orbitals can be separated into two, the occupied set $\{\phi\}$ and the virtual set $\{\phi'\}$. The idea is to construct two new sets of orbitals, $\{\psi\}$ and $\{\psi'\}$, which are useful for describing electronic transitions. We can form two, different, square matrices from $\mathbf{T}$ and obtain their eigenvalues and eigenvectors as

$$(\mathbf{TT}^{\dagger})\mathbf{U} = \mathbf{U}\boldsymbol{\Lambda}$$
$$(\mathbf{T}^{\dagger}\mathbf{T})\mathbf{V} = \mathbf{V}\boldsymbol{\Lambda}' \tag{4.24}$$

The dimension of $\mathbf{U}$ will be equal to the number of occupied orbitals, n_{occ}, and the dimension of $\mathbf{V}$ will be equal to the number of unoccupied orbitals, n_{virt}. The orbital sets $\{\psi\}$ and $\{\psi'\}$ are obtained from $\{\phi\}$ and $\{\phi'\}$ by transformation with $\mathbf{U}$ and $\mathbf{V}$, respectively,

$$\boldsymbol{\psi} = \boldsymbol{\phi}\mathbf{U}$$
$$\boldsymbol{\psi}' = \boldsymbol{\phi}'\mathbf{V} \tag{4.25}$$

The sets of eigenvalues, if ordered largest first, have the properties that: (i) $\Lambda_i = \Lambda_i'$; (ii) $0 \le \Lambda_i \le 1$ and (iii) $\sum_i^{n_{occ}} \Lambda_i = 1$, for $i \le n_{occ}$. The significance of the eigenvalues is that they are paired by property (i) above. The orbitals ψ_i and ψ_i' corresponding to the paired eigenvalues, $\Lambda_i = \Lambda_i'$, are the NTOs and they can be interpreted as the component of the excited state formed by excitation from ψ_i to ψ_i' with weight Λ_i. The principal ground-state orbitals of DMPO, which give rise to an intense absorption near 214 nm, are shown in Figure 4.8. These orbitals are the HOMO, LUMO and LUMO+1 levels. The TDDFT calculation tells us that in this excited state the HOMO→LUMO transition has a coefficient of 0.588 and the HOMO→LUMO+1 transition has a coefficient of 0.369. The first of these is simple to understand, as it is essentially a $\pi \rightarrow \pi^*$ transition. The second component, which has a very significant coefficient, is very difficult to interpret. If we form NTOs for this excited state we find the NTOs to look very similar to the HOMO and LUMO shown in Figure 4.8(a) and (b), and the corresponding eigenvalue is 0.980. This means that this transition between NTOs accounts for 98% of the character of this excited state, making the interpretation very simple.

4.6   Electronic Population Analysis

The idea of population analysis in molecules can provide useful insights into computed electronic wavefunctions. The natural orbitals and especially the natural bond orbitals provide information about occupancies of bonding and antibonding orbitals and also lone pairs. These notions can be tied in with the chemist's qualitative ideas on the interplay of different orbitals during chemical processes. Another useful idea is that of the partial charges on atoms in a molecule. These charges are not observables, since atoms in a molecule lose their individual electronic distributions and become part of the molecular electronic distribution. Methods do exist that partition the molecular space into atomic regions but we shall not deal with them here.

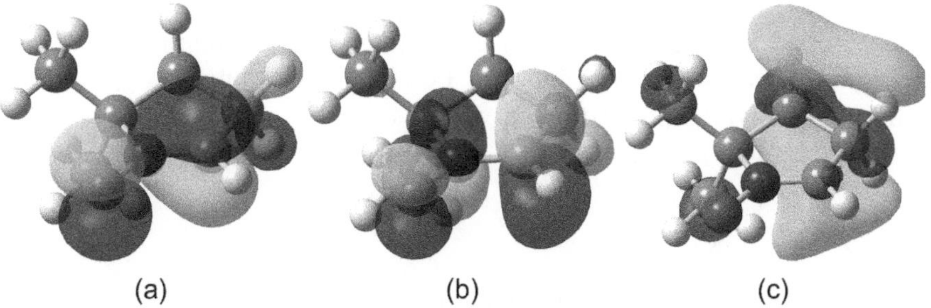

(a)   (b)   (c)

Figure 4.8   Canonical molecular orbitals of the ground state of DMPO. (a) HOMO, (b) LUMO and (c) LUMO+1. All orbitals are plotted at an isosurface value of 0.04 au.

The idea of atomic partial charges in a molecule again ties in well with the chemist's qualitative notions of electrophilicity and nucleophilicity that can determine very specific outcomes of reactions.

Returning to the electron density

$$\rho(\mathbf{r}) = \sum_{\mu\nu}^{m} P_{\mu\nu} \chi_\mu(\mathbf{r}) \chi_\nu(\mathbf{r}) \tag{4.26}$$

which is distributed through space, we can immediately see that were we to integrate over the spatial coordinate, $\mathbf{r}$, the integral of the density would simply return the number of electrons

$$\rho(\mathbf{r}) = \sum_{i}^{N} \sum_{\mu\nu}^{m} c_{\mu i} c_{\nu i} \chi_\mu(\mathbf{r}) \chi_\nu(\mathbf{r}) = \sum_{i}^{N} |\phi_i(\mathbf{r})|^2$$

$$\int \rho(\mathbf{r}) dr = \sum_{i}^{N} \int |\phi_i(\mathbf{r})|^2 = N \tag{4.27}$$

Written over basis functions this integral is

$$\int \rho(\mathbf{r}) d\mathbf{r} = \sum_{\mu\nu}^{m} P_{\mu\nu} S_{\mu\nu} = N \tag{4.28}$$

where $S_{\mu\nu}$ is the overlap integral between basis functions χ_μ and χ_ν. The Mulliken population scheme now partitions the summation in the equation above to

$$\sum_{\mu\nu}^{m} P_{\mu\nu} S_{\mu\nu} = \sum_{\mu}^{m} P_{\mu\mu} + 2 \sum_{\mu<\nu}^{m} P_{\mu\nu} S_{\mu\nu} = N \tag{4.29}$$

The factor of two and the restricted summation arise because the density and overlap matrices are both symmetric. For each basis function, χ_μ, we can define a gross *orbital* population, q_μ, as

$$q_\mu = P_{\mu\mu} + \sum_{\nu}^{m} P_{\mu\nu} S_{\mu\nu} \tag{4.30}$$

A gross *atomic* electron population, q_A, is defined as

$$q_A = \sum_{\mu \in A}^{m} q_\mu \tag{4.31}$$

This is quite arbitrary as a definition, since the overlap population $2 \sum\limits_{\mu<v}^{m} P_{\mu v} S_{\mu v}$ has been divided equally between the atoms on which χ_μ and χ_v are centred. This assumption is obviously valid if χ_μ and χ_v are centred on equivalent atoms, in equivalent chemical environments. However, when basis functions are centred on atoms in very different chemical environments, or on atoms of very different electronegativity, this assumption is a poor one. Provided we keep this concern in mind, we can go on to define the charge on an atom as

$$Q_A = Z_A - q_A \tag{4.32}$$

where Z_A is the atomic number of atom A. We can also define the bond order between atoms A and B as

$$B_{AB} = 2 \sum_{\mu \in A}^{m} \sum_{v \in B}^{m} P_{\mu v} S_{\mu v} \tag{4.33}$$

An alternative definition of the populations and charges is that due to Löwdin. The key difference to the Mulliken scheme is that the basis functions are symmetrically orthogonalised, see Appendix 2B, and the molecular orbital coefficients are transformed to this basis. The outcome is that instead of the product $\mathbf{P}^M = \mathbf{PS}$ being used to obtain the gross atomic populations and other quantities, $\mathbf{P}^L = \mathbf{S}^{\frac{1}{2}} \mathbf{PS}^{\frac{1}{2}}$ is used. $\mathbf{P}^L$ is the Löwdin population matrix and can also be used to obtain the Wiberg bond index (bond order)

$$B_{AB} = \sum_{\mu \in A}^{m} \sum_{v \in B}^{m} \left(P_{\mu v}^L \right)^2 \tag{4.34}$$

Table 4.1 shows the charges on the atoms of the N–O and O–H bonds in the nitroxy radical of DMPO as calculated by the Mulliken and Löwdin schemes.

Table 4.1 Atomic partial charges, calculated from Mulliken and Löwdin population analyses, for the atoms of the N–O and O–H bonds of the nitroxy radical.

Atom	Mulliken Charge	Löwdin Charge
>N–O		
N	−0.193	−0.004
O	−0.415	−0.237
–O–H		
O	−0.658	−0.254
H	0.339	0.163

The first thing to note is that the absolute magnitudes of the charges are predicted to be quite different between the Mulliken and Löwdin schemes. The Mulliken charges generally tend to be larger. This type of analysis is useful in providing a qualitative and comparative picture of molecular structure, but the absolute magnitudes of the charges should not be taken too literally.

4.7 Mayer Bond Orders and Valencies

We have already encountered the Mulliken and Wiberg definitions of bond order. To these we add the Mayer bond orders, which have been found to be particularly useful. Through many applications to organic and inorganic systems the Mayer bond orders and valencies have proved reliable in providing useful chemical pictures from a variety of wavefunctions.[5–6]

In Mayer's scheme, the bond order between atoms A and B is defined as

$$B_{AB}^{\text{Mayer}} = \sum_{\mu \in A}^{m} \sum_{\nu \in B}^{m} \left[(\mathbf{PS})_{\mu\nu}(\mathbf{PS})_{\nu\mu} + \left(\mathbf{P}^{\text{spin}}\mathbf{S}\right)_{\mu\nu}\left(\mathbf{P}^{\text{spin}}\mathbf{S}\right)_{\nu\mu} \right] \qquad (4.35)$$

where $\mathbf{P}^{\text{spin}} = \mathbf{P}^{\alpha} - \mathbf{P}^{\beta}$. For a closed-shell system, $\mathbf{P}^{\text{spin}} = 0$, and only the first term remains. The total valence for an atom is defined as

$$V_{A}^{\text{Mayer}} = 2q_A - \sum_{\mu \in A}^{m} \sum_{\nu \in B}^{m} (\mathbf{PS})_{\mu\nu}(\mathbf{PS})_{\nu\mu} \qquad (4.36)$$

and the free valence as

$$F_{A}^{\text{Mayer}} = V_{A}^{\text{Mayer}} - \sum_{B \neq A}^{M} B_{AB}^{\text{Mayer}} \qquad (4.37)$$

q_A in the Mayer scheme is the Mulliken gross atomic population given in eqn (4.31).

Table 4.2 compares the Wiberg and Mayer bond orders for the nitroxy radical, shown in Figure 4.1. The Wiberg bond orders tend to be larger than those obtained by the Mayer scheme. Mayer bond orders appear to make more sense from a qualitative chemistry point of view, and this perhaps explains their wide usage.

4.8 Electrostatic Potential

The molecular electrostatic potential, $E(\mathbf{r}_p)$, is a surface representing the energy of interaction of a unit point positive charge at some location in space, $\mathbf{r}_p$, with the electrons and nuclei in a molecule. $E(\mathbf{r}_p)$ is defined as

Table 4.2   Wiberg and Mayer bond orders for selected bonds in the nitroxy radical.

Bond	Wiberg bond order	Mayer bond order
N–O	1.704	1.342
O–H	1.037	0.867
(N)C–O(H)	1.222	0.889
(O)C–N(O)	1.025	0.837

$$E(\mathbf{r}_p) = \sum_{A}^{M} \frac{Z_A}{|\mathbf{r}_p - \mathbf{R}_A|} - \sum_{\mu v}^{m} P_{\mu v} \int \frac{\chi_\mu(\mathbf{r})\chi_v(\mathbf{r})}{|\mathbf{r}_p - \mathbf{r}|}\,\mathrm{d}\mathbf{r} \qquad (4.38)$$

Values of $\mathbf{r}_p$ are chosen over a grid and a map of the interaction is produced. Where the electrostatic potential is positive the charge will correspond to electrophilic regions and conversely, where negative, to nucleophilic regions. It is usual to map the electrostatic potential onto the total molecular density. Figure 4.9 shows the electrostatic potential of DMPO mapped onto the total electronic density. The map of the electrostatic potential is shaded to show the change in charge over the surface of the molecule. Accordingly, the nucleophilic region around the oxygen atom in DMPO appears much darker than the rest of the surface.

The electrostatic potential can also be used to derive more physically motivated definitions of atomic charges. These can be obtained by fitting the potential at the grid points using a set of atom-centred charges, q_A^{ESP},

$$E^{\mathrm{Fit}}(\mathbf{r}_p) = \sum_{A}^{M} \frac{q_A^{\mathrm{ESP}}}{|\mathbf{r}_p - \mathbf{R}_A|} \qquad (4.39)$$

The q_A^{ESP} values now fit the combination of the nuclear and electronic

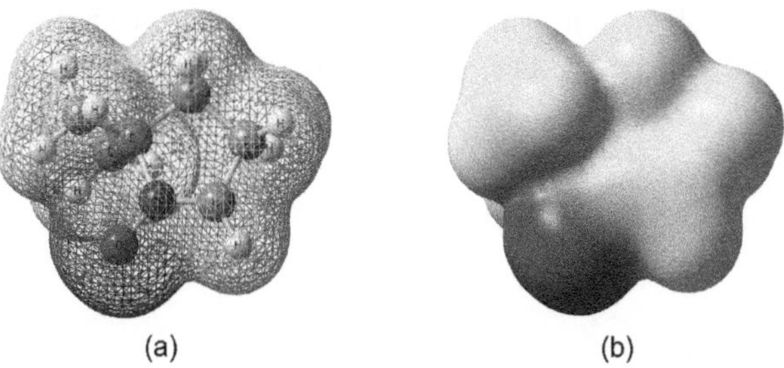

(a)   (b)

Figure 4.9   Molecular electrostatic potential of DMPO mapped on the total density shown as (a) a mesh and (b) a solid surface shaded by magnitude of charge. The electron-rich area around the oxygen atom appears much darker than the rest of the molecule. All surfaces are plotted at an isosurface value of 0.004 au.

interaction at each grid point. Some care should be exercised when using these electrostatic potential derived charges, since they depend on the choice of the grid. So in a sense they are no less arbitrary than the Mulliken and Löwdin definitions. However, experience shows these to provide more intuitive and reliable charges. Reviews can be found in refs. 7–8.

4.9 Energy Decomposition Analysis

It is desirable to be able to decompose a molecular energy into components that can be related to familiar chemical ideas. For example, if we have a ligand that can bind to a variety of different metals, we may wish to understand the binding trend in terms of electrostatic interactions, orbital interactions and the energy required to rearrange the free ligand and metals into the geometry of the metal–ligand complexes. This type of analysis can be performed *via* a number of computational schemes. Here we shall look at one scheme that can be applied to Hartree–Fock and density functional calculations.[9]

Consider a molecule made of two fragments, **A** and **B**, which we shall denote as **AB**. **A** and **B** can be atoms, for example C and O, in which case **AB** would be the diatomic carbon monoxide molecule. **A** and **B** can also refer to fragments, for example **A** could refer to two cyclopentadienyl rings and **B** to iron, in which case **AB** would be the ferrocene molecule. Equally **AB** could refer to the water dimer, see Figure 4.10. In each of the cases in Figure 4.10, **A** is bound to **B**, but the nature of the bonding is quite different. The binding energy, ΔE, can be separated into two components

$$\Delta E = \Delta E_{\text{Prep}} + \Delta E_{\text{Int}} \tag{4.40}$$

Starting with **A** and **B** as separated fragments, in their equilibrium electronic ground state and geometry, we must bring them together into the electronic and geometric configuration they adopt in **AB**. The preparative energetic requirement of this process is denoted as ΔE_{Prep}. Having brought the fragments together, the nuclei and electrons of each sub-system will interact with the other yielding the equilibrium electronic ground state of **AB**. The energy change accompanying this interaction of the fragments is denoted ΔE_{Int}. The interaction energy can be further decomposed into three terms:

$$\Delta E_{\text{Int}} = \Delta E_{\text{Elstat}} + \Delta E_{\text{Pauli}} + \Delta E_{\text{Orbital}} \tag{4.41}$$

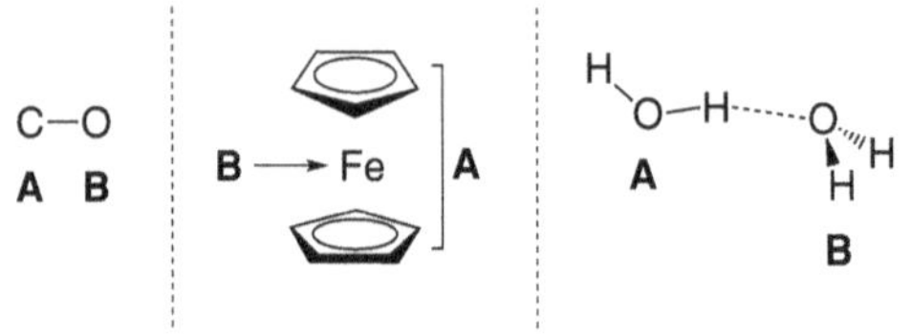

Figure 4.10   Partitioning of diverse systems into two fragments **A** and **B**.

The first of these, ΔE_{Elstat}, is the Coulomb interaction between the nuclei and electrons of one fragment with those of the other fragment. ΔE_{Pauli} is the steric repulsion between the two fragments. Typically ΔE_{Elstat} is attractive, while ΔE_{Pauli} is repulsive and arises from the Pauli principle requiring the wavefunction of **AB** to be properly antisymmetric and not the simple product of the wavefunctions of **A** and **B**. The final term, $\Delta E_{\text{Orbital}}$, is the energy accompanying the electronic orbital relaxation in **AB** which modifies the wavefunction from the antisymmetrised product of the fragment wavefunctions.

The necessary terms can be evaluated as follows. Let $E^{\text{A}}, \Psi^{\text{A}}, \rho^{\text{A}}$, be the optimised energy, wavefunction and density of the isolated fragment **A** and similarly, $E^{\text{B}}, \Psi^{\text{B}}, \rho^{\text{B}}$ for fragment **B**.

$$\left|\Psi^{\overline{\text{AB}}}\right\rangle = \hat{A}\left\{\left|\Psi^{\text{A}}\right\rangle\left|\Psi^{\text{B}}\right\rangle\right\} \tag{4.42}$$

where $\hat{A}$ implies antisymmetrisation. Thus $\left|\Psi^{\overline{\text{AB}}}\right\rangle$ is the properly antisymmetrised and renormalised product of the fragment wavefunctions $\left|\Psi^{\text{A}}\right\rangle$ and $\left|\Psi^{\text{B}}\right\rangle$. In practice this is simply obtained by forming a Slater determinant from the occupied spin–orbitals of the fragments **A** and **B**. The corresponding energy being

$$E^{\overline{\text{AB}}} = \left\langle\Psi^{\overline{\text{AB}}}\left|H\right|\Psi^{\overline{\text{AB}}}\right\rangle \tag{4.43}$$

Then

$$\begin{aligned}
\Delta E^{\overline{\text{AB}}} &= \Delta E_{\text{Elstat}} + \Delta E_{\text{Pauli}} \\
&= E^{\overline{\text{AB}}} - E^{\text{A}} - E^{\text{B}}
\end{aligned} \tag{4.44}$$

The electrostatic interaction between the fragments is given by

$$\begin{aligned}
\Delta E_{\text{Elstat}} = &-\int \sum_{A\in\text{A}}^{M} \frac{Z_A}{|\mathbf{r}-\mathbf{R}_A|}\rho^{\text{B}}(\mathbf{r})\,\mathrm{d}\mathbf{r} - \int \sum_{A\in\text{B}}^{M} \frac{Z_A}{|\mathbf{r}-\mathbf{R}_A|}\rho^{\text{A}}(\mathbf{r})\,\mathrm{d}\mathbf{r} \\
&+ \int \frac{\rho^{\text{A}}(\mathbf{r}_1)\rho^{\text{B}}(\mathbf{r}_2)\,\mathrm{d}\mathbf{r}_1\,\mathrm{d}\mathbf{r}_2}{\mathbf{r}_{12}} + \sum_{A\in\text{A}}^{M}\sum_{A\in\text{B}}^{M}\frac{Z_A Z_B}{|\mathbf{R}_B-\mathbf{R}_A|}
\end{aligned} \tag{4.45}$$

The first term is the coulombic attraction between the nuclei of **A** and the electrons of **B**. This is easily evaluated using the density matrix of **A** and nuclear attraction integrals in which the summation over the nuclei is restricted to those of fragment **A**. The second term is evaluated similarly. The third term is the classical coulombic repulsion between the electron clouds of **A** and **B**. It can be evaluated, as written, using numerical quadrature or equivalently over analytic integrals,

$$\int \frac{\rho^{\mathbf{A}}(\mathbf{r}_1)\rho^{\mathbf{B}}(\mathbf{r}_2)\,\mathrm{d}\mathbf{r}_1\,\mathrm{d}\mathbf{r}_2}{\mathbf{r}_{12}} = \sum_{i\in\mathbf{A}}\sum_{j\in\mathbf{B}}\left\langle \phi_i^{\mathbf{A}}(\mathbf{r}_1)\phi_i^{\mathbf{A}}(\mathbf{r}_1)\Big|\frac{1}{\mathbf{r}_{12}}\Big|\phi_j^{\mathbf{B}}(\mathbf{r}_2)\phi_j^{\mathbf{B}}(\mathbf{r}_2)\right\rangle$$

$$= \sum_{i\in\mathbf{A}}\sum_{j\in\mathbf{B}} J_{ij} \tag{4.46}$$

The final term in eqn (4.45) is the coulombic repulsion between the nuclei of $\mathbf{A}$ and those of $\mathbf{B}$. Given ΔE_{Elstat} and eqn (4.44), we can obtain ΔE_{Pauli} as

$$\Delta E_{\text{Pauli}} = E^{\overline{\mathbf{AB}}} - E^{\mathbf{A}} - E^{\mathbf{B}} - \Delta E_{\text{Elstat}} \tag{4.47}$$

The remaining term, $\Delta E_{\text{Orbital}}$, is due to the change of wavefunction from $\left|\Psi^{\overline{\mathbf{AB}}}\right\rangle$ to the optimum $\left|\Psi^{\mathbf{AB}}\right\rangle$ with energy

$$E^{\mathbf{AB}} = \left\langle \Psi^{\mathbf{AB}}\big|H\big|\Psi^{\mathbf{AB}}\right\rangle \tag{4.48}$$

which implies that

$$\Delta E_{\text{Orbital}} = E^{\mathbf{AB}} - E^{\overline{\mathbf{AB}}} \tag{4.49}$$

Equivalently, and more usually, this change can be expressed as a change in densities derived from $\left|\Psi^{\overline{\mathbf{AB}}}\right\rangle$ and $\left|\Psi^{\mathbf{AB}}\right\rangle$

$$\Delta\rho(\mathbf{r}) = \rho^{\mathbf{AB}}(\mathbf{r}) - \rho^{\overline{\mathbf{AB}}}(\mathbf{r})$$

$$= \sum_{\mu\nu}^{m}\left(P_{\mu\nu}^{\mathbf{AB}} - P_{\mu\nu}^{\overline{\mathbf{AB}}}\right)\chi_\mu(\mathbf{r})\chi_\nu(\mathbf{r}) \tag{4.50}$$

$$= \sum_{\mu\nu}^{m}\Delta P_{\mu\nu}\,\chi_\mu(\mathbf{r})\chi_\nu(\mathbf{r})$$

The elements of the density matrix corresponding to $\left|\Psi^{\mathbf{AB}}\right\rangle$ can be written as

$$P_{\mu\nu}^{\mathbf{AB}} = P_{\mu\nu}^{\overline{\mathbf{AB}}} + \Delta P_{\mu\nu} \tag{4.51}$$

The energy difference can be related to the idea of a "transition state" density matrix. The details of the derivation may be found in ref. 9, here we simply note that in addition to $P^{\overline{\mathbf{AB}}}$ and $P^{\mathbf{AB}}$ we need the transition state density matrix

$$P_{\mu v}^{TS} = \frac{1}{2}\left(P_{\mu v}^{\mathbf{AB}} + P_{\mu v}^{\overline{\mathbf{AB}}}\right) \tag{4.52}$$

From these density matrices we can form a transition state Fock matrix, or Kohn–Sham matrix, as

$$F_{\mu v}^{TS} = \frac{1}{6}F_{\mu v}\left(\mathbf{P}^{\mathbf{AB}}\right) + \frac{2}{3}F_{\mu v}\left(\mathbf{P}^{TS}\right) + \frac{1}{6}F_{\mu v}\left(\mathbf{P}^{\overline{\mathbf{AB}}}\right) \tag{4.53}$$

The orbital interaction term is then evaluated as

$$\Delta E_{\text{Orbital}} = \sum_{\mu v}^{m} \Delta P_{\mu v} F_{\mu v}^{TS} \tag{4.54}$$

over basis functions. It is possible to symmetry adapt the basis functions and so show the contribution to $\Delta E_{\text{Orbital}}$ from each irreducible representation. This expression should agree with the value obtained by $\left(E^{\mathbf{AB}} - E^{\overline{\mathbf{AB}}}\right)$ to high accuracy.

Table 4.3 shows the analysis at the Hartree–Fock and density functional levels for molecular nitrogen. From the discussion above, the binding energy in molecular nitrogen must be the net outcome of $\Delta E_{\text{Prep}} + \Delta E_{\text{Int}}$. In this example, $\Delta E_{\text{Prep}} = 0$ since the fragments are atoms. The bonding comes about from the electrostatic, steric and orbital interactions, $\Delta E_{\text{Int}} = \Delta E_{\text{Elstat}} + \Delta E_{\text{Pauli}} + \Delta E_{\text{Orbital}}$. As can be seen from Table 4.3, the energy decomposition analysis we have described reproduces the binding energy accurately at the Hartree–Fock and DFT levels. This allows us to understand the physical origin of the molecular binding.

Table 4.3 Energy decomposition analysis of molecular nitrogen, see text. Data taken from S. F. Vyboishchikov, A. Krapp and G. Frenking, *J. Chem. Phys.*, 2008, **129**, 144111.

Term	Hartree–Fock	DFT(BP86)
ΔE_{Int}	−454.4	−971.9
ΔE_{Pauli}	4472.7	3211.2
ΔE_{Elstat}	−1435.9	−1302.1
$\Delta E_{\text{Orbital}}$	−3491.1	−2881.1
ΔE_{Prep}	0.0	0.0
D_{e}	454.4	971.9

References

1. F. Neese, *J. Am. Chem. Soc.*, 2006, **128**, 10213.
2. A. E. Reed, L. A. Curtiss and F. Weinhold, *Chem. Rev.*, 1988, **88**, 899.
3. J. Pipek and P. G. Mezey, *J. Chem. Phys.*, 1989, **90**, 4916.
4. R. L. Martin, *J. Chem. Phys.*, 2003, **118**, 4775.
5. I. Mayer, *J. Comput. Chem.*, 2007, **28**, 204.
6. A. J. Bridgeman, G. Cavigliasso, L. R. Ireland and J. Rothery, *J. Chem. Soc., Dalton Trans.*, 2001, 2095.
7. *Molecular Electrostatic Potentials: Concepts and Applications*, ed. J. S. Murray and K. Sen, Elsevier Science, Amsterdam, 1996.
8. J. S. Murray and P. Politzer, *Wiley Interdiscip. Rev.: Comput. Mol. Sci.*, 2011, **1**, 153.
9. T. Ziegler and A. Rauk, *Theor. Chim. Acta*, 1977, **46**, 1.

CHAPTER 5

Relativistic Effects and Electronic Structure Theory

5.1 Relativistic Effects and Chemistry

The importance of relativistic effects on the quantum chemical description of molecular properties is now generally accepted. The early pioneers of the relativistic theory of the electron, including Dirac himself, were sceptical about their importance for chemistry. Their view was that chemistry was largely determined by the behaviour of the valence electrons, and since these electrons were in motion relatively far from the nucleus, their kinetic energies would be sufficiently small that relativistic effects should not influence them. Yet there are many periodic trends that appear to become anomalous as we descend the periodic table. For example, if we consider the bond lengths in dimers of the coinage metals we find: $R_e(Cu_2) = 222.0$ pm; $R_e(Ag_2) = 248.2$ pm; $R_e(Au_2) = 247.2$ pm. As we descend the group from Cu to Ag we obtain the expected increase in bond length, but proceeding on to Au we find a decrease in bond length! There are many other phenomena that do not follow the expected periodic trend. For example the colours of the metals: Cu, Ag and Au. If the Cu $\rightarrow$ Ag trend continued we would expect gold to be a pale-white substance. The explanation of the origin of these anomalies lies in the relativistic description of the atoms and molecules. Many other fascinating phenomena, such as the liquid form of mercury or the structural attraction between equivalently charged cations of gold, may also be attributed to the influence of relativistic effects.

To obtain some insight into what these relativistic effects might be, let us consider a one-electron (hydrogenic) atom with nuclear charge Z. How will the

RSC Theoretical and Computational Chemistry Series No. 5
Computational Quantum Chemistry: Molecular Structure and Properties *in Silico*
By Joseph J W McDouall
© Joseph J W McDouall 2013
Published by the Royal Society of Chemistry, www.rsc.org

velocity with which the electron moves change as Z increases? We can address this at the non-relativistic limit by using the virial theorem and the exact energy levels of the Schrödinger equation for the one-electron atom. The virial theorem tells us that a system with potential $V = r^n$ obeys the following relationship between its average kinetic, $\langle T \rangle$, and potential, $\langle V \rangle$, energies

$$2\langle T \rangle = n \langle V \rangle \tag{5.1}$$

In a hydrogenic atom, the Coulomb potential operates, that is $V = r^{-1}$. The total energy is the sum of the kinetic and potential terms, giving

$$E = \langle T \rangle + \langle V \rangle = \langle T \rangle - 2\langle T \rangle = -\langle T \rangle \tag{5.2}$$

The average kinetic energy has the form $\langle T \rangle = \frac{1}{2} m_e \langle v^2 \rangle$ (we shall briefly use SI units and then dispose of them once more). The exact energy levels for a hydrogenic atom, with quantum number n, are given as

$$E_n = -\frac{Z^2 h c R_H}{h^2} = -\frac{Z^2 \mu e^4}{32 \pi \varepsilon_0^2 \hbar^2 n^2} \tag{5.3}$$

R_H is the Rydberg constant for hydrogen and μ is the reduced mass of the electron and the nucleus

$$\mu = \frac{m_e \, m_{\text{nucleus}}}{m_e + m_{\text{nucleus}}} \tag{5.4}$$

Since $m_e < < m_{\text{nucleus}}$ we can take $\mu \approx m_e$. Using eqns (5.2)–(5.4) we can write

$$\frac{1}{2} m_e \langle v^2 \rangle = \frac{Z^2 m_e e^4}{32 \pi \varepsilon_0^2 \hbar^2 n^2} \tag{5.5}$$

For the 1s shell, $n = 1$, and so

$$\langle v^2 \rangle = \frac{Z^2 e^4}{(4\pi\varepsilon_0)^2 \hbar^2} \tag{5.6}$$

and the root-mean-square velocity is

$$\sqrt{\langle v^2 \rangle} = \frac{Ze^2}{4\pi\varepsilon_0 \hbar} \qquad \text{(in SI units)}$$
$$= Z \qquad \text{(in atomic units)} \tag{5.7}$$

or as a fraction of the speed of light ($c = 137.035999$ au)

$$\frac{\sqrt{\langle v^2 \rangle}}{c} \approx \frac{Z}{137} \tag{5.8}$$

We immediately see the effect of increasing Z on the (non-relativistic) motion of an electron in a hydrogenic atom. For gold, $Z = 79$, and in Au^{78+} the electron, in this model, will be moving at almost $\frac{3}{5}c$. The special theory of relativity assigns the mass of a body in motion as

$$m = \frac{m_0}{\sqrt{1 - (v^2/c^2)}} \tag{5.9}$$

where m_0 is the mass of the body at rest, which here is $m_0 = m_e$. We shall continue to denote this as m_0, even though in atomic units $m_0 = m_e = 1$. For gold this implies that the mass of the electron in the $1s$ orbital has increased to $m \approx 1.22 m_e$. The increased mass causes a contraction of the effective Bohr radius, which is the radius of the electron orbit in the Bohr model

$$a_0^{\mathrm{eff}} = \frac{4\pi\varepsilon_0\hbar}{me^2} \tag{5.10}$$

An increase in the mass of 22% ($m = 1.22 m_e$) will reduce the effective Bohr radius by about 19%. These numbers should be viewed only as qualitative indicators of the principles at play.

Contraction of a $1s$ orbital will affect all other s orbitals, since near the core region all s orbitals are orthogonal to each other. Consequently all s orbitals contract, so too do the p orbitals but to a much lesser extent. The contraction of the s orbitals provides a more effective screening of the nucleus from the more diffuse d and f orbitals. The d and f orbitals expand as a consequence. In terms of energy levels, the s and p levels are stabilised to lower energies while the d and f orbitals are destabilised to higher energies. This has important consequences on properties. For example, the colours of gold and silver arise from the absorption of visible light which induces a transition from the $4d$ to $5s$ levels in silver and from the $5d$ to $6s$ levels in gold. The relativistic effects are more pronounced in gold than in silver. The stabilisation of the $6s$ levels and the destabilisation of the $5d$ levels narrows the energy gap moving the absorption to a longer wavelength. Hence, gold is yellow rather than white.

From the preceding discussion it should be obvious that the proper treatment of relativistic effects is essential for heavier, larger Z, elements if we are to provide accurate descriptions of their molecular energies and properties.

5.2 Special Relativity and the Dirac Equation

The Principle of Special Relativity requires that all physical laws of mechanics and electrodynamics have the same mathematical form when expressed relative to any coordinate frame, that is they must be independent of the reference frame. Hence, if a physical object is described by a given system of coordinates and a set of physical laws, the same laws must hold given another system of coordinates which are moving at a constant velocity relative to the original frame of coordinates. In addition, the speed of light is constant and independent of the motion of the light source.

To meet these requirements, the coordinates of time and position in one system of coordinates, $(t, \mathbf{r})$, must be related to those in another system of coordinates, $(t', \mathbf{r}')$, by a strict set of four conditions. The transformation that meets these criteria is the Lorentz transformation, $T(v)$,

$$T(v) = \frac{1}{\sqrt{1-(v^2/c^2)}} \begin{bmatrix} 1 & -v^2/c^2 \\ -v & 1 \end{bmatrix}$$

$$\begin{pmatrix} t' \\ \mathbf{r}' \end{pmatrix} = \frac{1}{\sqrt{1-(v^2/c^2)}} \begin{bmatrix} 1 & -v^2/c^2 \\ -v & 1 \end{bmatrix} \begin{pmatrix} t \\ \mathbf{r} \end{pmatrix} \tag{5.11}$$

We shall not dwell on the many details and consequences of this transformation but simply note that the developments necessary for relativistic electronic structure theory arose out of the attempts to find an electronic equation which satisfied this type of transformation. Equations obeying this property are referred to as "Lorentz invariant".

A central result of the special theory of relativity is that the total mechanical energy of a moving mass, m, is the pythagorean sum of the constant rest-mass energy, m_0c^2, and the spatial quantity, pc, where p is the magnitude of the momentum $p = |\mathbf{p}|$, see Figure 5.1,

$$E^2 = m_0^2 c^4 + p^2 c^2 \tag{5.12}$$

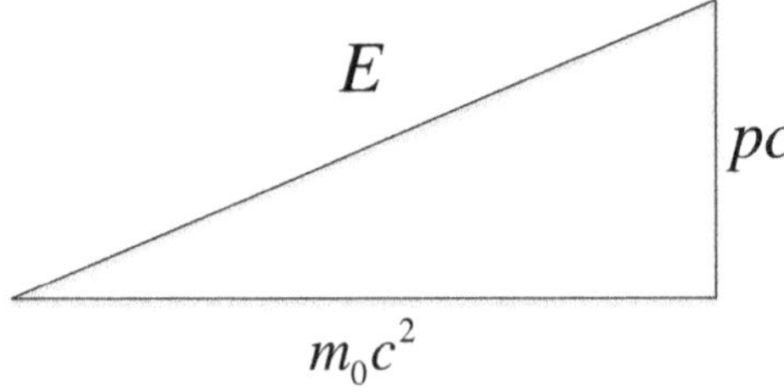

Figure 5.1   The energy–momentum relationship of special relativity, $E^2 = m_0^2 c^4 + p^2 c^2$.

Eqn (5.12) is known as the "energy–momentum relation" of special relativity and provides a first route to obtaining a relativistically consistent energy equation. Taking eqn (5.12) as our starting point we can apply the usual quantisation rules, in which we make the substitutions

$$E \rightarrow i\frac{\partial}{\partial t}$$

$$p_x \rightarrow -i\frac{\partial}{\partial x} \tag{5.13}$$

and introduce the wavefunction, $\Psi(\mathbf{r},t)$,

$$\left(-c^2\nabla^2 + m^2c^4\right)\Psi(\mathbf{r},t) = -\frac{\partial^2\Psi(\mathbf{r},t)}{\partial t^2} \tag{5.14}$$

This is known as the "Klein–Gordon equation". The shortcoming of the Klein–Gordon equation is that it does not take into account the fermionic nature of the electron and so is of little use in chemical problems, but it can be used successfully to treat systems which do not require an antisymmetric wavefunction, namely bosonic systems. Another point for consideration is that we started with an expression for E^2 and applied the quantisation rules. This has the very important property that there must be two solutions, one corresponding to a positive energy and one to a negative energy,

$$E = \pm\sqrt{p^2c^2 + m_0^2c^4} \tag{5.15}$$

We shall return to this 'square-root problem' when discussing solutions to the Dirac equation.

Dirac took a different approach and sought to find an equation of the form

$$\hat{h}^D \Psi(\mathbf{r},t) = i\frac{\partial\Psi(\mathbf{r},t)}{\partial t} \tag{5.16}$$

which satisfied the requirement of being Lorentz invariant. Since the right-hand side of the equation depends linearly on the first derivative of $\Psi(\mathbf{r},t)$ with respect to time, then $\hat{h}^D$ must depend linearly on the first derivative with respect to $\mathbf{r} = (x,y,z)$. Dirac obtained such an equation, and it takes the form

$$c\boldsymbol{\alpha}\cdot\mathbf{p}\,\Psi(\mathbf{r},t) + \boldsymbol{\beta}\,m_0c^2\,\Psi(\mathbf{r},t) = i\frac{\partial\Psi(\mathbf{r},t)}{\partial t} \tag{5.17}$$

The quantities $\boldsymbol{\alpha}$ and $\boldsymbol{\beta}$ are matrices

$$\alpha = \left\{ \begin{bmatrix} 0 & 0 & 0 & 1 \\ 0 & 0 & 1 & 0 \\ 0 & 1 & 0 & 0 \\ 1 & 0 & 0 & 0 \end{bmatrix} \begin{bmatrix} 0 & 0 & 0 & -i \\ 0 & 0 & i & 0 \\ 0 & -i & 0 & 0 \\ i & 0 & 0 & 0 \end{bmatrix} \begin{bmatrix} 0 & 0 & 1 & 0 \\ 0 & 0 & 0 & -1 \\ 1 & 0 & 0 & 0 \\ 0 & -1 & 0 & 0 \end{bmatrix} \right\}$$

(5.18)

$$\beta = \begin{bmatrix} 1 & 0 & 0 & 0 \\ 0 & 1 & 0 & 0 \\ 0 & 0 & -1 & 0 \\ 0 & 0 & 0 & -1 \end{bmatrix}$$

so that

$$c\alpha \cdot \mathbf{p} = c \left(\begin{bmatrix} 0 & 0 & 0 & 1 \\ 0 & 0 & 1 & 0 \\ 0 & 1 & 0 & 0 \\ 1 & 0 & 0 & 0 \end{bmatrix} \begin{bmatrix} 0 & 0 & 0 & -i \\ 0 & 0 & i & 0 \\ 0 & -i & 0 & 0 \\ i & 0 & 0 & 0 \end{bmatrix} \begin{bmatrix} 0 & 0 & 1 & 0 \\ 0 & 0 & 0 & -1 \\ 1 & 0 & 0 & 0 \\ 0 & -1 & 0 & 0 \end{bmatrix} \right) \cdot \begin{pmatrix} p_x \\ p_y \\ p_z \end{pmatrix}$$

$$= c \begin{bmatrix} 0 & 0 & p_z & (p_x - ip_y) \\ 0 & 0 & (p_x + ip_y) & -p_z \\ p_z & (p_x - ip_y) & 0 & 0 \\ (p_x + ip_y) & -p_z & 0 & 0 \end{bmatrix}$$

(5.19)

$$\beta m_0 c^2 = \begin{bmatrix} m_0 c^2 & 0 & 0 & 0 \\ 0 & m_0 c^2 & 0 & 0 \\ 0 & 0 & -m_0 c^2 & 0 \\ 0 & 0 & 0 & -m_0 c^2 \end{bmatrix}$$

Since the left-hand side of eqn (5.17) consists of a sum of 4×4 matrices, the solution, $\Psi(\mathbf{r},t)$, must be a four-component vector

$$\Psi(\mathbf{r},t) = \begin{bmatrix} \Psi_1(\mathbf{r},t) \\ \Psi_2(\mathbf{r},t) \\ \Psi_3(\mathbf{r},t) \\ \Psi_4(\mathbf{r},t) \end{bmatrix}$$

(5.20)

A more compact form can be obtained by introducing the 2×2 Pauli matrices

$$\sigma_x = \begin{bmatrix} 0 & 1 \\ 1 & 0 \end{bmatrix} \qquad \sigma_y = \begin{bmatrix} 0 & -i \\ i & 0 \end{bmatrix} \qquad \sigma_z = \begin{bmatrix} 1 & 0 \\ 0 & -1 \end{bmatrix} \qquad (5.21)$$

so that $\sigma = (\sigma_x, \sigma_y, \sigma_z)$ and

$$\alpha = \begin{bmatrix} 0 & \sigma \\ \sigma & 0 \end{bmatrix} \qquad \beta = \begin{bmatrix} \mathbf{I}_2 & 0 \\ 0 & -\mathbf{I}_2 \end{bmatrix} \qquad (5.22)$$

with $\mathbf{I_2}$ denoting the 2×2 identity matrix. The operator, $\hat{h}^D$, can now be written as

$$\hat{h}^{FE} = \begin{bmatrix} m_0 c^2 \mathbf{I}_2 & c\sigma \cdot \mathbf{p} \\ c\sigma \cdot \mathbf{p} & -m_0 c^2 \mathbf{I}_2 \end{bmatrix} \qquad (5.23)$$

The time-independent Dirac equation for a free electron is

$$\hat{h}^{FE} \begin{bmatrix} \Psi^{(+)} \\ \Psi^{(-)} \end{bmatrix} = E \begin{bmatrix} \Psi^{(+)} \\ \Psi^{(-)} \end{bmatrix} \qquad (5.24)$$

where

$$\Psi^{(+)} = \begin{bmatrix} \Psi_1 \\ \Psi_2 \end{bmatrix} \qquad \Psi^{(-)} = \begin{bmatrix} \Psi_3 \\ \Psi_4 \end{bmatrix} \qquad (5.25)$$

$\Psi^{(+)}$ refers to electronic solutions with positive energies and $\Psi^{(-)}$ to positronic solutions with negative energies. Solutions to the Dirac equation are also solutions to the Klein–Gordon equation. The positive energy states have energies close to $+m_0 c^2$, while negative energy states are close to $-m_0 c^2$. The $\Psi^{(+)}$ and $\Psi^{(-)}$ solutions occur symmetrically. To understand the behaviour of the positronic solutions, the second row of eqn (5.24) can be expanded to give

$$c\sigma \cdot \mathbf{p}\Psi^{(+)} - m_0 c^2 \mathbf{I}_2 \Psi^{(-)} = E\Psi^{(-)} \qquad (5.26)$$

which can be rearranged to give

$$\Psi^{(-)} = \frac{c\sigma \cdot \mathbf{p}}{m_0 c^2 + E} \Psi^{(+)} \qquad (5.27)$$

From this equation we can establish that $\Psi^{(-)}$ is significantly smaller than $\Psi^{(+)}$. Accordingly, $\Psi^{(+)}$ and $\Psi^{(-)}$ are often referred to as the large

$\left(\Psi^L = \Psi^{(+)}\right)$ and small $\left(\Psi^S = \Psi^{(-)}\right)$ components of the Dirac equation. From eqn (5.20) we can see that the large and small components consist of two parts, each referring to one of the spin orientations, $\uparrow$ or $\downarrow$.

To proceed we must consider the case of a bound electron, as in the hydrogen atom. To $\hat{h}^D$ in eqn (5.17) we now add the electron–nucleus potential $-Z/r$ to obtain

$$\hat{h}^D = c\boldsymbol{\alpha}\cdot\mathbf{p} + m_0 c^2 \boldsymbol{\beta} - \frac{Z}{r} \tag{5.28}$$

and $\hat{h}^D \Psi = E\Psi$ remains Lorentz invariant, the matrix form being

$$\begin{bmatrix} (m_0 c^2 - Z/r)\mathbf{I}_2 & c\boldsymbol{\sigma}\cdot\mathbf{p} \\ c\boldsymbol{\sigma}\cdot\mathbf{p} & (-m_0 c^2 - Z/r)\mathbf{I}_2 \end{bmatrix} \begin{bmatrix} \Psi^L \\ \Psi^S \end{bmatrix} = E \begin{bmatrix} \Psi^L \\ \Psi^S \end{bmatrix} \tag{5.29}$$

The extension to many electrons can be achieved by introducing the Coulomb inter-electron term $\sum\limits_{i>j} \dfrac{1}{r_{ij}}$

$$\hat{H}^{DC}(i) = c\boldsymbol{\alpha}\cdot\mathbf{p}_i + m_0 c^2 \boldsymbol{\beta} - \sum_A^M \frac{Z_A}{r_{iA}} + \sum_{i<j} \frac{1}{r_{ij}} \tag{5.30}$$

The resulting Dirac–Coulomb hamiltonian for electron i is *not* Lorentz invariant, but has been found to give accurate results in many cases. An improved description of the inter-electronic interaction is provided by the Breit operator, $\hat{H}^{Breit}$,

$$\hat{H}^{Breit}(i,j) = \frac{1}{r_{ij}} - \frac{1}{2r_{ij}} \left[\boldsymbol{\alpha}_i \boldsymbol{\alpha}_j + \frac{(\boldsymbol{\alpha}_i\cdot r_{ij})(\boldsymbol{\alpha}_j\cdot r_{ij})}{r_{ij}^2} \right] \tag{5.31}$$

This is not exactly Lorentz invariant, but is more accurate than the Coulomb description. The additional terms introduce inter-electron magnetic interactions and account for the finite speed of light.

The energy spectrum of the Dirac equation is rather different to that used in non-relativistic theory, see Figure 5.2. The Dirac equation energies can be shifted by $-m_0 c^2$ to match the familiar non-relativistic scale. This is easily accomplished by modifying the definition of β to

$$\beta' = \begin{bmatrix} \mathbf{0}_2 & \mathbf{0}_2 \\ \mathbf{0}_2 & -2\mathbf{I}_2 \end{bmatrix} \tag{5.32}$$

where $\mathbf{0}_2$ is the 2×2 null matrix.

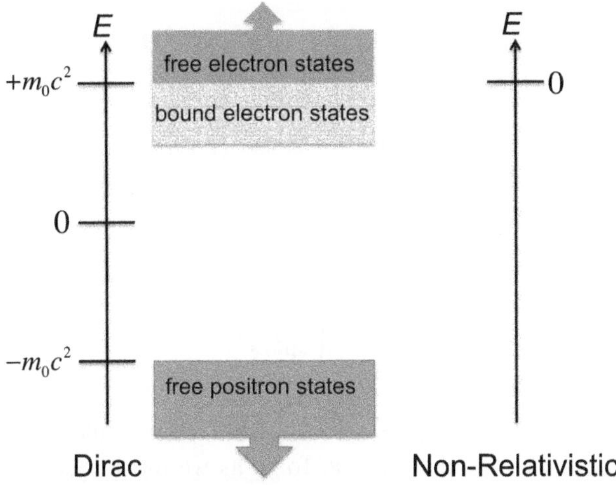

Figure 5.2 Comparison of the energy spectrum of the Dirac equation and that of the conventional non-relativistic energy.

The problem that remains is that simple application of a variational principle to the Dirac equation must yield the positronic states as the global minimum of the energy. Hence, various constraints must be imposed to prevent any numerical procedure from this variational collapse. One such scheme, used with basis set expansions, is known as "kinetic balance". It gives a prescription for obtaining the small component basis functions as derivatives of the large component basis functions. Such calculations are very demanding since the kinetic balance requirement means that the number of basis functions required to describe the small component is twice the number of basis functions used for the large component. An alternative strategy is to decouple the large and small components in eqn (5.29), by eliminating the off-diagonal terms. If this is achieved, it enables the problem to be reduced to dealing with just the electronic block of eqn (5.29).

5.3 Elimination of the Small Component: The Pauli Approximation

The time-independent Dirac equation with potential $V = -Z/r$ can be written as

$$h^{\mathrm{D}}\Psi^{\mathrm{D}} = \varepsilon\,\Psi^{\mathrm{D}} \tag{5.33}$$

where

$$\hat{h}^D = c\boldsymbol{\alpha}\cdot\mathbf{p} + m_0c^2\boldsymbol{\beta}' + \left(V - m_0c^2\right)\mathbf{I}_4$$

$$= \begin{pmatrix} V & c\boldsymbol{\sigma}\cdot\mathbf{p} \\ c\boldsymbol{\sigma}\cdot\mathbf{p} & V - 2m_0c^2 \end{pmatrix} \tag{5.34}$$

and

$$\Psi^D = \begin{pmatrix} \Psi^L \\ \Psi^S \end{pmatrix} \tag{5.35}$$

We shall ignore magnetic fields for now, but they could be included by substituting the kinetic momentum, $\boldsymbol{\pi}$, for $\mathbf{p}$ as we did in Section 3.6.2. This equation contains the energy shift of $-m_0c^2$ as previously discussed. In matrix form eqn (5.33) is

$$\begin{pmatrix} V & c\boldsymbol{\sigma}\cdot\mathbf{p} \\ c\cdot\mathbf{p} & V - 2m_0c^2 \end{pmatrix} \begin{pmatrix} \Psi^L \\ \Psi^S \end{pmatrix} = \varepsilon \begin{pmatrix} \Psi^L \\ \Psi^S \end{pmatrix} \tag{5.36}$$

Expanding this yields two coupled equations

$$V\Psi^L + c\boldsymbol{\sigma}\cdot\mathbf{p}\Psi^S = \varepsilon\,\Psi^L$$
$$c\boldsymbol{\sigma}\cdot\mathbf{p}\Psi^L + \left(V - 2m_0c^2\right)\Psi^S = \varepsilon\,\Psi^S \tag{5.37}$$

We can solve the second equation for Ψ^S

$$\Psi^S = \frac{c\boldsymbol{\sigma}\cdot\mathbf{p}}{\left(\varepsilon - V + 2m_0c^2\right)}\Psi^L \tag{5.38}$$

Substituting into the first equation above gives

$$V\,\Psi^L + c\boldsymbol{\sigma}\cdot\mathbf{p}\left(\varepsilon - V + 2m_0c^2\right)^{-1}c\boldsymbol{\sigma}\cdot\mathbf{p}\,\Psi^L = \varepsilon\,\Psi^L \tag{5.39}$$

We now have an equation from which Ψ^S has been eliminated. The term in parentheses can be factored as

$$\left(\varepsilon - V + 2m_0c^2\right)^{-1} = \frac{1}{2m_0c^2}\left(1 + \frac{\varepsilon - V}{2m_0c^2}\right)^{-1}$$
$$= \frac{1}{2m_0c^2}K(\varepsilon) \tag{5.40}$$

How the $K(\varepsilon)$ term is represented determines the nature of any approximation. $K(\varepsilon)$ can be expanded in the geometric series

$$K(\varepsilon) = \sum_{\kappa=0}^{\infty} \left(\frac{V-\varepsilon}{2m_0c^2}\right)^{\kappa} \tag{5.41}$$

and the Pauli approximation is obtained by retaining only the first two terms of the series.

$$K(\varepsilon) \approx 1 - \frac{\varepsilon-V}{2m_0c^2} \tag{5.42}$$

Provided that $(\varepsilon-V)<<2m_0c^2$ this will be a valid representation. The difficulty arises when the electron approaches the region of the nucleus, since as $r\to0$, the potential $-Z/r\to-\infty$. However, proceeding with the expansion in eqn (5.42), and substituting into eqn (5.39) we obtain

$$\frac{1}{2m_0c^2}\left[c\boldsymbol{\sigma}\cdot\mathbf{p}\left(1+\frac{V-\varepsilon}{2m_0c^2}\right)c\boldsymbol{\sigma}\cdot\mathbf{p}\right] \tag{5.43}$$

After considerable rearrangement, and renormalisation of Ψ^L, eqn (5.43) yields the Pauli equation.

$$\left[\frac{\mathbf{p}^2}{2m_0}+V-\frac{\mathbf{p}^4}{8m_0^3c^2}+\frac{Z\,\mathbf{s}\cdot\boldsymbol{l}}{2m_0^2c^2r^3}+\frac{Z\pi}{2m_0^2c^2}\delta(r)\right]\Psi^L=\varepsilon\,\Psi^L \tag{5.44}$$

The first two terms of the Pauli equation are the non-relativistic kinetic and potential energies. The remaining three terms are known as

(i) $-\dfrac{1}{8m_0^3c^2}\mathbf{p}^4$ the "mass–velocity correction"

(ii) $\dfrac{Z\pi}{2m_0^2c^2}\delta(r)$ the "Darwin correction"

(iii) $\dfrac{Z}{2m_0^2c^2}\dfrac{\mathbf{s}\cdot\boldsymbol{l}}{r^3}$ the "spin–orbit interaction"

Correction (i) is due to the velocity dependence of the relativistic mass. The Darwin term is associated with the rapid oscillation of the electron about its mean position, which is referred to as *Zitterbewegung*. Both the mass–velocity and Darwin terms have no dependence on spin and so are referred to as "scalar relativistic corrections". The spin–orbit term arises from the interaction of spin-angular momentum of the electron with its orbital angular momentum. We have included the Pauli approximation since it is historically important and it illustrates some useful principles. However the Pauli hamiltonian is not bounded from below since as the electron approaches the nucleus, $r\to0$ and

$V \to - \infty$. The consequence of this is that the Pauli hamiltonian cannot be used in any variational procedure, given that the global minimum must correspond to $V \to -\infty$! The Pauli hamiltonian can be used in first-order perturbation theory to recover relativistic effects. When the magnitude of such relativistic effects is large, first-order perturbation theory and the Pauli hamiltonian must be expected to be inadequate.

5.4 Elimination of the Small Component: Regular Approximations

To address the problems associated with the lack of a variational bound in the Pauli hamiltonian a family of methods were developed based on regular approximations to the $(\varepsilon - V + 2m_0 c^2)^{-1}$ term.[1] Instead of factoring $\frac{1}{2} m_0 c^2$ from this inverse, as in eqn (5.40), the operator $(2m_0 c^2 - V)$ is factored out as

$$(\varepsilon - V + 2m_0 c^2)^{-1} = \frac{2m_0 c^2}{2m_0 c^2 - V} \left[1 + \frac{\varepsilon}{2m_0 c^2 - V} \right]^{-1} \tag{5.45}$$

Employing an expansion for the inverse on the right-hand side now gives

$$(\varepsilon - V + 2m_0 c^2)^{-1} = \frac{2m_0 c^2}{2m_0 c^2 - V} \sum_{\kappa=0}^{\infty} \left(\frac{\varepsilon}{V - 2m_0 c^2} \right)^{\kappa} \tag{5.46}$$

The factor $(2m_0 c^2 - V)$ will always be positive since V is always negative. So $[2m_0 c^2 / (2m_0 c^2 - V)]$ will always be positive and less than unity. The expansion term $[\varepsilon / (2m_0 c^2 - V)]$ will also always be less than unity for all bound electronic states since these have energies less than $m_0 c^2$. If we apply this regular expansion to eqn (5.39) and retain only the $\kappa = 0$ component we obtain the zero-order regular approximation (ZORA) hamiltonian

$$\hat{h}^{\mathrm{ZORA}} = V + \frac{1}{2m_0} \boldsymbol{\sigma} \cdot \mathbf{p} \left(\frac{2m_0 c^2}{2m_0 c^2 - V} \right) \boldsymbol{\sigma} \cdot \mathbf{p} \tag{5.47}$$

The great advantage of $\hat{h}^{\mathrm{ZORA}}$ is that it is variationally stable and has found wide use. At zeroth-order the regular approximation is energy independent and relatively easy to apply. Higher orders contain energy-dependent terms, which complicates their computational implementation. The dependence on the potential means that the kinetic energy in the ZORA approximation must be evaluated using numerical quadrature or possibly some form of density fitting. A useful idea due to Filatov[2] avoids both these strategies by providing a set of working equations that can be evaluated using analytic integrals over basis functions. The idea is to introduce a unit operator to eliminate the potential dependence. Accordingly the method is referred to as the ZORA-RI (resolution of the identity) method. The kinetic energy operator in the ZORA approach is

$$\hat{T}^{\text{ZORA}} = (\boldsymbol{\sigma}\cdot\mathbf{p})K(\boldsymbol{\sigma}\cdot\mathbf{p}) \tag{5.48}$$

where

$$K = \frac{1}{2m_0}\left(\frac{2m_0c^2}{2m_0c^2 - V}\right) = \frac{1}{2m_0}\frac{1}{\left(1 - \dfrac{V}{2m_0c^2}\right)} \tag{5.49}$$

Multiplying both sides of the final expression by $\left(1 - V/2m_0c^2\right)$ and rearranging gives the identity

$$K = \frac{1}{2m_0} + \frac{VK}{2m_0c^2} \tag{5.50}$$

which can be inserted into eqn (5.48) to give

$$\hat{T}^{\text{ZORA}} = \frac{(\boldsymbol{\sigma}\cdot\mathbf{p})(\boldsymbol{\sigma}\cdot\mathbf{p})}{2m_0} + \frac{(\boldsymbol{\sigma}\cdot\mathbf{p})VK(\boldsymbol{\sigma}\cdot\mathbf{p})}{2m_0c^2} \tag{5.51}$$

In scalar relativistic treatments we ignore the $\boldsymbol{\sigma}$ operator. This approximation is not essential in the development of the ZORA-RI method but we shall adopt it here for simplicity. The kinetic energy operator now reads

$$\hat{T}^{\text{ZORA}} = \frac{\mathbf{p}^2}{2m_0} + \frac{\mathbf{p}VK\mathbf{p}}{2m_0c^2} \tag{5.52}$$

Note that the first term is simply the non-relativistic kinetic energy operator $\hat{T}^{\text{NR}}$, $(\mathbf{p} = -i\nabla)$. Filatov defined an identity operator as

$$\hat{I} = \Phi\Phi^+$$

$$= \frac{1}{2m_0}\sum_{\lambda\sigma\tau}\mathbf{p}|\chi_\lambda\rangle T^{\text{NR}\,-\frac{1}{2}}_{\lambda\tau} T^{\text{NR}\,-\frac{1}{2}}_{\tau\sigma}\langle\chi_\sigma|\mathbf{p} \tag{5.53}$$

$$= \frac{1}{2m_0}\sum_{\lambda\sigma}\mathbf{p}|\chi_\lambda\rangle T^{NR\,-1}_{\lambda\sigma}\langle\chi_\sigma|\mathbf{p}$$

In this expression, $\{\chi\}$, represents the set of basis functions to be used in the calculation. The identity operator is inserted between V and K in the second term of eqn (5.52), the resulting expression is then pre-multiplied by χ_μ and post-multiplied by χ_ν and integrated

$$\langle \chi_\mu | \hat{T}^{ZORA} | \chi_v \rangle = \langle \chi_\mu | \hat{T}^{NR} | \chi_v \rangle +$$

$$\frac{1}{2m_0} \frac{1}{2m_0 c^2} \sum_{\lambda\sigma} \langle \chi_\mu | \mathbf{p} V \mathbf{p} | \chi_\lambda \rangle T_{\lambda\sigma}^{NR\,-1} \langle \chi_\sigma | \mathbf{p} K \mathbf{p} | \chi_v \rangle \qquad (5.54)$$

In matrix form, this can be written

$$\mathbf{T}^{ZORA} = \mathbf{T}^{NR} + \mathbf{W}_0 \mathbf{T}^{NR\,-1} \mathbf{T}^{ZORA} \qquad (5.55)$$

where

$$(W_0)_{\mu v} = \frac{1}{4m_0^2 c^2} \langle \chi_\mu | \mathbf{p} V \mathbf{p} | \chi_v \rangle \qquad (5.56)$$

$\mathbf{T}^{ZORA}$ can be represented as a sum of the non-relativistic kinetic energy and a relativistic correction

$$\mathbf{T}^{ZORA} = \mathbf{T}^{NR} + \mathbf{W} \qquad (5.57)$$

with

$$\mathbf{W} = \mathbf{W}_0 \mathbf{T}^{NR\,-1} \mathbf{T}^{ZORA} \qquad (5.58)$$

substituting the previous equation into this gives

$$\mathbf{W} = \mathbf{W}_0 \mathbf{T}^{NR\,-1} \left(\mathbf{T}^{NR} + \mathbf{W} \right)$$
$$= \mathbf{W}_0 + \mathbf{W}_0 \mathbf{T}^{NR\,-1} \mathbf{W} \qquad (5.59)$$

So $\mathbf{W}$ depends on $\mathbf{W}$! Post-multiplying by $\mathbf{W}^{-1}$ gives

$$\mathbf{I} = \mathbf{W}_0 \mathbf{W}^{-1} + \mathbf{W}_0 \mathbf{T}^{NR\,-1} \qquad (5.60)$$

Then pre-multiplying by $\mathbf{W}_0^{-1}$

$$\mathbf{W}_0^{-1} = \mathbf{W}^{-1} + \mathbf{T}^{NR\,-1} \qquad (5.61)$$

or more usefully

$$\mathbf{W}^{-1} = \mathbf{W}_0^{-1} - \mathbf{T}^{NR\,-1} \qquad (5.62)$$

Hence the ZORA kinetic energy operator, which is dependent on V, has been reduced to a form involving analytic integrals, $\mathbf{T}^{NR}$ and $(\mathbf{p} V \mathbf{p})$. Note that here

V refers to the one-electron nuclear attraction potential only. In principle, it should include the electron–electron potential as well, but this approximation does not appear to be too severe. The integrals of the form $\langle \chi_\mu | \mathbf{p} V \mathbf{p} | \chi_\nu \rangle$ are readily evaluated in standard quantum chemistry programs. Recalling that $\mathbf{p} = -i\nabla$, and allowing $\mathbf{p}$ to act on the basis functions (turn over rule)

$$\langle \chi_\mu | \mathbf{p} V \mathbf{p} | \chi_\nu \rangle = \langle p_x \chi_\mu | V | p_x \chi_\nu \rangle + \langle p_y \chi_\mu | V | p_y \chi_\nu \rangle + \langle p_z \chi_\mu | V | p_z \chi_\nu \rangle \tag{5.63}$$

Hence $\langle \chi_\mu | \mathbf{p} V \mathbf{p} | \chi_\nu \rangle$ can be evaluated in terms of nuclear attraction integrals in which the basis functions have been differentiated, see eqn (3.37).

The principal disadvantage of the ZORA method is that it is gauge dependent, that is a constant shift applied to V does not produce the same shift in the energy. This has serious consequences in molecular calculations since the ZORA energy will not depend correctly on the molecular structure and will produce erroneous geometries. This can be circumvented approximately by using the scaled ZORA energy,[3] which includes a renormalisation term. The gauge dependence can be eliminated rigorously by applying the ZORA correction only within atomic (one-centre) blocks. The relativistic correction then becomes constant for each atom in the molecule and there is no structural dependence. This can also facilitate the evaluation of analytic gradients of the energy.[4] The use of atomic corrections may seem a huge approximation but is in fact quite effective, as well as efficient. The largest component of the relativistic corrections arises from within atomic blocks, with the two-centre terms being of much smaller magnitude.

5.5 Elimination of the Small Component: Unitary Decoupling of the Dirac Equation

An alternative approach to decoupling the electronic and positronic solutions of the Dirac equation is to seek a unitary transformation that will block diagonalise the Dirac hamiltonian.

$$\mathbf{U}^\dagger \mathbf{h}^D \mathbf{U} = \begin{pmatrix} \mathbf{h}^L & 0 \\ 0 & \mathbf{h}^S \end{pmatrix} \tag{5.64}$$

If $\mathbf{U}$ can be found then the electronic solutions can be found from

$$\mathbf{h}^L \Psi^L = \varepsilon^L \Psi^L \tag{5.65}$$

Applying $\mathbf{U}$ to Ψ^D will yield the electronic solution

$$\mathbf{U}^\dagger\Psi^D = \mathbf{U}^\dagger\begin{pmatrix}\Psi^L\\ \Psi^S\end{pmatrix} = \begin{pmatrix}\Psi^L\\ 0\end{pmatrix} \tag{5.66}$$

The transformation $\mathbf{U}$ was obtained by Foldy and Wouthuysen[5] for the free electron Dirac hamiltonian, $\hat{h}^{FE}$. We shall refer to this transformation as $\mathbf{U}_0$, and it has the form

$$\mathbf{U}_0 = A_i\begin{pmatrix} \mathbf{I} & \boldsymbol{\sigma}\cdot\mathbf{p}_i \\ -\boldsymbol{\sigma}\cdot\mathbf{p}_i & \mathbf{I} \end{pmatrix} \tag{5.67}$$

where

$$A_i = \sqrt{\frac{E_i + m_0 c^2}{2E_i}}$$
$$p_i = \frac{c\mathbf{p}}{E_i + m_0 c^2} \tag{5.68}$$
$$E_i = \sqrt{p_i^2 c^2 + m_0^2 c^4}$$

It should be noted there are many different, but equivalent, forms of the Foldy–Wouthuysen transformation, $\mathbf{U}_0$, used in the literature. The outcome of this transformation is

$$\mathbf{U}_0^\dagger\mathbf{h}^{FE}\mathbf{U}_0 = \beta(E_i - m_0 c^2)$$
$$= \begin{pmatrix} E_i - m_0 c^2 & 0 & 0 & 0 \\ 0 & E_i - m_0 c^2 & 0 & 0 \\ 0 & 0 & -(E_i - m_0 c^2) & 0 \\ 0 & 0 & 0 & -(E_i - m_0 c^2) \end{pmatrix} \tag{5.69}$$

If we move from $\hat{h}^{FE}$ to $\hat{h}^D$ or $\hat{H}^{DC}$, the Foldy–Wouthuysen transformation will *not* achieve a similar block diagonalisation of the hamiltonian. However, we can look for a further transformation, $\mathbf{U}_1$, to complete the block diagonalisation,

$$\mathbf{U}_1^\dagger\mathbf{U}_0^\dagger\mathbf{h}^D\mathbf{U}_0\mathbf{U}_1 = \begin{pmatrix} \mathbf{h}^L & 0 \\ 0 & \mathbf{h}^S \end{pmatrix} \tag{5.70}$$

This was the approach developed by Douglas and Kroll[6] and made practical by Hess.[7] The approach can be extended to higher orders by using a general transformation, **U**,

$$\mathbf{U} = \ldots \mathbf{U}_3\mathbf{U}_2\mathbf{U}_1\mathbf{U}_0 \tag{5.71}$$

We shall briefly discuss some of the details related to implementing the Douglas–Kroll–Hess method at second order (DKH2). The hamiltonian we shall consider will be spin-free (scalar relativistic) with the inter-electron interaction represented by the Coulomb operator, $1/r_{ij}$. We shall only consider the one-electron terms of the DKH2 transformation. In this approximation the resultant hamiltonian appears deceptively complicated but can be evaluated using the familiar matrices of integrals: $\mathbf{T}^{\text{NR}}$, $\mathbf{V}$ and $(\mathbf{pVp})$, that we have already met.

$$\hat{h}^{\text{DKH2}} = \sum_i^N \varepsilon_0^+(i) + \sum_{ij}^N \varepsilon_1^+(i,j) + \sum_{ijk}^N \varepsilon_2^+(i,j,k)$$

$$\varepsilon_0^+(i) = E_i - m_0c^2$$

$$\varepsilon_1^+(i,j) = A_iV_{ij}A_j + A_i(pVp)_{ij}A_j$$

$$\varepsilon_2^+(i,j,k) = \frac{1}{2}\left\{\begin{array}{l} -A_i\overline{(pVp)}_{ij}A_jA_jV_{jk}A_k + A_i\overline{(pVp)}_{ij}A_j\frac{1}{p_j^2}A_j(pVp)_{jk}A_k \\[2ex] +A_i\overline{V}_{ij}A_jp_j^2A_jV_{jk}A_k - A_i\overline{V}_{ij}A_jA_j(pVp)_{jk}A_k \\[2ex] -A_i(pVp)_{ij}A_jA_j\overline{V}_{jk}A_k + A_i(pVp)_{ij}A_j\frac{1}{p_j^2}A_j\overline{(pVp)}_{jk}A_k \\[2ex] +A_iV_{ij}A_jp_j^2A_j\overline{V}_{jk}A_k - A_iV_{ij}A_jA_j\overline{(pVp)}_{jk}A_k \end{array}\right\} \tag{5.72}$$

with

$$\overline{V}_{ij} = \frac{V_{ij}}{E_i + E_j}$$

$$\overline{(pVp)}_{ij} = \frac{(pVp)_{ij}}{E_i + E_j} \tag{5.73}$$

In order to evaluate these terms Hess[7] suggested using a basis in which the $\mathbf{p}^2$ operator is diagonal. Recalling that

$$\hat{T}^{\text{NR}} = \frac{\mathbf{p}^2}{2m_0} \tag{5.74}$$

for which integrals are readily available in any quantum chemistry program, we can diagonalise $\mathbf{T}^{NR}$

$$\mathbf{t} = \Omega^{\dagger} \mathbf{T}^{NR} \Omega \tag{5.75}$$

$\mathbf{t}$ is a diagonal matrix with eigenvalues $t_i = p_i/2m_0$. This immediately provides us with

$$p_i = 2m_0 t_i \tag{5.76}$$

which in turn enables us to evaluate E_i and A_i in eqn (5.68). The integral matrices $\mathbf{V}(\chi)$ and $(\mathbf{pVp})(\chi)$ that are evaluated over basis functions, $\{\chi\}$, are also transformed to the $\mathbf{p}^2$ basis

$$\begin{aligned}
\left[\mathbf{V}(\mathbf{p}^2)\right] &= \Omega^{\dagger}\left[\mathbf{V}(\chi)\right]\Omega \\
\left[(\mathbf{pVp})(\mathbf{p}^2)\right] &= \Omega^{\dagger}\left[(\mathbf{pVp})(\chi)\right]\Omega
\end{aligned} \tag{5.77}$$

The matrices $\overline{\mathbf{V}}$ and $\overline{(\mathbf{pVp})}$ are scaled element by element in the $\mathbf{p}^2$ basis.

With the various integral matrices transformed to the $\mathbf{p}^2$ basis, it is a straightforward task to assemble $\hat{h}^{DKH2}$. Having done so, $\hat{h}^{DKH2}$ must be back-transformed to the coordinate (basis function) space

$$\mathbf{h}^{DKH2}(\chi) = \Omega\, \mathbf{h}^{DKH2}(\mathbf{p}^2)\Omega^{\dagger} \tag{5.78}$$

$\mathbf{h}^{DKH2}(\chi)$ can now replace the non-relativistic one-electron operator $\mathbf{h}$ in any Hartree–Fock or Kohn–Sham procedure. Since $\mathbf{h}/\mathbf{h}^{DKH2}$ do not depend on the molecular orbitals, or density, this process need only be carried out once at the beginning of the self-consistent-field process. The use of $\hat{h}^{DKH2}$ is variationally stable, does not suffer from gauge dependence, and has become a very successful tool in relativistic computational quantum chemistry.

5.6   Elimination of the Small Component: The Picture Change Transformation of Operators

The transformation of the Dirac hamiltonian to two-component form has an important consequence for the calculation of expectation values of property operators. The decoupled electronic wavefunction contains only the upper component of Ψ^D

$$\mathbf{U}^{\dagger}\Psi^D = \mathbf{U}^{\dagger}\begin{pmatrix} \Psi^L \\ \Psi^S \end{pmatrix} = \begin{pmatrix} \Psi^L \\ 0 \end{pmatrix} \tag{5.79}$$

In the Dirac theory, all property operators are four-component quantities. If the expectation value of an operator, say $\hat{M}$, is evaluated as

$$\langle \Psi^L | M | \Psi^L \rangle = \bar{m} \tag{5.80}$$

the picture change of the operator $\hat{M}$ has been neglected. The correct evaluation of the expectation value of $\hat{M}$ must involve transformation of its four component structure by **U**

$$\langle \Psi^L | \mathbf{U}^\dagger M \mathbf{U} | \Psi^L \rangle = m$$
$$= \langle \Psi^D | \mathbf{U}\mathbf{U}^\dagger M \mathbf{U}\mathbf{U}^\dagger | \Psi^D \rangle \tag{5.81}$$

The difference $m - \bar{m}$ is called the "picture change error". For properties that depend on the core region of heavy nuclei, for example hyperfine coupling constants, electric field gradients or nuclear quadrupole moments, the picture change error can be large. In such cases it is essential to apply the picture change transformation to the property operator. This applies to any two-component scheme such as the ZORA method or the DKH2 method.

5.7 Spin–Orbit Coupling

We discussed the role of the spin–orbit coupling in determining spin-dependent properties in Section 3.6.2. In relativistic theory the description of the electron's angular momentum in terms of the orbital angular momentum, l, and the spin angular momentum, s, is no longer valid. Rather, it becomes necessary to take the vector sum of s and l to yield the quantum number, j. For each electron

$$j = l \pm s \tag{5.82}$$

The vector sum of the one electron quantum numbers j, yields the overall quantum number J. For example, consider the OH radical that, in the non-relativistic treatment, has molecular orbitals of π and σ type formed from the interaction of the $1s$ orbital on H and the $2p$ orbitals on OH, see Figure 5.3(a). In the relativistic treatment, the $2p$ electrons in oxygen couple their orbital and spin angular momentum to produce $j = \frac{3}{2}, \frac{1}{2}$. Hence the $2p$ levels in atomic oxygen are split, as are the resultant π molecular orbitals. See Figure 5.3(b). The magnitude of the splitting in OH is approximately 140 cm^{-1}, but in heavy elements this can rise to a few electron volts (1 eV $= 8065.5$ cm^{-1}).

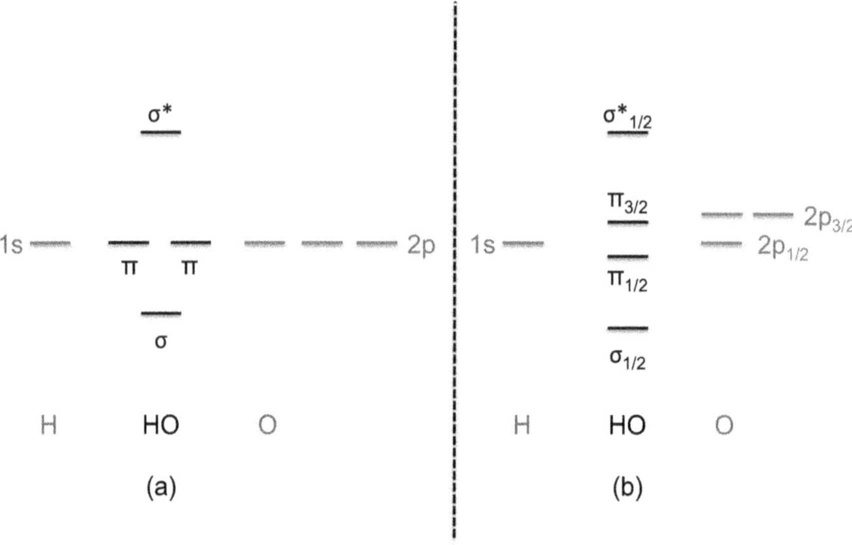

Figure 5.3   (a) Non-relativistic atomic orbitals of O and H and the molecular orbitals of OH. (b) Relativistic orbitals of O, H and OH corresponding to spin–orbit coupled states $j = l \pm s$.

The ZORA and DKH2 methods that we have discussed are able to treat spin–orbit coupling, provided that all spin-dependent components are retained. This complicates their implementation quite significantly. Alternatively, spin–orbit coupling can be included using simpler operators, such as the Pauli spin–orbit operator, eqn (5.44), or the Breit–Pauli spin–orbit operator, eqn (3.107). (The Breit–Pauli spin–orbit operator is obtained by applying the Foldy–Wouthuysen transformation to the Dirac–Breit hamiltonian.) The defect in both the Pauli and Breit–Pauli operators is that they are not variationally bound and so cannot be used in variational methods. However, they can be used in a first-order perturbation-type treatment. Typically a set of electronic states are solved for in some variety of CI method. These states are then coupled using the spin–orbit operator. Each CI state, label it P, is a linear combination of determinants, or configuration state functions

$$\Psi_P = \sum_M C_M^P \Psi_M \tag{5.83}$$

The CI hamiltonian matrix is diagonal

$$H_{PQ} = E_P^{CI} \delta_{PQ} \tag{5.84}$$

where P,Q label the CI states. The spin–orbit operator is included as

$$H_{PQ}^{SO} = E_P^{CI}\delta_{PQ} + \langle \Psi_P|\hat{H}^{SO}|\Psi_Q\rangle(1-\delta_{PQ}) \tag{5.85}$$

The spin–orbit coupled states are obtained by diagonalising $\mathbf{H}^{SO}$. The evaluation of $\langle \Psi_P|H^{SO}|\Psi_Q\rangle$ is quite simple in a determinantal basis

$$\langle \Psi_P|H^{SO}|\Psi_Q\rangle = \sum_{MN} C_M^P C_N^Q \langle \Psi_M|H^{SO}|\Psi_N\rangle \tag{5.86}$$

The one- and two-electron spin–orbit operators have the form given in eqn (3.107). Diagonal terms vanish, $\langle \Psi_M|H^{SO}|\Psi_M\rangle = 0$. For one spin–orbital difference, for example if $|\Psi_M\rangle$ and $|\Psi_N\rangle$ differ by ϕ_i and ϕ_j, the matrix element is

$$\langle \Psi_M(\ldots i \ldots)|H^{SO}|\Psi_N(\ldots j \ldots)\rangle =$$

$$\langle \phi_i|h^{SO}|\phi_j\rangle + \frac{1}{2}\sum_{k\neq i,j} n_k \left\{ \begin{array}{l} \langle i(1)k(2)|g^{SO}|j(1)k(2)\rangle \\ -\langle i(1)k(2)|g^{SO}|k(1)j(2)\rangle \\ -\langle k(1)i(2)|g^{SO}|j(1)k(2)\rangle \end{array} \right\} \tag{5.87}$$

where n_k denotes the occupancy of the orbitals ϕ_k that are the same in both determinants. For two spin–orbital differences, for example if $|\Psi_M\rangle$ and $|\Psi_N\rangle$ differ by ϕ_i, ϕ_k and ϕ_j, ϕ_l, the matrix element is

$$\langle \Psi_M(\ldots i \ldots k \ldots)|H^{SO}|\Psi_N(\ldots j \ldots l \ldots)\rangle =$$

$$\frac{1}{2}\left\{ \begin{array}{l} \langle i(1)k(2)|g^{SO}|j(1)l(2)\rangle \\ +\langle k(1)i(2)|g^{SO}|l(1)j(2)\rangle \\ -\langle i(1)k(2)|g^{SO}|l(1)j(2)\rangle \\ -\langle k(1)i(2)|g^{SO}|j(1)l(2)\rangle \end{array} \right\} \tag{5.88}$$

These expressions are written over spin–orbitals and the spin integration has yet to be performed. The spin integration is a little more complicated than we have previously met due to the presence of the spin operators. The additional terms arise because the two-electron spin–orbit operator does not possess permutational symmetry between the two electrons. It is necessary to use a symmetrised operator of the form $\hat{g}_{12}^{SO} = \frac{1}{2}(\hat{g}_{12}^{SO} + \hat{g}_{21}^{SO})$. The two-electron spin–orbit integrals are numerous and the mean field spin–orbit (SOMF) operator discussed in Section 3.6.2 provides an efficient means of dealing with them. Since the SOMF operator is an effective one-electron operator, the matrix

Table 5.1 Equilibrium bond distances (R_e), bond dissociation energies (D_e) and harmonic vibrational wavenumbers ($\tilde{\nu}_e$) obtained using non-relativistic and relativistic computational methods. "4C" refers to the full four-component Dirac–Coulomb hamiltonian.

Method	R_e / Å	D_e / eV	$\tilde{\nu}_e$ / cm^{-1}
HF[a]	2.930	0.384	97
HF-4C[a]	2.594	0.895	159
HF-DKH2[b]	2.596	—	183
MP2[a]	2.701	1.566	140
MP2-4C[a]	2.449	2.544	205
CCSD(T)-DKH2[b]	2.484	2.24	206
ZORA/BP86[c]	2.511	2.31	178
SR-ZORA/BP86[c]	2.517	2.25	177
Experiment[d]	2.472	2.29	191

[a]R. Wesendrup, J. K. Laerdahl and P. Schwerdtfeger, *J. Chem. Phys.*, 1999, **110**, 9457. [b]T. Fleig and L. Visscher, *Chem. Phys.*, 2005, **211**, 113. [c]E. van Lenthe, J. G. Snijders and E. J. Baerends, *J. Chem. Phys.*, 1996, **105**, 6505. [d]G. A. Bishea and M. D. Morse, *J. Chem. Phys.*, 1991, **95**, 5646.

element in eqn (5.87) is trivially evaluated once the SOMF integrals are available.

The implementation of spin–orbit operators is an advanced matter, with a range of additional complications that are absent in non-relativistic methods. The contracted quasi-degenerate perturbation scheme we have outlined above is only one of many strategies available. The interested reader should consult the reviews in refs. 8 and 9 for more details.

5.8 Summary

Returning to our opening discussion on the properties of gold dimer, we list in Table 5.1 some results obtained using the methods we have discussed. Two very clear conclusions can be drawn from these calculations: (1) the inclusion of relativistic effects is essential for the proper description of molecular properties, (2) the effects of electron correlation can sometimes be as large as the effects of relativity on molecular properties.

In this chapter we have touched on some of the key ideas in relativistic quantum chemistry, but there is much that we have omitted. A detailed description of the background theory can be found in ref. 10 and a survey of theory and consequences relevant to chemists can be found in ref. 11.

References

1. E. van Lenthe, E. J. Baerends and J. G. Snijders, *J. Chem. Phys.*, 1993, **99**, 4597.

2. M. Filatov, *Chem. Phys. Lett.*, 2002, **365**, 222.
3. E. van Lenthe, E. J. Baerends and J. G. Snijders, *J. Chem. Phys.*, 1994, **101**, 9783.
4. J. H. van Lenthe, S. Faas and J. G. Snijders, *Chem. Phys. Lett.*, 2000, **328**, 107.
5. L. L. Foldy and S. A. Wouthuysewn, *Phys. Rev.*, 1950, **78**, 29.
6. M. Douglas and N. M. Kroll, *Ann. Phys.*, 1974, **82**, 89.
7. B. A. Hess, *Phys. Rev. A*, 1985, **32**, 756.
8. C. M. Marian, *Reviews in Computational Chemistry*, ed. K. B. Lipkowitz and D. B. Boyd, Wiley, New York, 2001, vol. **17**, pp. 99.
9. B. A. Hess, C. M. Marian and S. D. Peyerimhoff, *Modern Electronic Structure Theory*, ed. D. R. Yarkony, World Scientific, Singapore, 1995, vol. **1**, pp. 152–278.
10. M. Reiher and A. Wolf, *Relativistic Quantum Chemistry*, Wiley, Weinheim, 2009.
11. Relativistic Methods for Chemists. Challenges and Advances in Computational Chemistry and Physics, ed. M. Barysz and Y. Ishikawa, Springer, New York, 2010, vol. **10**.

Subject Index

References to tables are given in **bold** type. References to figures are given in *italic* type.